# 全国大学生工业设计大赛优秀作品集·2022

Collection of Award-winning Works of
China Universities Industrial Design Competition
2022

裴悦舟 陈 江 何人可 主编

中国文联出版社

图书在版编目（CIP）数据

全国大学生工业设计大赛优秀作品集. 2022 / 裴悦舟，陈江，何人可主编. -- 北京 : 中国文联出版社，2024. 9. -- ISBN 978-7-5190-5604-9

Ⅰ. TB47

中国国家版本馆CIP数据核字第20241XB352号

主　　编　裴悦舟　陈　江　何人可
责任编辑　胡世勋　张　甜
责任校对　燕城实业
装帧设计　陈泳存
翻　　译　杨洪玉　石梓熹　黄怡雯

出版发行　中国文联出版社有限公司
社　　址　北京市朝阳区农展馆南里10号　邮编　100125
电　　话　010-85923025（发行部）　010-85923091（总编室）
经　　销　全国新华书店等
印　　刷　廊坊佰利得印刷有限公司

开　　本　889毫米 × 1194毫米　1/16
印　　张　20
字　　数　100千字
版　　次　2024年9月第1版第1次印刷
定　　价　278.00元

**指导单位：**

教育部高等教育司

**主办单位：**

广东省教育厅

全国工业设计一流专业建设协同创新平台

广东省本科高校工业设计专业教学指导委员会

**承办单位：**

广州美术学院

**协办单位：**

广东省工业设计协会

全国各赛区承办单位：北方工业大学　天津商业大学　河北工业大学　太原科技大学　内蒙古科技大学　沈阳航空航天大学　长春建筑学院　哈尔滨工业大学　上海理工大学　南京工业大学　浙江理工大学　安徽工业大学　福州大学　九江学院　青岛理工大学　郑州轻工业大学　武汉理工大学　湖南大学　广州美术学院　桂林电子科技大学　三亚学院　重庆大学　西华大学　贵州大学　昆明理工大学　西安美术学院　西藏民族大学　兰州理工大学　北方民族大学　新疆大学

**2022年全国大学生工业设计大赛组委会秘书处**

秘 书 长：吴华明

副秘书长：陈　江　涂渊

成　　员：张　兵　杨　杰　林　婕　黄筠志　裴悦舟　王　涛　彭圣芳　安　娃　田　顺　吴东杰　王倩蕾　钟　君　杨晓姗　袁　苑　郑　冰　庄　婕　关佩芳　李少娟　黄永华　张　政　付　皓　齐　媛　朱　荻

# 前言

全国大学生工业设计大赛是由全国工业设计一流专业建设协同创新平台、广东省教育厅、广东省工业和信息化厅联合主办的全国大学生学科竞技类赛事，每两年一届，自2012年至今已成功举办五届。2022年（第六届）全国大学生工业设计大赛于2022年4月正式启动。

大赛旨在贯彻落实《制造业设计能力提升专项行动计划（2019-2022年）》《国务院关于大力推进大众创业万众创新若干政策措施的意见》《国务院关于深化高等学校创新创业教育改革的实施意见》《教育部关于加快建设高水平本科教育全面提高人才培养能力的意见》和《教育部关于深化本科教育教学改革全面提高人才培养质量的意见》等文件精神，深入推进高校工业设计人才培养模式改革。大赛以工匠精神为核心，引导参赛者树立崇高的理想和良好的职业道德，致力于培养崇尚劳动、敬业守信、精益求精、敢于创新的文化创意及制造业的领军人才。

本届大赛主题为“跨界共融·设计同行”。在分析机遇与挑战、展望社会和行业未来发展前景的时候，工业设计将如何发挥跨学科、跨领域的特点，将对人类社会及经济发展、社群健康与福祉、科技与人的关系等这些重大议题的关注转化为切实可行的产品或服务，是当下设计学科的责任。



本届大赛收到来自全国750所高等院校共4万余件参赛作品。经各分赛区组委会初评，共有1281件作品入围全国总决赛。大赛在复评、终评等环节中遵照“符合本次大赛主题、可用、易用、环保、审美、经济”等评价标准，对参赛作品进行综合评审。经分赛区初评、全国复评、全国终评，本届大赛评选出金奖10名、银奖40名、铜奖80名、优秀奖165名，以及最佳人气奖1名、优秀指导教师奖10名、优秀组织奖10名、最佳协办单位奖6名。参赛作品充分展示了年轻的中国设计师们已经走出课堂，更多地关注国家的发展，敏锐地发现当前生活中的民生需求，用创造性的思维提出优良的解决方案，利用优秀传统文化及现代科技优势，创新未来生活方式。

大赛旨在以赛促学、以赛促教，引导参赛者关注社会重大议题，夯实基础技能，结合文化传统及科学技术，激发创造潜能，培养创新思维、协作意识、实践能力与社会责任感。大赛将继续发挥品牌与平台作用，以多种形式为高校及企业搭建教育成果与经验的交流平台，向全社会展示新时代高校工业设计的面貌，扶持、培育优秀的新锐设计人才，进一步推动工业设计教育事业发展，激发工业设计教育及产业动力。

陈江
教育部高等学校工业设计专业教学指导分委员会　副主任
2022年全国大学生工业设计大赛评审委员会　副主任
广东省本科高校工业设计专业教学指导委员会　主任
广州美术学院工业设计学院　院长 教授

2023年5月18日

# PREFACE

China Universities Industrial Design Competition is a biennial national university students subject competition jointly organized by the National Industrial Design First Class Professional Construction Collaborative Innovation Platform, Guangdong Provincial Department of Education, and Guangdong Provincial Department of Industry and Information Technology. It has been successfully held for ten years and a total of five sessions since 2012. The 6th National Industrial Design Competition for University Students was officially launched in April 2022.

The competition aims to implement the spirit of documents such as the Special Action Plan for Enhancing Manufacturing Design Ability (2019-2022), the Opinions of the State Council on Vigorously Promoting Mass Entrepreneurship and Innovation, the Implementation Opinions of the State Council on Deepening the Reform of Innovation and Entrepreneurship Education in Higher Education Institutions, the Opinions of the Ministry of Education on Accelerating the Construction of High level Undergraduate Education and Improving Talent Training Ability, and the Opinions of the Ministry of Education on Deepening the Reform of Undergraduate Education and Improving the Quality of Talent Training, and to deepen the reform of industrial design talent training models in universities. The competition advocates an industrial spirit centered on the spirit of craftsmanship, guiding participants to establish lofty ideals and good professional ethics, and is committed to cultivating leading talents in the cultural, creative, and manufacturing industries who value labor, are dedicated to work, keep improving, and dare to innovate.



The theme of this competition is "Cross-industry Integration going along with Design". When analyzing opportunities and challenges, and looking forward to the future development prospects of society and industry, it is the responsibility of the current design discipline to leverage the interdisciplinary and interdisciplinary characteristics of industrial design, and transform the focus on major issues such as human social and economic development, community health and well-being, and the relationship between technology and people into practical and feasible products or services.

This competition received over 40000 entries from 750 higher education institutions across the country. After preliminary evaluation by the organizing committees of each division, a total of 1281 works were nominated for the national finals. The competition will conduct a comprehensive evaluation of the entries in accordance with evaluation criteria such as "meeting the theme, usability, ease of use, environmental protection, aesthetics, and economy" during the reevaluation and final evaluation stages. After preliminary evaluation in different regions, national reevaluation, and national final evaluation, this competition has selected 1 special prize, 10 gold awards, 40 silver awards, 80 bronze awards, 165 excellent awards, as well as 1 best popularity award, 10 excellent guidance teacher awards, 10 excellent organization awards, and 6 best coorganizing unit awards. The entries fully demonstrate that young Chinese designers have stepped out of the classroom, paid more attention to the development of the country, keenly discovered the current needs of people's livelihoods, proposed excellent solutions with creative thinking, and utilized the advantages of excellent traditional culture and modern technology to create new ways of life for the future.

The competition aims to promote learning and teaching through competitions, guide participants to pay attention to major social issues, consolidate basic skills, combine cultural traditions and scientific technology, stimulate creative potential, cultivate innovative thinking, collaborative awareness, practical ability, and social responsibility. The competition will continue to play the role of a brand and platform, providing various forms of exchange platforms for educational achievements and experiences for universities and enterprises, showcasing the face of industrial design in universities in the new era to the whole society, and supporting and cultivating excellent emerging design talents. Further promote the development of industrial design education, stimulate industrial design education and industrial motivation.



Chen Jiang
Deputy Director of the Teaching Guidance Sub Committee for Industrial Design Majors in Higher Education Institutions of the Ministry of Education
Deputy Director of the Evaluation Committee for the 2022 National Universities Student Industrial Design Competition
Director of the Teaching Guidance Committee for Industrial Design in Under-graduate Universities in Guangdong Province
Dean of the Industrial Design School of Guangzhou Academy of Fine Arts

May 18, 2023

# 目录
# CONTENTS

# 全国大学生工业设计大赛
# 优秀作品集·2022

Collection of Award-winning Works of
China Universities Industrial Design Competition
2022

# 金奖

## Gold Award

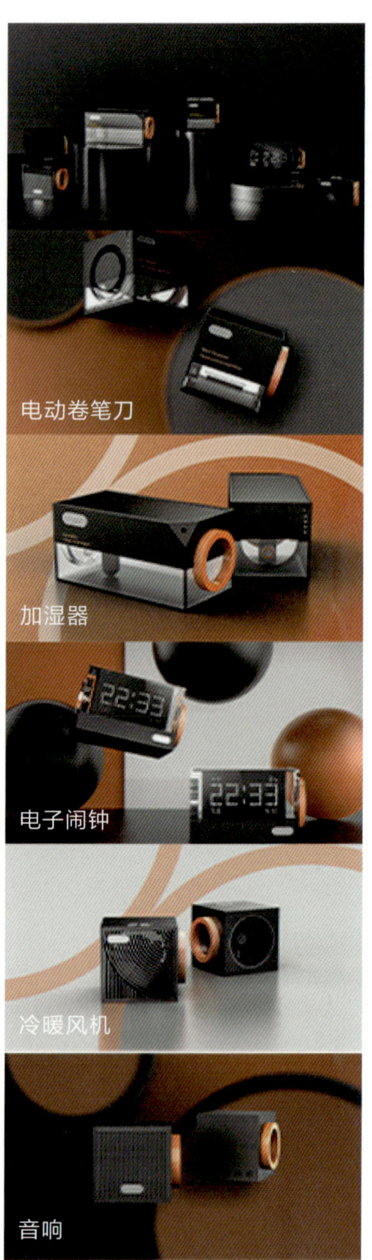

电动卷笔刀

加湿器

电子闹钟

冷暖风机

音响

# TAILLESS——无尾书桌生态设计

## TAILLESS——Intelligent Tailless Desk

作　　者：王鸣杰
指导老师：定　律
所在院校：上海理工大学

### 设计说明

桌面办公用品正在朝着智能化、生态化的方向发展，与此同时电子产品搭配的电源线往往导致现在的书桌十分杂乱，因而“无尾书桌”这一概念应运而生。“无尾书桌”概念是基于松耦合磁共振技术、摒弃桌面繁杂的电源线和实现桌面智能办公用品无线充电的构想。为使桌面收纳达到一个极致简洁的效果，将系列办公用品进行模块化设计，融入无尾书桌生态，形成极简的视觉效果。

“无尾书桌”使用场景图

折叠后的宽度为1110mm
女生也可以轻松携运

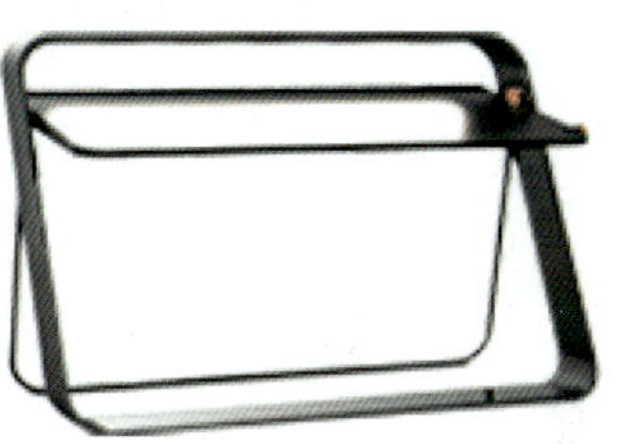
书桌顶部开灯后的效果
桌板呈现出一个倾斜的角度

## Design notes

Desktop office supplies are developing in the direction of intelligence and ecology, and the power cord of desktop electronic products has led to the fact that the desk is now very messy, so the concept of "tailless desk" came into being. The "tailless desk" is based on loosely coupled magnetic resonance technology, abandoning the complicated power cord on the desktop and realizing the concept of wireless charging of desktop smart office supplies. In order to achieve the ultimate storage and concise effect of the desktop, the series of office supplies are molded and designed, integrated into the tailless desk ecology, and forms a minimalist visual effect.

# GREAT WALK——可穿戴式足下垂辅助行走设备



## GREAT WALK —— Wearable Foot Drop Assistive Walking Equipment

作　　者：闫显龙　邹雪靖
指导老师：谌　涛
所在院校：上海理工大学

### 设计说明

Great Walk 是专为腰椎椎管狭窄引起的神经压迫症状患者而设计的户外行走辅助设备。这种病症会导致足下垂，患者在走路时因脚部控制神经受压，难以自主抬起脚尖，造成脚尖先触地，这显著增加了患者跌倒和受伤的风险。Great Walk 的设计注重佩戴便捷、操作简单和耐用性，能有效帮助足下垂患者调整步态，并减轻行走过程中的不适症状，从而减少行走时的伤害和风险。该产品采用非固定性结构设计，带动踝关节和胫骨肌肉积极参与运动，有助于患者的健康恢复。在整体形态上，它巧妙地与鞋子结合，既确保了辅助行走的功能，又降低了产品佩戴时的负担，为用户带来轻松舒适的体验。

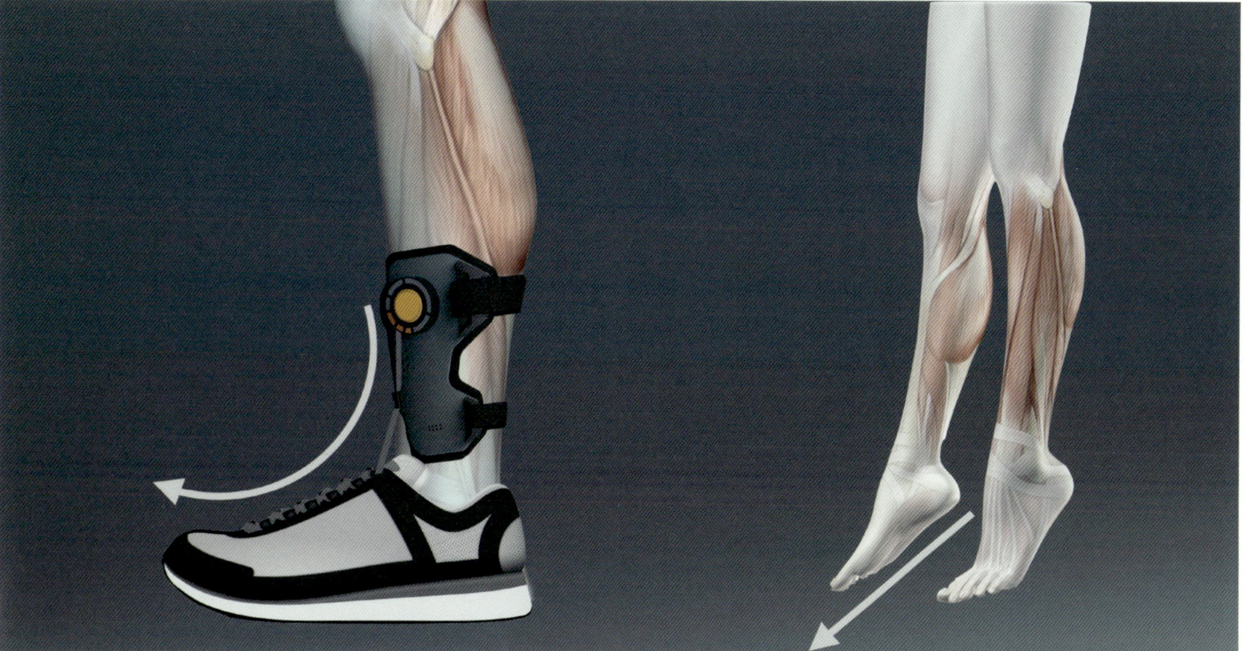

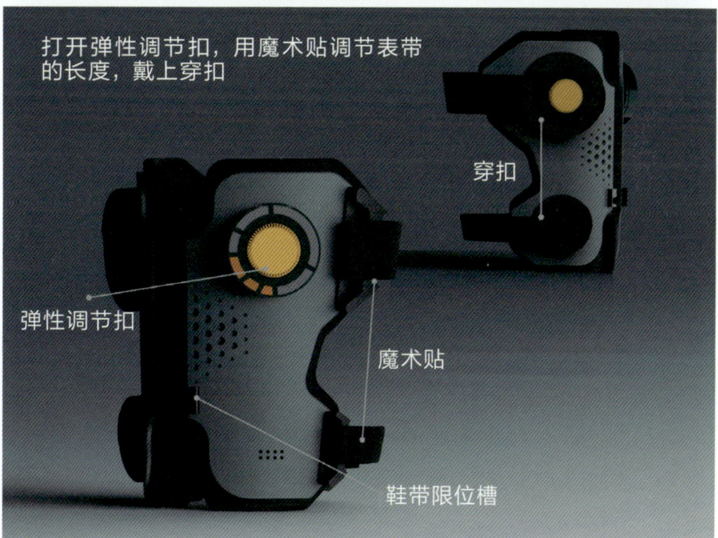


## Design notes

The Great Walk is an outdoor walking aid designed for symptoms of nerve compression caused by lumbar spinal stenosis. This condition causes foot drop, and patients have difficulty lifting their toes voluntarily when walking due to compression of the controlling nerves in their feet, causing their toes to touch the ground first, which significantly increases the risk of falls and injuries. Designed with a focus on ease of wear, simple operation and durability, Great Walk can effectively help patients with foot drop adjust their gait and reduce discomfort during walking, thereby reducing injuries and risks while walking. The product is designed with a non-fixed structure to actively participate in the ankle joint and tibial muscles, which helps patients recover healthily. In terms of overall form, it is cleverly combined with shoes, which not only ensures the function of assisted walking, but also reduces the burden when wearing the product, bringing users a relaxed and comfortable experience.

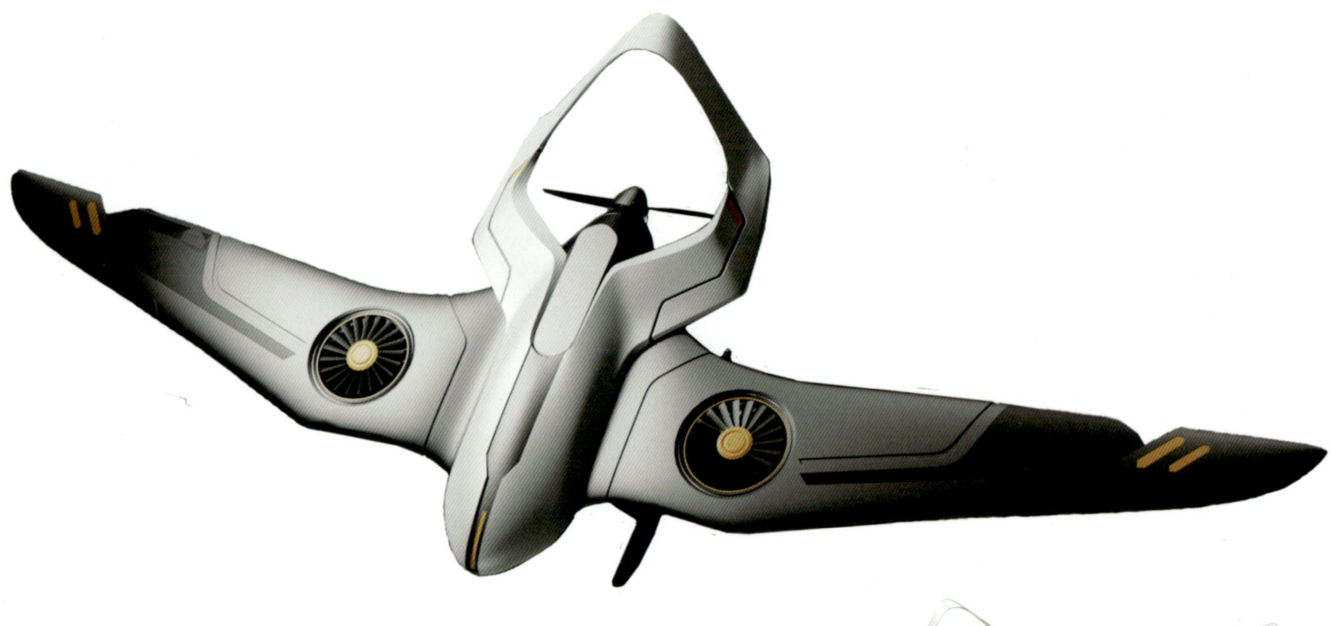

# 极地鹰科考无人机

## Polar Eagle Expedition Drone



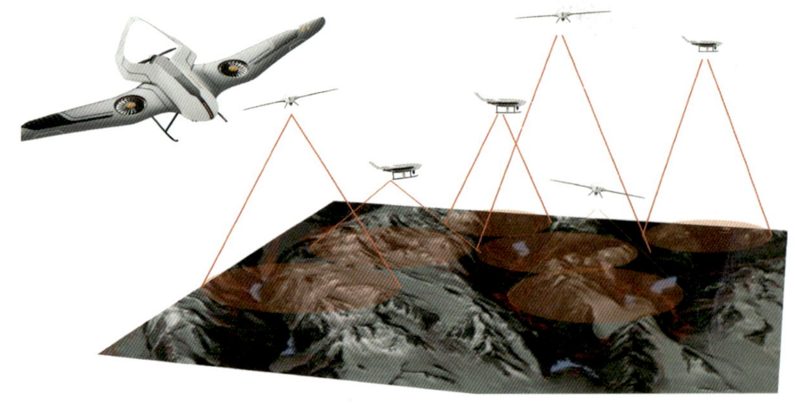

作　　者：娄素琦
指导老师：陈安全
所在院校：南京航空航天大学

### 设计说明

南北极的考察活动对于国家在政治、领土、资源能源以及气候环境等多个方面具有深远的战略意义。鉴于这些地区极端恶劣的环境条件和高度的安全风险，人类直接进行的勘测活动受到了显著的限制。因此，需要通过创新手段来拓展极地科考的时空领域，提升考察的效率和安全性。为此，我们研发了一款名为“极地鹰”的无人机，该无人机集成了采样器、存储仓、多种传感器、太阳能板以及高精度摄像头等功能，能够执行科学勘测、数据收集和样品采集等任务。它具备在复杂地形中进行自主导航和路线规划的能力，极大地减少了因极地严酷自然环境对科考人员可能造成的危险。此外，其机翼上的太阳能板设计使其能够在飞行过程中进行充电，有效节省能源。“极地鹰”无人机不仅提高了极地科考的安全性和效率，还为未来在这片神秘而重要的区域开展研究工作提供了新的技术手段。

## Design notes

Expedition activities in the Arctic and Antarctic regions have far-reaching strategic significance for the country in many aspects such as politics, territory, resources and energy, climate and environment. Given the extremely harsh environmental conditions and high levels of risk in these areas, direct human exploration activities are significantly limited. Therefore, it is necessary to expand the space and time field of polar scientific exploration through innovative means to improve the efficiency and safety of the expedition. To this end, a drone called the "Polar Eagle" has been developed, which integrates a sampler, storage bin, multiple sensors, solar panels, and high-precision cameras to perform tasks such as scientific surveys, data collection, and sample collection. It has the capability of autonomous navigation and route planning in complex terrain, which greatly reduces the danger that may be caused by the harsh natural environment of the polar region. In addition, the design of solar panels on its wings enables it to be charged during flight, effectively saving energy. The "Polar Eagle" UAV not only improves the safety and efficiency of polar research, but also provides a new technical means for future research in this mysterious and important region.

# 左与右——现代公用桌面办公系列设计

## LEFT AND RIGHT —— Modern Public Desktop Office Series Design



作　　者：陈子禾　焦奕博　颜廷旻
指导老师：崔天剑　陈　绘
所在院校：东南大学

### 设计说明

为了适应和满足广大办公场所用户的使用习惯和需求，精心设计了一系列适合左右手通用的桌面办公用品，包括经过再设计的键盘、鼠标和剪刀，旨在为不同惯用手的办公人员提供更舒适、更安全的使用体验。通过这些细致入微的考量，在设计中致力于提升办公环境的人类工程学标准，确保每一位用户都能在享受工作的同时，得到必要的舒适性和安全性保障。

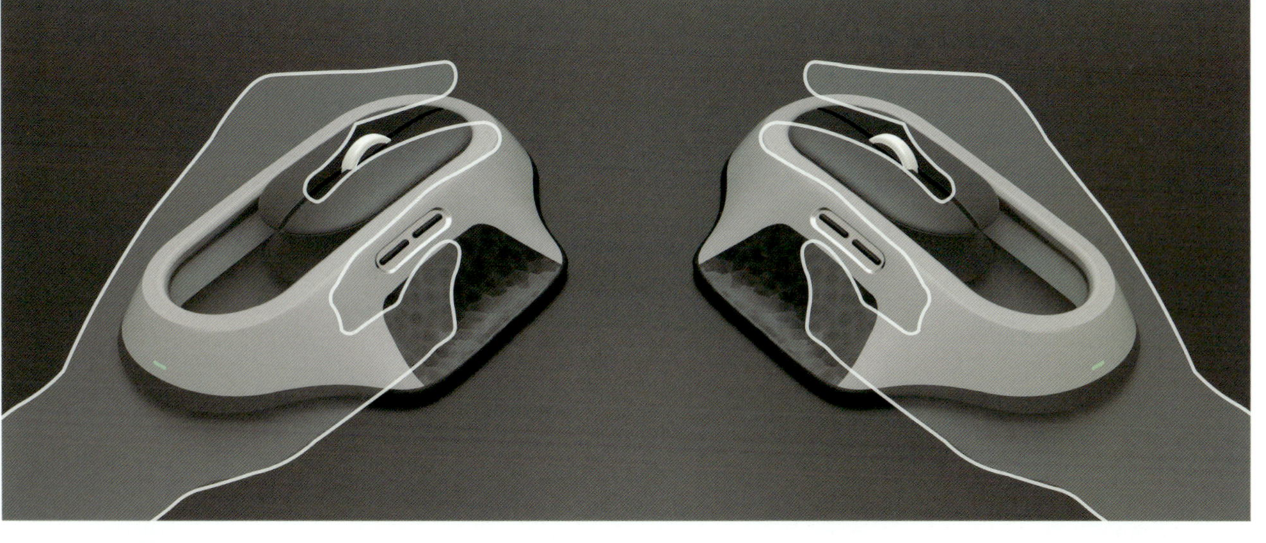

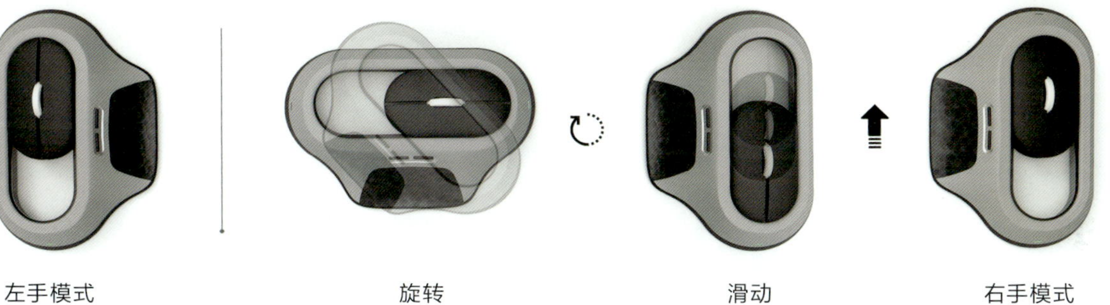

左手模式　　旋转　　滑动　　右手模式

## Design notes

In order to meet the habits and needs of the majority of office users, a series of desktop office supplies suitable for left and right hands have been carefully designed. The collection includes a redesigned keyboard, mouse, and scissors designed to provide a more comfortable and secure experience for office workers with different handers. Through these meticulous considerations, the design aims to improve the ergonomics of the office environment, ensuring that every user can enjoy their work while receiving the necessary comfort and safety.

金奖

银奖

铜奖

优秀奖

# 研思高效智能桌椅学习系统

## Intelligent Learning System



作　　者：孔　媛　王成宝　唐文祝
指导老师：傅晓云　冯　迪
所在院校：浙江工业大学

智能操作系统

## 设计说明

当前的儿童学习桌设计往往让书架占据了桌面的大部分空间，并且通常缺乏合理的收纳功能，导致桌面容易变得杂乱无章。这样的环境不仅使孩子们容易分心，还有可能影响他们的学习效率。针对这一问题，我们推出了一款创新设计的学习桌，它能够智能切换多种使用场景，以保持桌面的整洁，减少干扰，并提高孩子们的专注力和学习效率。此外，这款学习桌配备了自适应灯光系统，能够保护青少年的视力健康。座椅采用半包围式的头枕设计和柔性吸音材料，有效降低周围环境的噪音，能为孩子们创造一个便于集中精力而高效的学习环境。座椅还配有自适应充气腰靠，通过压力开关启动智能追踪功能，全面支撑青少年的背部和腰部，有助于他们的骨骼健康生长。在使用过程中，用户可以通过桌面上的操作界面以及家长专用的 APP 来控制整个学习系统。操作界面包含了语音提醒、高度调节、书架隐藏和灯光调整等功能；而 APP 端则提供了桌椅当前状态查看、学习专注度监测等多种功能，使得管理和了解孩子的学习环境变得更加便捷和直观。

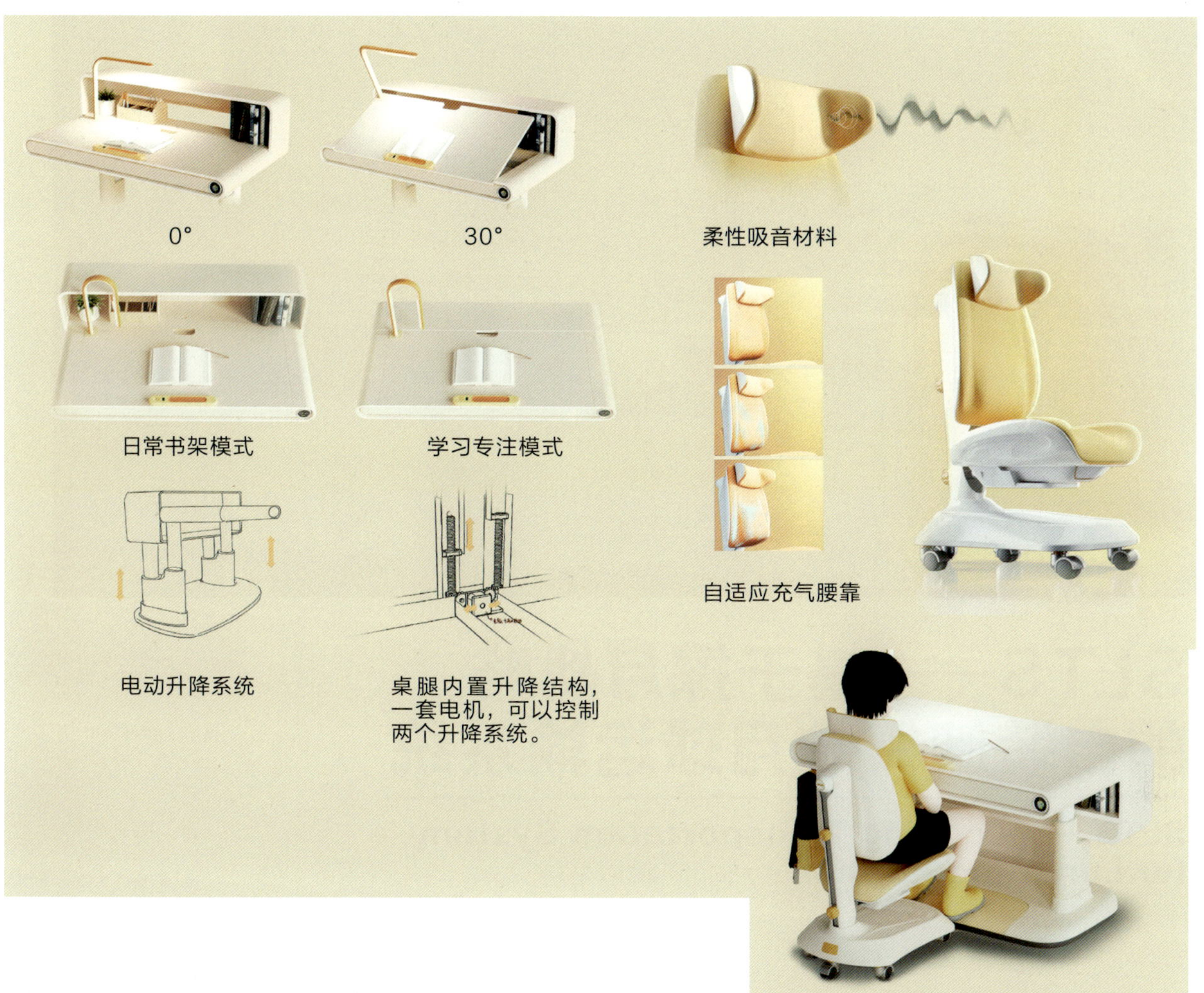




## Design notes

The current children's study table design often allows the bookshelf to occupy most of the space on the desktop, and often lacks reasonable storage functions, resulting in the desktop easily becoming cluttered. Such an environment not only makes children easily distracted, it may also affect their learning efficiency. To address this problem, we have introduced an innovative design of the study desk, which can intelligently switch between multiple use scenarios to keep the desk clean, reduce distractions, and improve children's focus and learning efficiency. In addition, the study table is equipped with an adaptive lighting system to protect the visual health of teenagers. The seat uses a semi-enclosed headrest design and soft sound-absorbing materials to effectively reduce ambient noise and create a focused and efficient learning environment for children. The seat is also equipped with adaptive inflatable lumbar support, which activates the intelligent tracking function through the pressure switch, and fully supports the back and waist of teenagers, helping their healthy bone growth. In the process of use, the user can control the entire learning system through the operation interface on the desktop and the APP for parents. The operation interface includes voice reminder, height adjustment, bookshelf hiding and lighting adjustment. The APP side provides a variety of functions such as the current status of tables and chairs, learning concentration monitoring, etc., making the management and understanding of children's learning environment more convenient and intuitive.

# SLTS——基于探月战略的自装卸概念月球运输系统

## SLTS —— Loading Transportation System for Lunar Exploration

作　　者：胡雨晗　梁焯涵　谭义晖　程门雪
指导老师：段正洁
所在院校：浙江理工大学

### 设计说明

SLTS 是一个针对未来探月任务而设计的月面货物运输和月面、地面间货物转运系统的设计概念，其设计定位于2040年及以后的中国探月战略需求。该运输系统采用了仿生学和模块化的设计理念，以高效率的货物转运能力和卓越的机动性满足多样化的月球探索任务需求。SLTS 系统由可重复使用的载人航天器、可变形的月球车以及模块化的运输舱（包含人员、物资和充气式基地模块）3个核心组件构成。这一系统的独特之处在于它无需依赖外部设备即可完成货物的交换与转运，这得益于其自身的结构设计，SLTS 能够独立实现各接口间的货物交接。此外，系统的可逆交接功能允许进行临时充气式基地的迁移和快速部署等操作。在外观设计上，SLTS 倾向于科幻风格的硬表面造型，强调了其作为工程设备的机械感。整体而言，SLTS 的设计不仅为未来的探月工程提供了一个货物转运高效化的解决方案，同时也为月面基地活动提供了前所未有的机动性和灵活性。

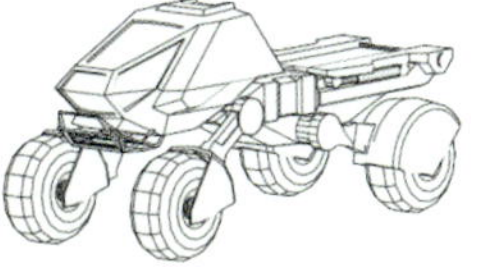

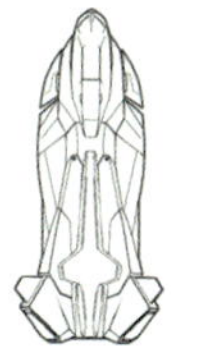

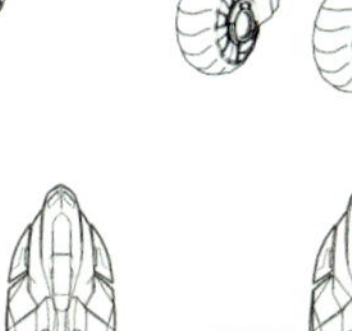

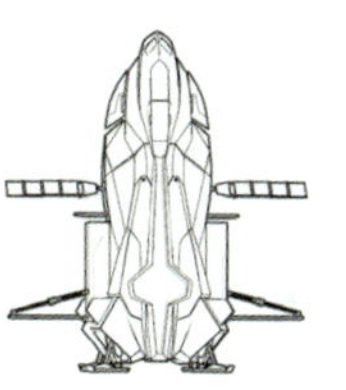

## Design notes

SLTS is a conceptual design for a lunar surface cargo transport and lunar to Earth cargo transport system designed for future lunar exploration missions, which is designed to meet China's strategic needs for lunar exploration in 2040 and beyond. The transportation system uses bionic and modular design concepts to meet the needs of diverse lunar exploration missions with efficient cargo transfer capabilities and superior mobility. The SLTS system consists of three core components: a reusable manned spacecraft, a deformable lunar rover, and a modular transport module containing personnel, supplies, and inflatable base modules. The unique feature of this system is that it does not need to rely on external equipment to complete the exchange and transfer of goods, which thanks to its own structural design, SLTS can independently realize the transfer of goods between the interfaces. In addition, the system's reversible handover function allows operations such as temporary inflatable base migration and rapid deployment. In terms of exterior design, the SLTS leans towards sci-fi style hard surface styling, emphasizing its mechanical feel as an engineering device. Overall, the SLTS design not only provides an efficient solution for cargo transfer for future lunar exploration projects, but also provides unprecedented mobility and flexibility for lunar base activities.

金奖

银奖

铜奖

优秀奖

# 生存式科考车

## Survivable Research Vehicle

作　　者：马圣涛　方　浩　张艺瀚
指导老师：艾　萍
所在院校：青岛理工大学

### 设计说明

本产品是一款专为极端环境科考勘测设计的大型生存式全地形车。它配备了宽敞的生活和科研车厢，能够通过深入分析不同的自然地形和气候条件，确保在多种恶劣自然环境下的适用性，并为科研人员提供优越的生活和研究空间。这款全地形车为全天候水陆两用型多功能车辆，可根据不同地形需求轻松切换驾驶模式，使其能够在雪地、沙漠、滩涂、戈壁、山地及水域等各类复杂地形中行驶。此外，采用模块化设计的全地形车能够进行勘测、领航、驻扎等多种功能的快速转换，从而更全面地满足科研人员在野外作业中的多样化需求。

支腿模块展示　　履带模块展示　　麦克纳姆轮展示

## Design notes

This product is a large survivable all-terrain vehicle designed for extreme environmental research and exploration. It is equipped with spacious living and research carriages, which can ensure suitability in a variety of harsh natural environments through in-depth analysis of different natural terrain and climatic conditions, and provide superior living and research space for researchers. This all-terrain vehicle is an all-weather dual-purpose vehicle, which can easily change the driving mode according to different terrain needs, so that it can drive in various complex terrain such as snow, desert, beach, Gobi, mountain and water. In addition, the modular design of the all-terrain vehicle can be quickly transformed into a variety of functions such as surveying, piloting, and stationing, so as to more comprehensively meet the diverse needs of scientific researchers in field operations.

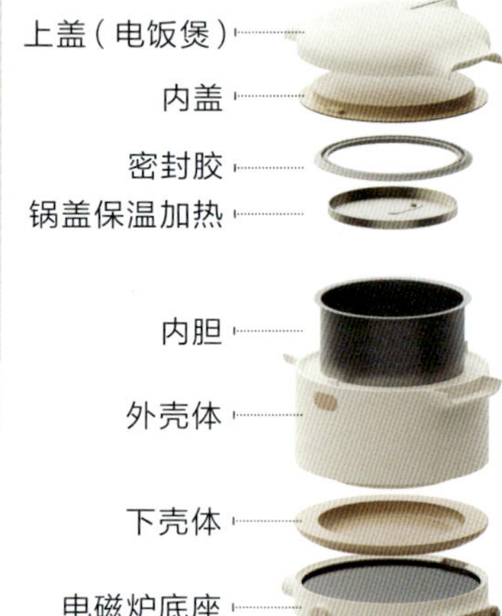


锅盖
烤盘
锅
烤盘
IH 面板
电磁线圈
壳体
下壳体



# MODULAR MULTI——面向都市单身合租青年群体的模块化多功能厨电

## MODULAR MULTI —— Functional Kitchen Appliances for Urban Single-Rental Youth Groups

作　　者：张开淇
指导老师：汤　军　许　昌
所在院校：武汉理工大学

### 设计说明

multi++ 是一套创新的模块化厨房电器产品，通过不同的模块组合实现多样化的烹饪功能。该产品最初的设计构想是将 IH 电饭煲中的电磁加热元件与锅体分离，使得电磁线圈能够独立作为小型电磁炉使用。这样的设计不仅实现了多功能一体化，还显著减少了厨房中电器所占的空间，并便于用户收纳。为了扩展其功能性，设计中增加了多个新模块，包括2个加热底座、1个电饭煲锅体、2个烤盘、1个小型锅以及1个连接件。用户可以根据需要自由组合这些模块，以适应不同的烹饪任务。在不使用时，这些模块可以便捷地堆叠收纳，节省空间。考虑到目标消费群体“一人食”的饮食习惯，multi++ 的产品容积被精心设计以适应1~3人的日常食量。

收纳模式

将所有元件进行堆叠放置，节省空间

电饭煲 + 电煮锅

日常做饭、饮食常见的组合。
蒸米饭、炒菜、煲汤等

电烤盘 + 电煮锅

适合应用于与一两个好友在家中聚会的场景，烧烤与火锅同时进行

电饼铛

除了烙饼以外，也可以用于制作一些特殊的美食，如比萨、蛋卷等

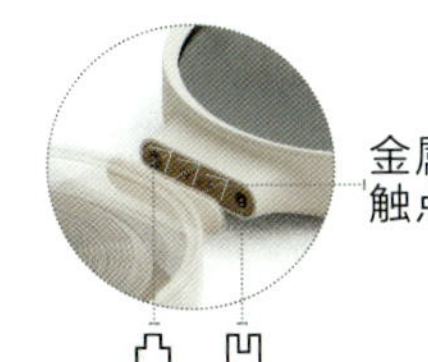


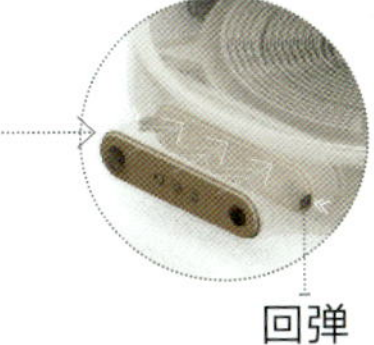


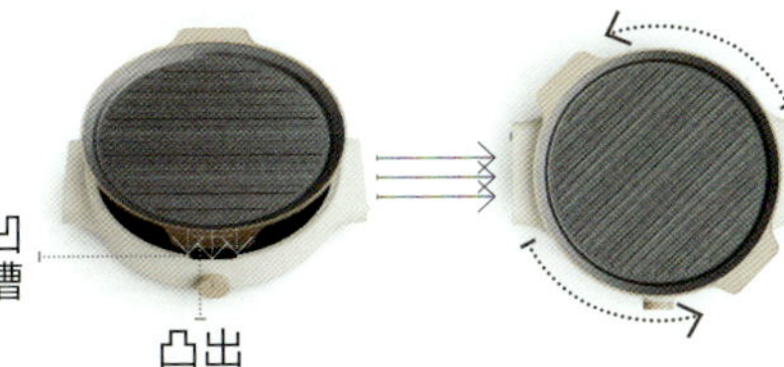


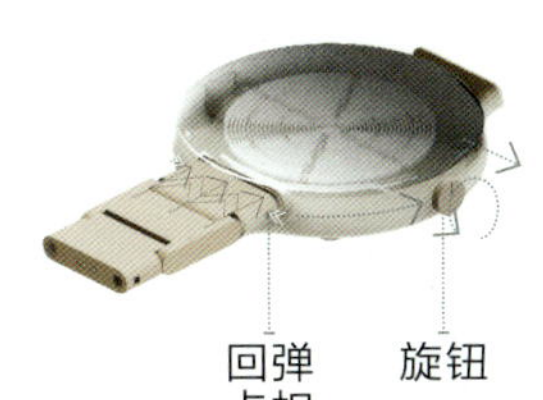


## Design notes

multi++ is an innovative set of modular kitchen appliances with diverse cooking functions through a combination of different modules. The initial design concept of the product is to separate the electromagnetic heating element in the IH rice cooker from the pot body, so that the electromagnetic coil can be used independently as a small induction cooker. This design not only achieves multi-functional integration, but also significantly reduces the space occupied by appliances in the kitchen and facilitates user storage. To extend its functionality, several new modules were added to the design, including two heating bases, a rice cooker pot body, two baking pans, a small pot and a connector. Users can freely combine these modules as needed to suit different cooking tasks. When not in use, these modules can be easily stacked and stored, saving space. Taking into account the "one for one" eating habits of the target consumer group, the product volume of multi++ has been carefully designed to accommodate the daily food intake of 1 to 3 people.

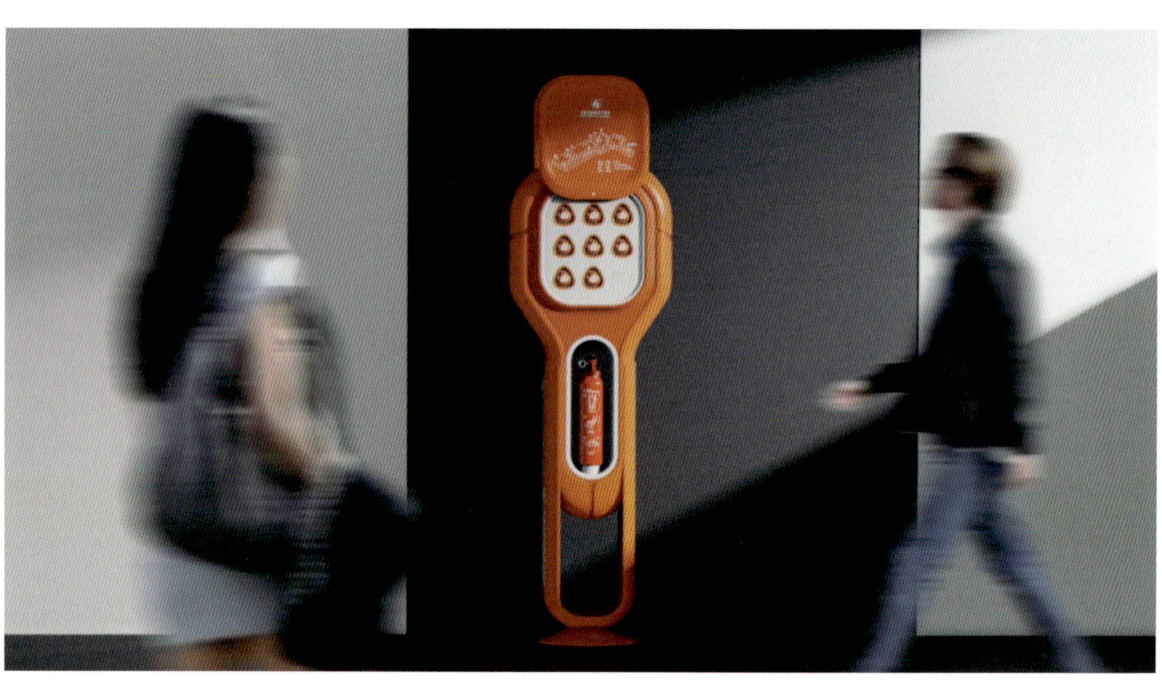

# 5G 物联网——INTEGRATOR 火灾救援集成设备

## INTEGRATOR 5G-IoT Integrated Instrument of Fire Rescue

作　　者：孙华杰　刘奕辰　秦祖淦　周亿博
指导老师：黄国梁　胡永攀
所在院校：四川美术学院

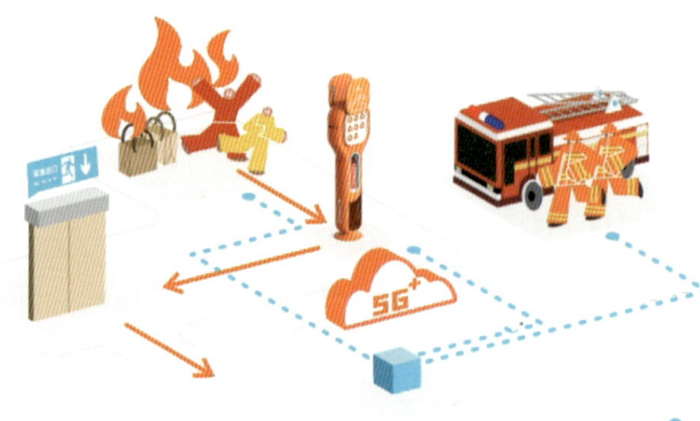


### 设计说明

本设计为公共场所的火灾预防和救援工作提供了一种创新的解决方案。它与市场上现有的单一功能救援设备（如口罩、灭火器）和灭火设施（如消防栓、喷淋系统等）不同，INTEGRATOR 设备配置有多个口罩和灭火器。更重要的是，它通过5G 物联网技术整合了现有的公共安全防火设备，实现了资源的集中管理和高效调配，从而提升了火灾防控的能力。此外，该设备内置的传感器可以实时监测火场环境状况，并将数据实时传输给消防机构。这一功能使消防员能够根据精确的环境信息有针对性地制定救援计划，并有效部署救援力量以快速扑灭火灾。

## Design notes

This design provides an innovative solution for fire prevention and rescue work in public places. Unlike single-function rescue equipment (such as masks, fire extinguishers) and fire-fighting facilities (such as fire hydrants, sprinkler systems, etc.) available on the market, INTEGRATOR equipment is equipped with multiple masks and fire extinguishers. More importantly, it integrates the existing public safety fire prevention equipment through 5G Internet of Things technology, and realizes the centralized management and efficient deployment of resources, thus improving the ability of fire prevention and control. In addition, the device's built-in sensors can monitor the environmental conditions of the fire site in real time and transmit the data to the fire squadron in real time. This feature allows firefighters to tailor rescue plans based on accurate environmental information and effectively deploy rescue forces to quickly extinguish fires.

# BAMBOO ATTORNEY——数字赋能的竹编工具与计算设计平台

## BAMBOO ATTORNEY —— Digitally Empowered Bamboo Weaving Tools and Computational Design Platform

作　　者：高培中　杨彦彬　侍建宇　李　晋
指导老师：刘震元　刘　胧　樊　中　宋善威　郑康奕
所在院校：同济大学设计创意学院

### 设计说明

随着数字化进程的不断加速，传统工艺传承正面临数字化转型与升级的挑战，由静态转向动态发展。本设计旨在将先进的数字技术与传统的竹编工艺相结合，推动非物质文化遗产的发展与创新。通过这种融合创新设计，不仅增强了传统手工艺的魅力，还降低了竹编工艺品的制作难度，有助于振兴手工艺市场，并促进边远地区的经济增长。设计的核心在于开发一种新的数字工艺形式——利用计算机算法生成数字化的竹编结构，并进行力学模拟生成辅助编织的支撑框架，然后通过3D 打印制造出来，从而创造出更加多样化的竹编产品。与此同时，配套的数字平台成为用户与工匠共同协作参与定制产品的桥梁。

木质素聚乳酸复合材料用于替代经篾的支撑功能，可以辅助编织出更复杂的造型

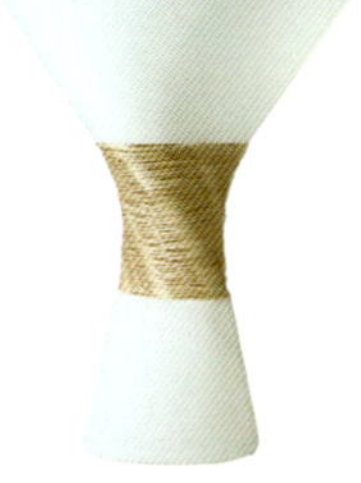

灵感来自竹丝扣瓷工艺，由算法生成三维的支撑框架

网状支撑结构作为对抗相互作用力的承载者

通过算法生成三角面形成复杂的几何体





交互界面展示

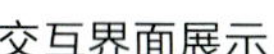

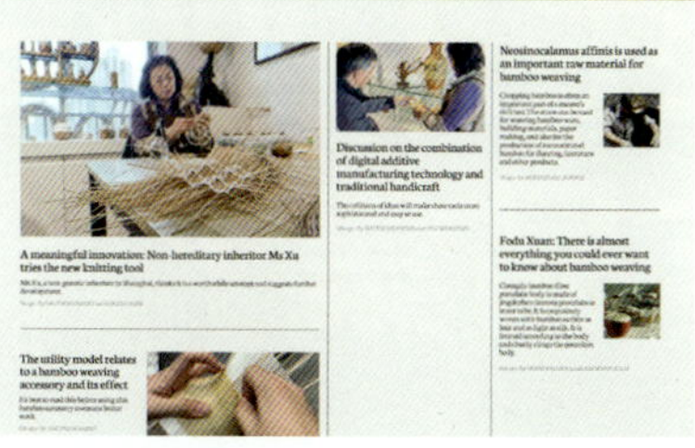
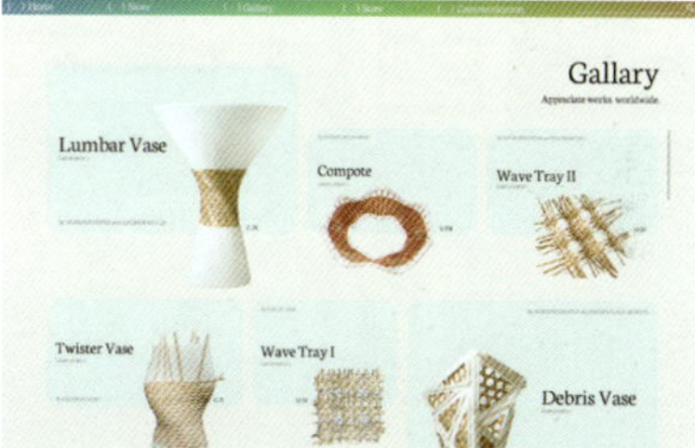


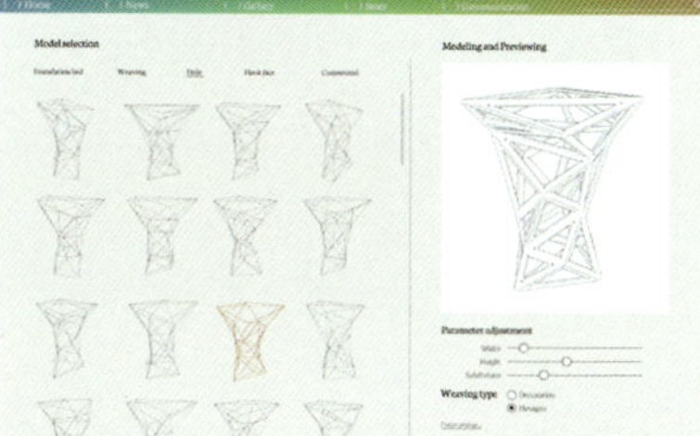
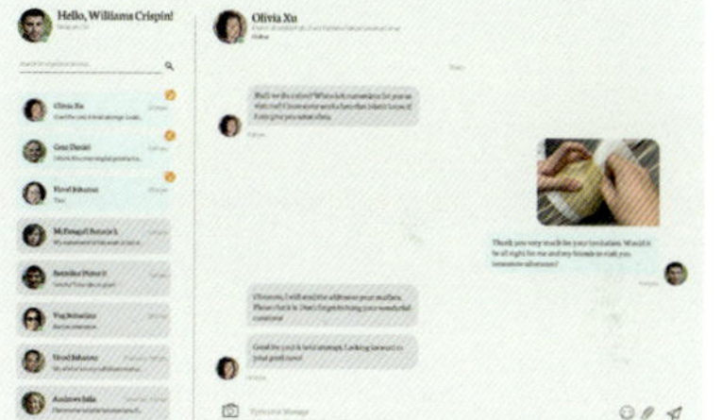

## Design notes

With the continuous acceleration of the digital process, the traditional process inheritance is facing the challenge of digital transformation and upgrading, from static to active development. The design aims to combine advanced digital technology with traditional bamboo weaving techniques to promote the development and innovation of intangible cultural heritage. Through this integrated and innovative design, it not only enhances the charm of traditional handicrafts, but also reduces the difficulty of making bamboo woven handicrafts, which helps to revitalize the handicraft market and promote economic growth in remote areas. The core of the design is to develop a new digital process form —— the use of computer algorithms to generate digital bamboo fabric structure, and mechanical simulation to generate auxiliary woven support frame, and through 3D printing to create a more diversified bamboo woven products. At the same time, the supporting digital platform serves as a bridge for users and artisans to collaborate and participate in customized products.

# 全国大学生工业设计大赛
# 优秀作品集·2022

Collection of Award-winning Works of
China Universities Industrial Design Competition
2022

# 银奖

# Silver Award

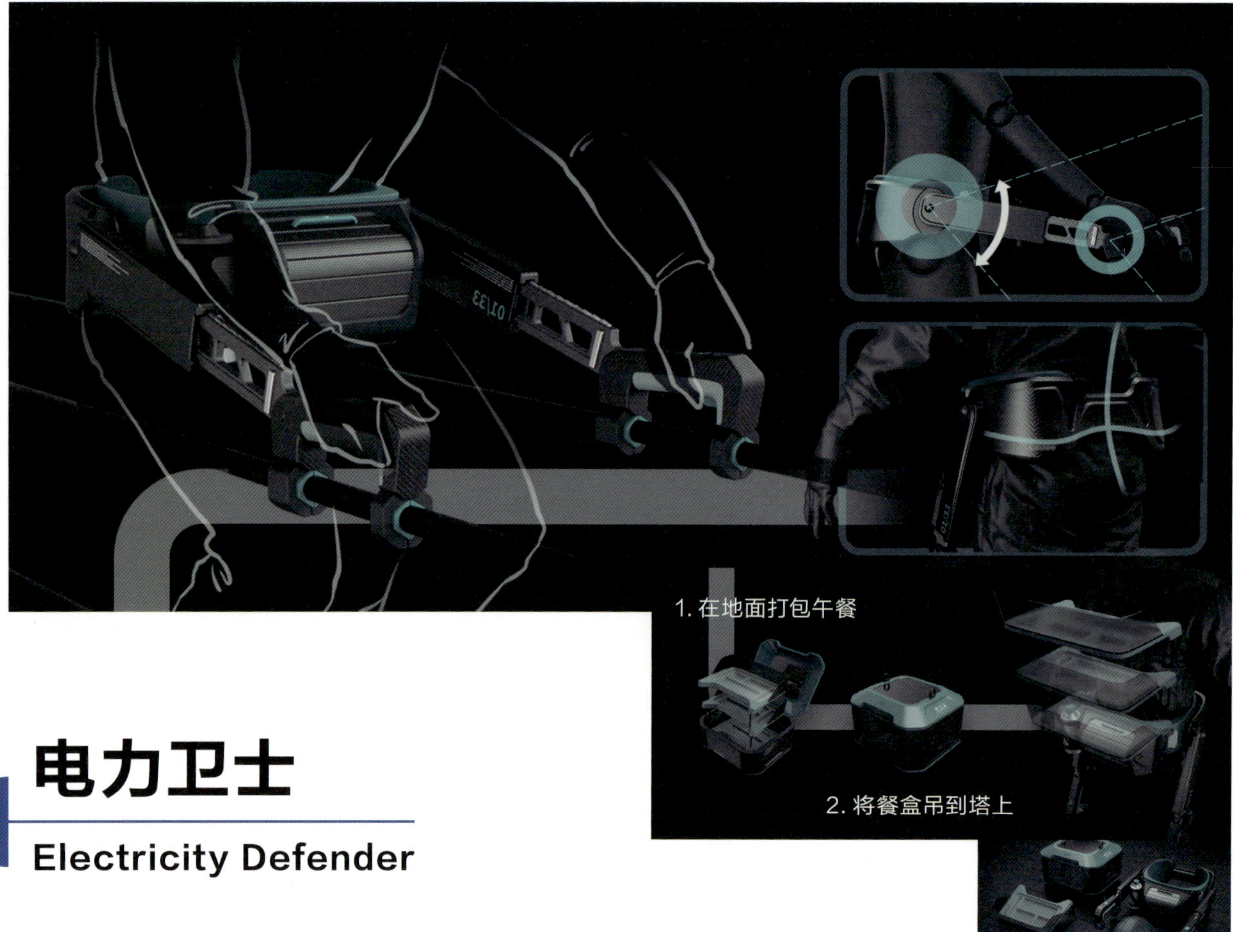


# 电力卫士

## Electricity Defender



作　　者：胡若晨　杨梦石
指导老师：杨　新
所在院校：北京理工大学

## 设计说明

辛勤工作的电网工人为社会带来光明，但他们的安全很少被注意。针对这一群体的特定需求，设计了一款高空作业设备，旨在为电网工人提供更安全、舒适和体面的工作环境。设计中为了确保使用者在操作过程中的灵活性与安全性，装备采用了一种三段式结构设计，其中包括腰部和肘关节的连接，这使得工作人员在电缆上的活动更为自如。人类工程学背板则是基于人体髋骨的自然形态演化而成，紧密贴合腰部肌肉，提供必要的保护和支撑。装置还配备了基于导线的坚固支撑系统，极大地降低了因高空作业时遭遇强风或操作失误而导致坠落的风险，从而确保使用者的安全。

## Design notes

Hard-working grid workers bring light to society, but their safety is rarely noticed. In response to the specific needs of this group, a high-altitude working equipment was designed to provide a safer, more comfortable and decent working environment for grid workers. In order to ensure the flexibility and safety of the user during the operation, the equipment adopts a three-stage structural design, which includes the connection of the waist and elbow joints, which allows the staff to move more freely on the cable. The ergonomic backplane is based on the natural shape of the body's hip bones and closely fits the waist muscles to provide necessary protection and support. The unit is also equipped with a robust wire-based support system, which greatly reduces the risk of falling due to strong winds or operational errors while working at altitude, thus ensuring user safety.

# 夫诸号火星车

## FU ZHU Mars Vehicle

作　　者：杨梦石　胡若晨
指导老师：徐　悬　冯　明
所在院校：北京理工大学

### 设计说明

工业设计不仅要解决身边的实际问题，更要为人类的未来规划蓝图。夫诸号火星车为探索火星的先驱者们保驾护航，满足不同场景下的栖息和科研需求。夫诸号火星车由牵引车和移动实验室两部分组成：前者是主要的生活场所，并能提供强劲动力；后者配备模块化实验柜，用于科学研究。同时，夫诸号配套的无人驾驶车可以代替人力完成大部分危险的探索和测绘工作。“夫诸”是中国传统神话中水神共工的坐骑，寓意其能够征服火星的恶劣环境，开创人类的未来。

### Design notes

Industrial design is not only to solve the practical problems around us, but also to plan the blueprint for the future of mankind. The Fuzhu rover has served as an escort for pioneers exploring Mars, meeting their habitat and scientific needs in different scenarios. The Fuzhu consists of two parts: a tractor, which is the main living space and can provide strong power, and a mobile laboratory, which is equipped with a modular experimental cabinet for scientific research. At the same time, the Fuzhu unmanned vehicle can replace human manpower to complete most of the dangerous exploration and mapping work. "Fu Zhu" is the mount of Gonggong, the water god in Chinese traditional mythology, which means he can conquer the harsh environment of Mars and create the future of mankind.

# SKYNEX——单人飞行载具系统



## SKYNEX —— Single Person Flying Vehicle System

作　　者：胡　翔
指导老师：苏　艺　山　娜
所在院校：北京服装学院

### 设计说明

随着经济全球化的发展，城市化的进程不断加快。人类为此向自然界过度索取资源，在城市面积增加的同时，导致人类与自然界的距离逐渐增加，对自然的了解逐渐减少。人类之间出现信任危机，自私主义的兴起导致人与人之间的隔阂不断增加。本人通过设计低空飞行载具系统的方式，以“机”为纽带将人与人、人与自然的关系重新连接起来，达成人与人互帮互助，人与自然和谐共生的最终目的。

### Design notes

With the development of economic globalization, the process of urbanization is accelerating. In order to achieve this goal, human beings are demanding excessive resources from the natural world. As the area of cities increases, the distance between human beings and the natural world gradually increases and the understanding of nature gradually decreases. Meanwhile the crisis of trust between human beings occurs, the rise of selfishness leads to the growing gap between people. Through the design of low-altitude flight vehicle system, I take "Machine" as a link to reconnect the relationship between man and man, man and nature, to achieve mutual help, the ultimate goal of harmonious coexistence between human and nature.

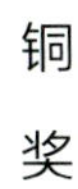

前往除沙区域

除沙车上轨

连接火车

自动除沙

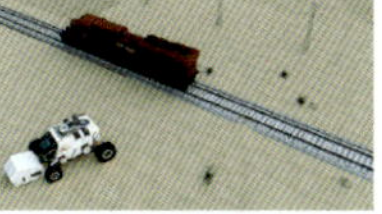

离开轨道

返回基站充电

# 沙漠地区铁路除沙系统设计

# Desert Railroad Sand Removal System Design

作　　者：冯汉禹　魏　玮　殷海洋
指导老师：张芳兰
所在院校：燕山大学

## 设计说明

本设计为沙漠地区轨道提供一种高效的除沙系统。该系统主要由DS轨道除沙车、DS生态维护车、DS车辆管理基地以及DS系统管理终端4大核心组件构成。首先，DS轨道除沙系统的主要功能是在沙尘暴来袭时，及时发现轨道埋沙并前往清理，减少人工操作，提高除沙效率。其次，本设计还秉持可持续设计与绿色理念，致力于探索沙害防治方案的新可能。

## Design notes

This design provides an efficient sand removal system for track in desert area. The system is mainly composed of DS track sand removal vehicle, DS ecological maintenance vehicle, DS vehicle management base and DS system management terminal four core components. Firstly, the main function of the DS track sand removal system is to find the track sand in time and clean it when the dust storm hits, reducing manual operation and improving the efficiency of sand removal. Secondly, the design also upholds the concept of sustainable design and green, and is committed to exploring new possibilities of sand damage control schemes.

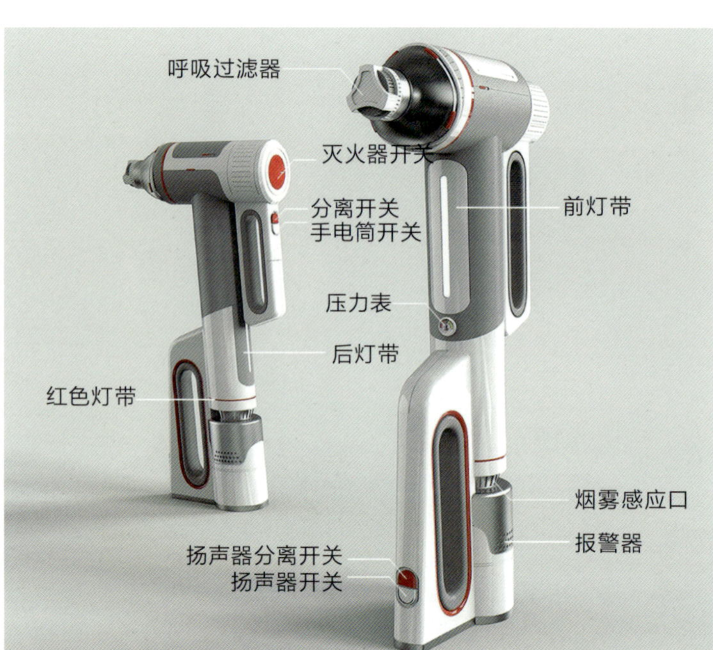


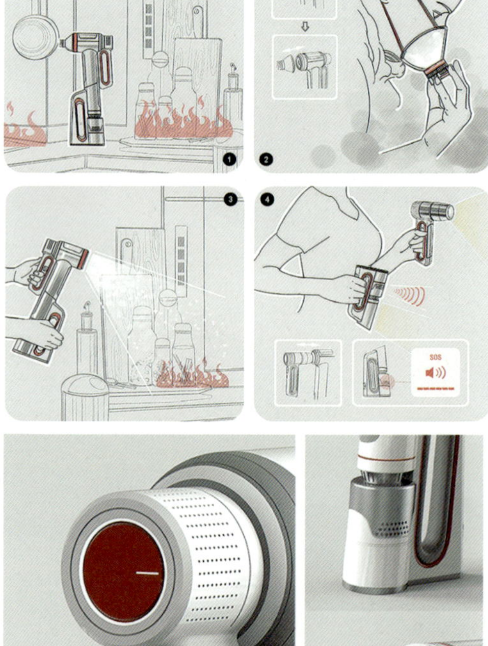

# FLEXO——多功能家用小型灭火器设计

## FLEXO —— Design of Multifunctional Household Small Fire Extinguisher



作　　者：王伊桐
指导老师：孙　兵
所在院校：鲁迅美术学院

### 设计说明

FLEXO 灭火器的设计以家居环境为基点，其核心功能是以提高灭火效率而保障人们安全。该灭火器还集成了照明、报警和过滤等多项辅助功能，这些功能使得灭火过程更加安全、便捷和高效。具体来说，该灭火器的上半部把手可以与瓶身分离，转变为手电筒，以帮助用户在紧急情况下更安全地逃生，而下半部把手也可脱离瓶身，变身为扬声器，这有助于救援人员快速准确地确定被困人员的位置。此外，灭火器前端配备的过滤面罩能够确保用户在灭火时呼吸顺畅，从而在火场中实现更安全高效的灭火和

### Design notes

The FLEXO fire extinguisher is designed with the home environment in mind, and its core function is to keep people safe by improving fire extinguishing efficiency, in addition to integrating multiple auxiliary functions such as lighting, alarm and filtration. These features make the fire fighting process safer, easier and more efficient. Specifically, the upper handle of the extinguisher can be detached from the bottle and converted into a flashlight to help the user escape more safely in an emergency. The lower part of the handle can also be removed from the bottle and turned into a speaker, which helps rescue workers quickly and accurately locate the trapped person. In addition, the filter mask at the front of the extinguisher ensures that the user can breathe easily while extinguishing the fire, thus achieving a safer and more efficient fire extinguishing and escape in the fire site.

# OUREA——新立体交通系统下的概念交通工具设计

## OUREA——A Design of Vehicle Bared on the Concept of New Grade Separation Traffic System

作　　者：王家霖
指导老师：范乐明
所在院校：上海大学

### 设计说明

在2040年的中国，人口和建筑密度极度增加，汽车数量也大幅提高。为了解决越来越拥堵的大城市交通问题，城市高架和快速路将以一种纵置的形态建造，以便在有限的地面空间上设置更多的车道。车辆通过磁力吸附在特殊高架的两侧行驶，车辆座椅也会单独旋转，保证每位乘客始终处于水平状态。

### Design notes

The project is under the background that by 2040, the traffic system in big cities will have changed to reduce congestion. The viaducts and urban expressways are in an upright position, and cars run on both sides by magnetic attraction. The purpose of these kinds of roads is to set more lanes on limited ground area. When the car runs on the upright road, the seats will rotate independently to keep all the passengers horizontal.

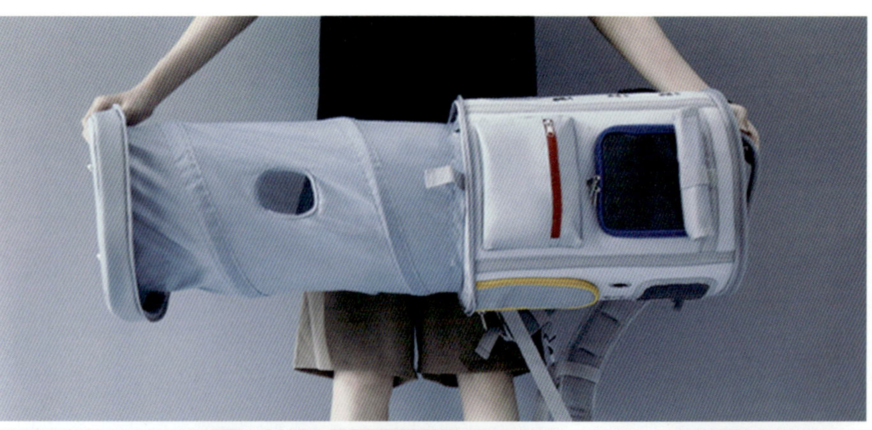

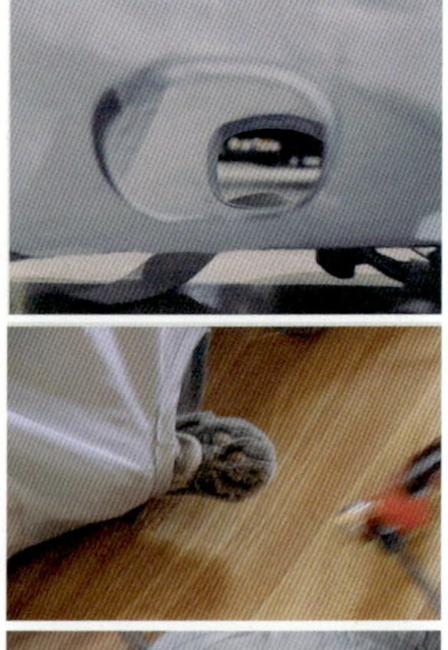

拉开底部拉链

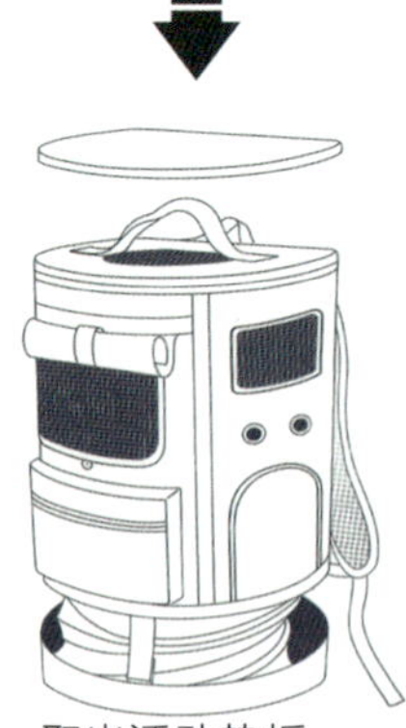

# TUTOY ——宠物猫背包

## TUTOY —— Pet Cat Backpack

作　　者：田　源
指导老师：韩　挺
所在院校：上海交通大学

### 设计说明

猫背包底部的可扩展弹簧袋与包身的多入口设计，闲置时作为猫咪玩具；扩展底袋在外也是活动空间的扩展。

### Design notes

The expandable spring bag at the bottom of the cat backpack and the multi-entry design of the bag body can be used as a cat toy when idle; expanding the bottom pocket is also an extension of the activity space.

# 智能仓库 AGV 搬运机器人

## Intelligent Warehouse AGV Handling Robot

作　　者：牛金子　童莹莹　李　瑶
指导老师：张　昆
所在院校：中国矿业大学

### 设计说明

本设计主要用于无人智慧仓库中的入库、出库、理货等物料搬运，代替传统需要人工操作的叉车和搬运车的工作，而且 AGV 机器人还可以通过自动巡航更换电池，实现全天超长时间工作。搭载 WMS 智慧仓库的系统，可实现从终端进行调度管理，并将 AGV 机器人工作状态与任务进程可视化，提高仓库管理效率，使仓库工作有序进行。

### Design notes

It is mainly used for material handling in the unmanned intelligent warehouse, such as warehouse in, warehouse out, tally, etc., instead of the traditional manual forklift and carrier. In addition, the AGV robot can also change the battery through automatic cruise, so as to work for a long time throughout the day. The system equipped with WMS intelligent warehouse realizes the scheduling management from the terminal, visualizes the working state and task process of AGV robot, improves the efficiency of warehouse management, and makes the warehouse work orderly.

# AIPO-6智能农用环保钻探机系统

## AIPO-6 Intelligent Agricultural Environmental Protection Drilling Machine System

作　　者：李熔华
指导老师：沙春发
所在院校：江苏大学

### 设计说明

AIPO-6智能农用钻探机系统是探索产品转型升级的可行性，探索机械装备多样化、智能化和信息化趋势下综合解决问题的设计方案。AIPO-6智能钻探机明确以钻探勘测一体、智能交互、地质数据集成化管理为核心创意，力求搭建地质探测的全流程系统。

### Design notes

AIPO-6 intelligent agricultural drilling machine system is a design scheme to explore the feasibility of product transformation and upgrading, and to explore a comprehensive solution to problems under the trend of diversification, intelligence and informatization of mechanical equipment.The care creative idea of AIPO-6 Intelligent drilling machine is to realize the integration of drilling and survey, the intelligent interaction and integrated management of geology data, and set up the system of geology survey with all-process.

# SA-REVB 移动式紧急道路救援机械

## Design of SA-REVB Mobile Emergency Road Rescue Machine



作　　者：孟忠涛　巢文博　陈　榕
　　　　　孙　超　卢凌宇
指导老师：张　凯　孙宁娜
所在院校：江苏大学

### 设计说明

这款移动式紧急道路救援机械装置专为应对地震、塌方、泥石流等自然灾害，以及火灾、矿山事故和危险化学品安全生产事故等灾难救援而设计。上述情况下，可能因道路损毁、边坡垮塌或桥梁损坏而导致交通中断，短时间内难以修复，从而阻碍了紧急救援工作的进行。针对这一挑战，提出了一种通过搭载伸缩式桥架的全地形智能机械，能够迅速铺设临时抢险道路的创新解决办法。该移动式紧急道路救援机械的工作原理是升起钢索杆，拉出钢索，并展开伸缩桥架。同时，设备每侧配备的8根液压杆可用来调节桥架角度，确保快速有效地铺设临时道路，以便及时恢复交通，保障救援车辆和人员顺畅通行。

### Design notes

This mobile emergency road rescue mechanism is designed to respond to natural disasters such as earthquakes, landslides and mudslides, as well as disaster sites such as fires, mine accidents and hazardous chemical safety accidents. In these cases, road damage, slope collapse or bridge damage may interrupt traffic and make it difficult to repair in a short period of time, thus hampering emergency relief efforts. Aiming at this challenge, an innovative solution is proposed to quickly lay temporary emergency roads through all-terrain intelligent machinery equipped with retractable bridge. The working principle of the mobile emergency road rescue machine is to raise the steel cable rod, pull out the steel cable, and deploy the telescopic bridge. At the same time, eight hydraulic rods on each side of the equipment can be used to adjust the Angle of the bridge to ensure that the temporary road is quickly and effectively laid in order to restore traffic in time and ensure the smooth passage of rescue vehicles and personnel.

# ARCTO 城市自然连接系统



## ARCTO Urban-natural Connection System

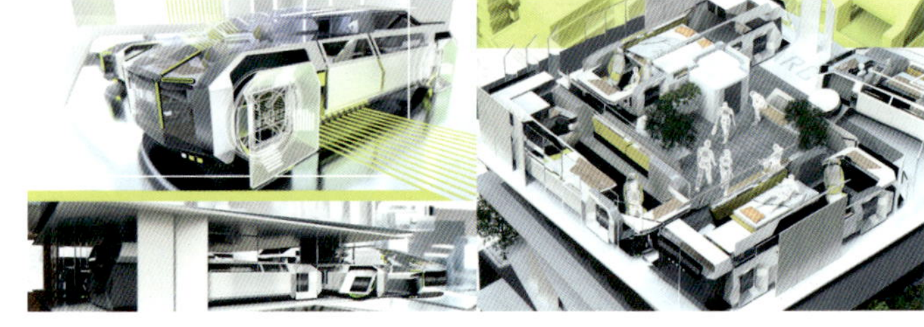

作　　者：薛立言
指导老师：刘　华
所在院校：南京理工大学

## 设计说明

ARCTO，作为未来派新型交通工具，通过其两款车型 X-ARC 和 ARCTO 001，实现城市生态与自然融合。X-ARC 拥有4种独特的模式，使其成为连接城市与自然的桥梁——飞行模式（可在城市或大自然中穿梭）、驾驶模式（可在道路上行驶）、露营模式（X-ARC 变形为小型露营基地）、自然模式（X-ARC 变形为融入自然的小型建筑），让人们在享受自然之美的同时，也不失舒适体验。

## Design notes

ARCTO, a brand of futuristic new vehicles, proposes a new way of integrating urban ecology and nature with its two models, X-ARC and ARCTO 001. X-ARC has four unique modes that make it a bridge between the city and nature —— flight mode (which can shuttle through the city or nature), driving mode (which can drive on the road), camping mode (X-ARC transforms into a small camping base), and natural mode (X-ARC transforms into a small building that blends into nature). Let people enjoy the beauty of nature at the same time, without losing the comfortable experience.

# 单腿坐姿滑行助行器

## Single Leg Sitting Sliding Walker

作　　者：杨　良
指导老师：于东玖
所在院校：南京工业大学

### 设计说明

单腿坐姿滑行助行器是一种以单腿、坐姿的方式进行滑行的助行器械，适合各个年龄段的侧小腿及以下受伤人群使用，填补市场上此类人群助行器械的空缺。产品可拆卸，尺寸规格较小便于携带；采用坐姿滑行的方式，使出行更加省力。同时车身配备减震结构，让出行更加舒适。

### Design notes

Single-leg sitting sliding walker is a kind of walking aid that can glide in the way of single-leg support and sitting posture. It is suitable for the use of injured people of lower leg and below at all ages. It fills the gap of walking aids in the market. The product can be disassembled, and its size is small and easy to carry; the way of sitting and sliding makes travel more labor-saving. At the same time, the car body is equipped with shock absorption structure to make travel more comfortable.

金奖

**银奖**

铜奖

优秀奖

# 自行式模块运输车及其夹具系统设计

## Design of Self-Propelled Modular Transport Vehicle and Its Fixture System

作　　者：董　轩　王英钒　马冰洁
张海龙　施新婕
指导老师：张乐凯　唐智川　李文杰
所在院校：浙江工业大学

### 设计说明

现有自行式模块化运输车的夹具系统仅提供转向功能，无法保证运输的管状物体在通过具有高度差的山路转角路段时处于水平状态，从而可能造成一定的损坏与财产损失。本产品提供了一种创新的夹具系统，它同时具有升降和旋转的功能，可以通过前后至少两个夹具系统的协同调节，使得管状运输物实时保持水平状态。

### Design notes

The existing fixture system of self-propelled modular transport vehicle only provides steering function, which can not ensure that the transported tubular objects are in a horizontal state when passing through the corner section of mountain road with height difference, which may cause certain damage and property loss. Provides an innovative fixture system, which has the functions of lifting and rotating at the same time. Through the coordinated adjustment of at least two fixture systems before and after, the tubular transport can be kept in a horizontal state in real time.

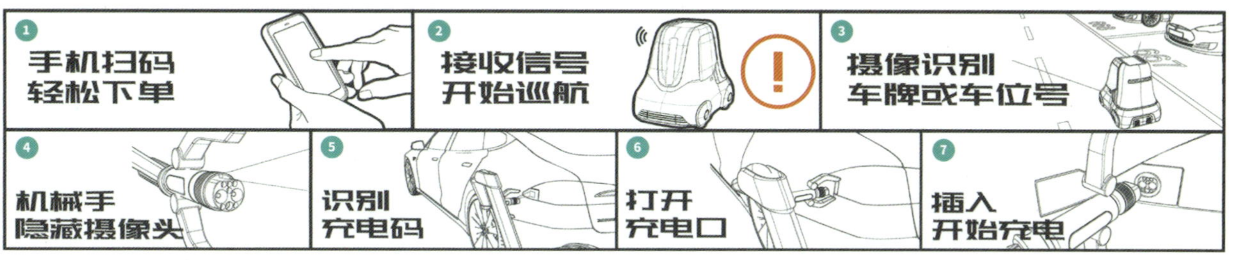


# SUPER-P ERRANT 分布式智能汽车充电系统

## SUPER-P ERRANT Distributed Intelligent Vehicle Charging System

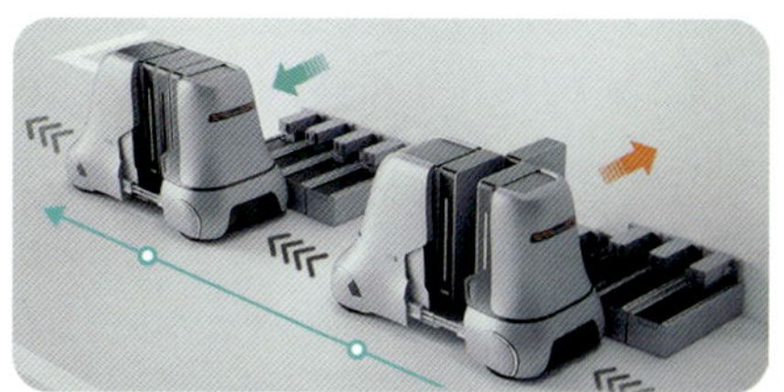

作　　者：王碧凌　赖志枨　袁宗炜
　　　　　范梦琳　黄之胤
指导老师：张祖耀
所在院校：浙江理工大学

### 设计说明

SUPER-P ERRANT 分布式智能汽车充电系统采用模块化设计，能够灵活适应不同场景的需求。该系统能够自动导航至指定的车位，并为已预约的电动车辆提供充电服务，有效解决了用户在找桩、用桩时面临的难题，提供了一个便捷的一站式解决方案。

### Design notes

The SUPER-P ERRANT distributed intelligent vehicle charging system adopts a modular design and can flexibly adapt to the needs of different scenarios. The system can automatically navigate to the designated parking space, and provide charging services for electric vehicles that have been booked, effectively solving the difficulties faced by users in building piles, finding piles and using piles, and providing a convenient one-stop solution.

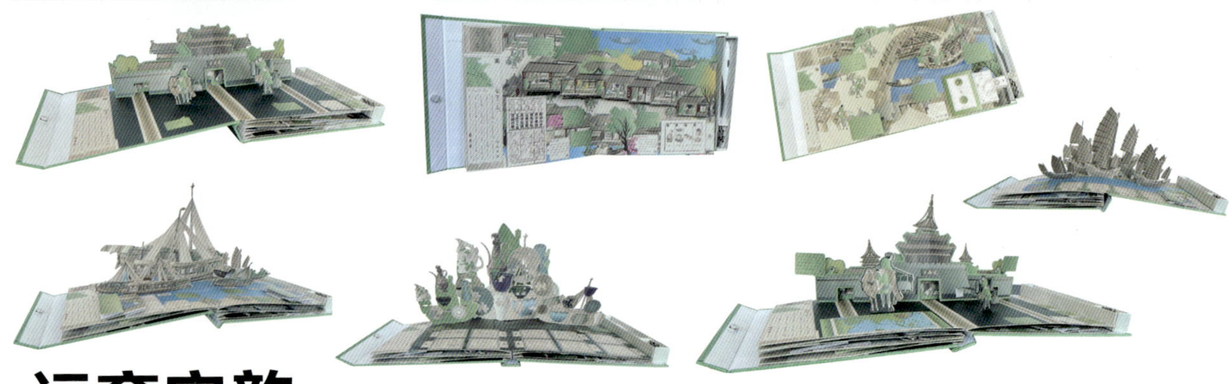



# 运育宋韵——立体书

## YUNYU SONGYUN —— Stereoscopic Book

作　　者：翁佳玲
指导老师：俞　凯
所在院校：浙江财经大学东方学院

### 设计说明

“运育宋韵”立体书文创产品汲取了千年宋韵及中国大运河深厚的文化底蕴，采用“书＋笔”的形式进行创意设计。本产品通过提炼两“Yun”文化中的“韵”与“运”来寻找文化间的融合。它以宋代的运河商贸文化作为故事主线，将宋代运河所孕育出的风景、民俗和风采之美巧妙融入。结合立体纸艺工艺、磁性拼图、手工DIY、寻宝游戏以及点睛之笔等互动功能，为读者提供充满趣味、互动性和陪伴感的新型阅读体验。通过这种创新方式，不仅展现了大宋时期因运河而繁荣的景象，还彰显了中华民族“开放”与“包容”的精神。

### Design notes

"Yun Yu Song Yun" stereo book creative products draw from the millennium Song Yun and China's Grand Canal profound cultural heritage, using the form of "book + pen" for creative design. This product seeks the integration between cultures by refining the "rhyme" and "yun" in the "Yun" culture (the pronunciation of the two words are "Yun" in Chinese). It takes the canal trade culture of the Song Dynasty as the main line of the story, and cleverly integrates the scenery, folk customs and beauty of style nurtured by the canal of the Song Dynasty. Combined with three-dimensional paper art, magnetic puzzles, DIY, scavenger hunt and other interactive functions, to provide readers with a new reading experience full of fun, interactive and accompanying sense. Through this innovative approach, it not only shows the prosperity of the Great Song Dynasty due to the canal, but also highlights the spirit of "openness" and "inclusiveness" of the Chinese nation.

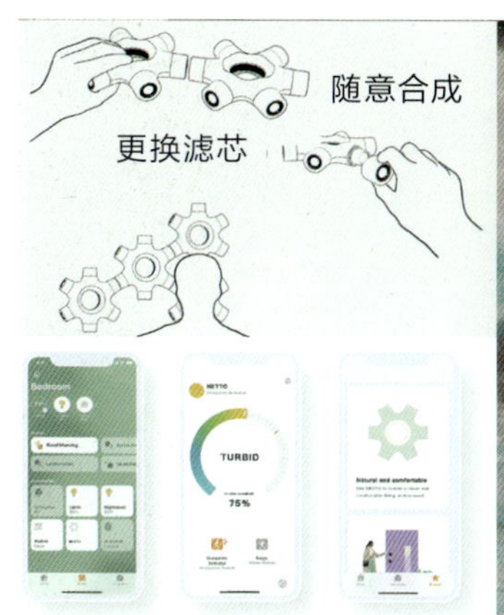

# NETTO——未来数智家居植物产品创新设计

# NETTO —— Innovative Design of Future Smart Home Plant Products

作　　者：陈晶晶　姜俊豪　倪子涵　张　烨
指导老师：包德福　沈丹妮　虞宇翔
所在院校：浙江理工大学

## 设计说明

NETTO 是一款探讨未来室内环境调理的数智家居情感化植物产品，其设计理念源于植物自我调节的天然机制，目的在于加深人们对植物作用的认知与理解。产品设计围绕“净”这一核心概念展开，拟用海藻在遭遇污染时颜色变化和净化能力的特性，通过磁性拼接的设计，用户可以根据实际环境和空间大小调整产品组合的数量。内置于产品的是用户检测模块、环境颗粒物监测模块及空气净化模块。空气由产品的“角”部分吸入，再从中心部分释放，同时系统能够识别空气中的污染物并将污染程度反馈到 RGB 灯带上，以此调节净化效率。NETTO 致力于利用技术简化用户交互，当激光传感器检测到用户接近时，产品会通过光的律动来引导用户并缓解他们的情绪。为了在保持界面简洁的同时提供必要信息，产品通过配套的 APP 应用和平台向需要详细数据的用户展示更多信息，实现了简约与实用的完美平衡。

## Design notes

NETTO is a digital home emotional plant product that explores the future indoor environment conditioning. Its design concept is derived from the natural mechanism of plant self-regulation, and aims to deepen people's cognition and understanding of the role of plants. The product design revolves around the core concept of "clean," using the characteristics of seaweed's color change and purification ability when exposed to pollution. Through the design of magnetic splicing, users can adjust the number of product combinations according to the actual environment and space size. Built into the product are the user detection module, environmental particulate monitoring module and air purification module. Air is inhaled from the "corner" part of the product and released from the center part, while the system is able to identify pollutants in the air and feed the pollution level back to the RGB lamp strip to adjust the purification efficiency. NETTO is committed to using technology to simplify user interaction. When the laser sensor detects the user's approach, the product guides the user through the rhythm of light and relives their mood. In order to provide the necessary information while keeping the interface simple, the product presents more information to users who need detailed data through supporting APP applications and platforms, achieving a perfect balance of simplicity and practicality.

# NUMEROSO——户外野营灯具

## NUMEROSO —— Outdoor Camping Lighting Fixtures

作　　者：屈靖琳　陆　野　钱佳蕊
指导老师：张乐凯
所在院校：浙江工业大学

### 设计说明

Numeroso 译为“众”，是一款针对户外野营的情感化氛围灯具，单独的 Numeroso 户外野营灯造型形似汉字“人”。当夜晚来临时，围坐在一起的朋友将 Numeroso 户外野营灯堆叠成一团“篝火”，又形似汉字“众”。 我们将野营过程中的团队合作精神融入到产品设计中，使产品更有归属感和意义。此外，Numeroso 户外野营灯能有效减少碳排放，达到碳中和，减少山林火灾的悲剧发生。

### Design notes

Numeroso, which translates as" 众 " ,is an emotional lighting fixture for outdoor camping.the single Numeroso outdoor camping light is shaped like the Chinese character for "人". When night comes, the friends sitting together stack the Numeroso outdoor camping lights into a "bonfire" , which also looks like the Chinese character for " 众 ". We integrate the team spirit in the process of camping into the product design, so that the product has a sense of belonging and significance.In addition, the Numeroso outdoor camping light is effective in reducing carbon emissions, achieving carbon neutrality and reducing the tragic occurrence of bushfires.

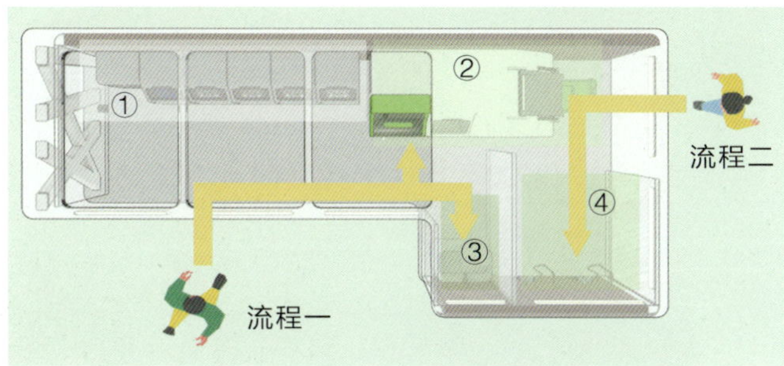


① 垃圾分类区
（不含厨余垃圾）

② 堆肥出肥区
（含厨余垃圾投放口）

③ 洗手区

④ 垃圾箱冲洗区

# Wisoil 社区智能垃圾分类服务系统

## Wisoil Community Intelligent Garbage Classification Service System

作　　者：余明洁　管伊朵
指导老师：张露芳
所在院校：浙江工业大学

### 设计说明

近年来，随着垃圾分类处理和资源化利用的不断推广，绿色环保理念已经逐渐深入人心。但是在城市环境中，厨余垃圾因其易腐烂、易变质和散发恶臭的特性，常常难以得到有效的资源化管理。针对这一问题，Wisoil 采用了先进的堆肥技术，打造了一套专注于厨余垃圾分散处理的社区垃圾分类服务系统。该系统能够实现厨余垃圾的就地堆肥化，并将所产生的营养土进行合理分配与应用，既减少了垃圾处理的环境负担，又提高了资源的循环利用率。

### Design notes

In recent years, with the continuous promotion of garbage classification and resource utilization, the concept of green environmental protection has gradually been deeply rooted in people's hearts. However, in the urban environment, kitchen waste is often difficult to get effective resource management because of its perishable, metamorphic and odorous characteristics. In response to this problem, Wisoil adopted advanced composting technology to create a community waste sorting service system focused on the decentralized disposal of kitchen waste. The system can realize the composting of food waste in situ, and distribute and apply the nutritive soil reasonably, which not only reduces the environmental burden of waste disposal, but also improves the recycling rate of resources.

# 气压平衡农用雨靴

## Air-Pressure Balanced Rainboat

作　　者：李　政
指导老师：张　萍
所在院校：合肥工业大学

### 设计说明

这是一款方便农民兄弟在水田工作中使用的雨靴，以便他们在泥泞的环境中更轻松地工作。在传统雨靴的使用过程中，尤其在水田等含淤泥的地方，靴子经常会被淤泥紧紧“吸附”，导致穿靴者在行走时非常费力。针对这一问题，设计时在鞋身外侧和鞋底内部增设了气管结构，这些气管从鞋口延伸至鞋底。当靴子从淤泥中拔出来的时候，外部空气可以通过这些气管进入鞋底，平衡鞋内外的压力差，消除真空区域，从而减少阻力，让使用者在淤泥中能够行走自如。

### Design notes

It is a rain boot that is convenient for the farmer brothers to use in the paddy field, so that they can work more easily in the muddy environment. In the use of traditional rain boots, especially in muddy places such as paddy fields, boots are often tightly "adsorbed" by silt, resulting in very laborious walking. To solve this problem, air pipes were added on the outside of the shoe body and inside the sole during the design, and these air pipes extended from the shoe mouth to the sole. When the boots are pulled out of the mud, the outside air can enter the sole through these air pipes, balancing the pressure difference between the inside and outside of the shoe, eliminating the vacuum area, thereby reducing resistance and allowing the user to walk freely in the mud.

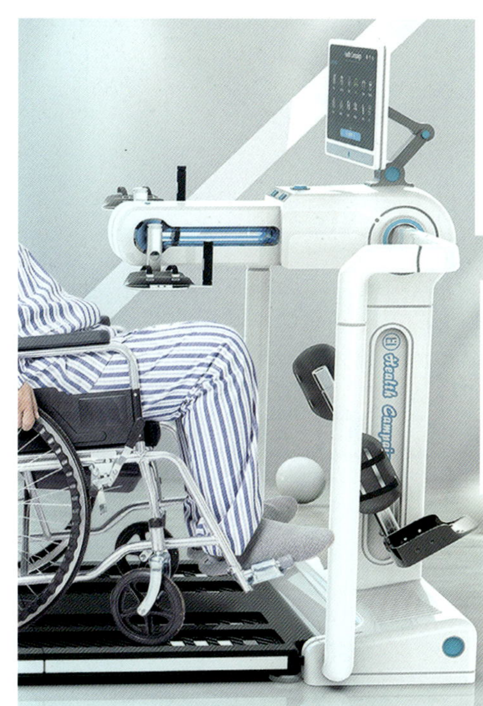

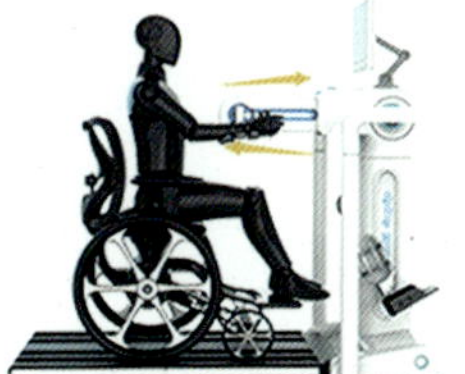

拉伸运动——肩关节

上旋运动——手关节

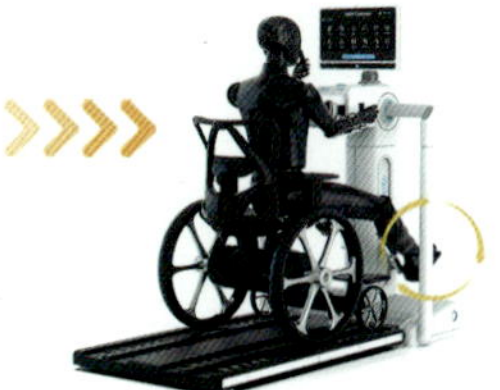

下旋运动——腿关节

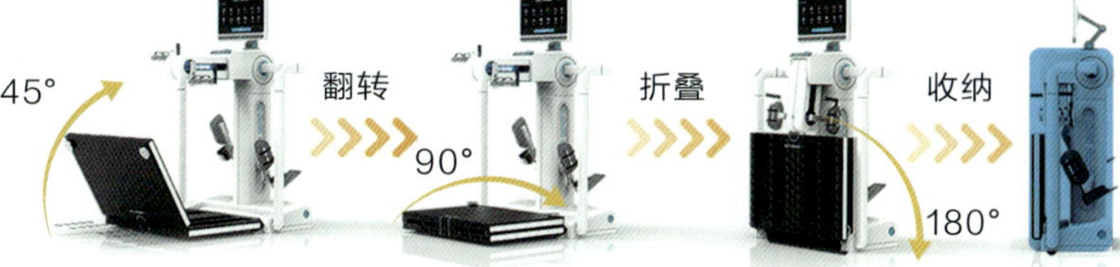


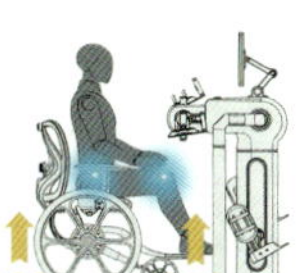

康复准备
肌肉放松

康复前期
手脚功能

结构转换
肩关节

康复后期
数据反馈

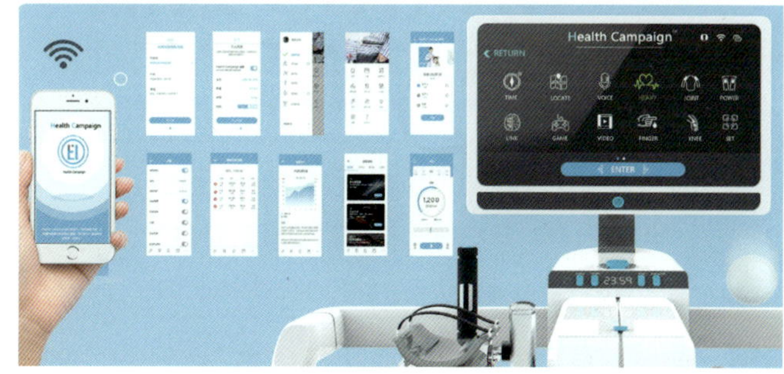


# HEALTH——公共康复综合锻炼器材设计

## HEALTH —— Design of Public Rehabilitation Comprehensive Exercise Equipment

作　　者：黄梓锋　洪浚栊　张小敏
指导老师：林　伟
所在院校：福州大学

### 设计说明

Health——公共康复综合锻炼器材设计，完善了残疾患者完整的康复流程，包括前期放松、主动与被动运动、后期数据反馈。通过一个简易的结构转化，使患者可以完成上旋下旋、前伸和后缩等运动，配合虚拟现实技术可重复练习，设置音乐、画面、文字以及语音提示功能，增加患者康复训练的乐趣。

### Design notes

Health —— the design of public rehabilitation comprehensive exercise equipment, which improves the complete rehabilitation process of patients, including early relaxation, active + passive exercise and later data feedback. It enables patients to complete up-down rotation, forward extension, backward contraction and other movements, cooperate with virtual reality technology to provide repeated practice, increase music, pictures, words and voice prompts to stimulate feedback, and increase the fun of patients' rehabilitation training. It has comprehensive functions, vivid and interesting.

金奖
**银奖**
铜奖
优秀奖

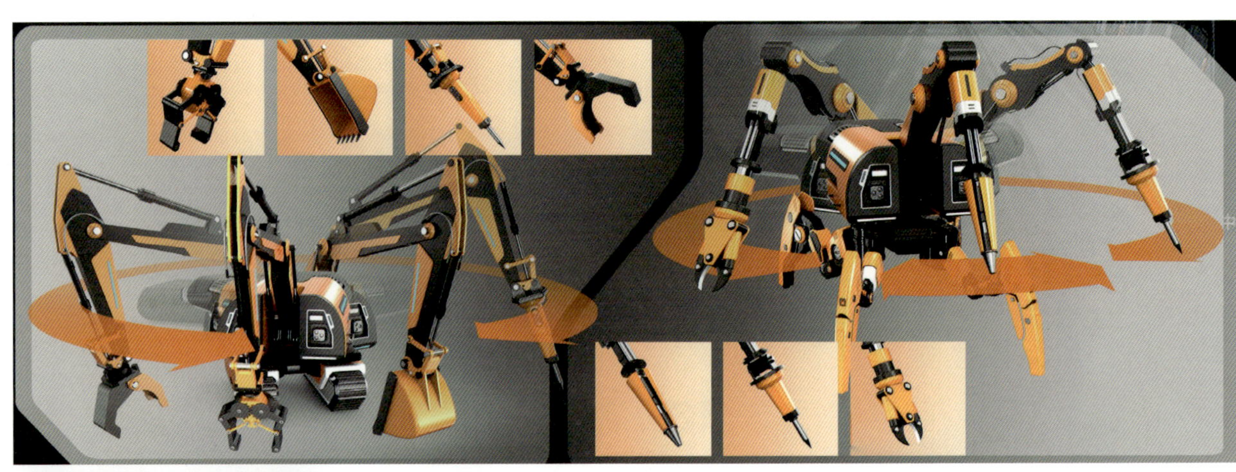

# G-Mantis 灾后救援破拆机器人

## G-Mantis Post Disaster Rescue and Demolition Robot


作　　者：谭启轲　朱俊衡　王玲玮

指导老师：宋玉凤

所在院校：山东科技大学


### 设计说明

针对传统灾后救援费时费力、危险性大等问题，采用模块化理念设计救援破拆机器人，通过搭载不同的行进模块与作业模块，适应不同的救援环境与作业需求，实现灵活高效的救援。路面较为平坦时，对外围建筑物进行破拆，采用履带行进模块搭配重工机械臂，完成如破拆、切割、物品搬运、解体等作业。在环境为坍塌的墙体废墟时，选用四足行进模块搭配轻型机械臂模块，便于机器人在坍塌物上行进，通过生命勘探装置定位受灾人员并展开施救作业，可配备金属剪、分裂器、侧摆装置、支撑器等工具头，可在狭小的空间灵活作业，完成钢筋剪切、分离与搬除构筑物、搭建支撑等作业，实现高效安全救援。此外，机器人还配备摄录像系统与照明系统，可进行远程操控。

### Design notes

In response to the traditional post-disaster rescue problems such as time-consuming, labor-intensive and dangerous, the modular concept is used to design the rescue and demolition robot, which can adapt to different rescue environments and operational needs by carrying different travel modules and operational modules to achieve flexible and efficient rescue. When the road is relatively flat, the peripheral buildings are broken, and the sandal belt travel module with heavy mechanical arm is used to complete operations such as breaking, cutting, item handling and disintegration. When the environment is collapsed wall ruins, the quadruped travel module with light mechanical arm module is used to facilitate the robot to travel on the collapsed objects, locate the affected people through the life exploration device and start rescue operations, and can be equipped with metal shears, splitters, side swing devices, supporters and other tool heads, which can work flexibly in a small space to complete steel shearing, separating and removing structures, building support and other operations to achieve efficient safe rescue. In addition, the robot is also equipped with video camera system and lighting system, and can be controlled remotely.

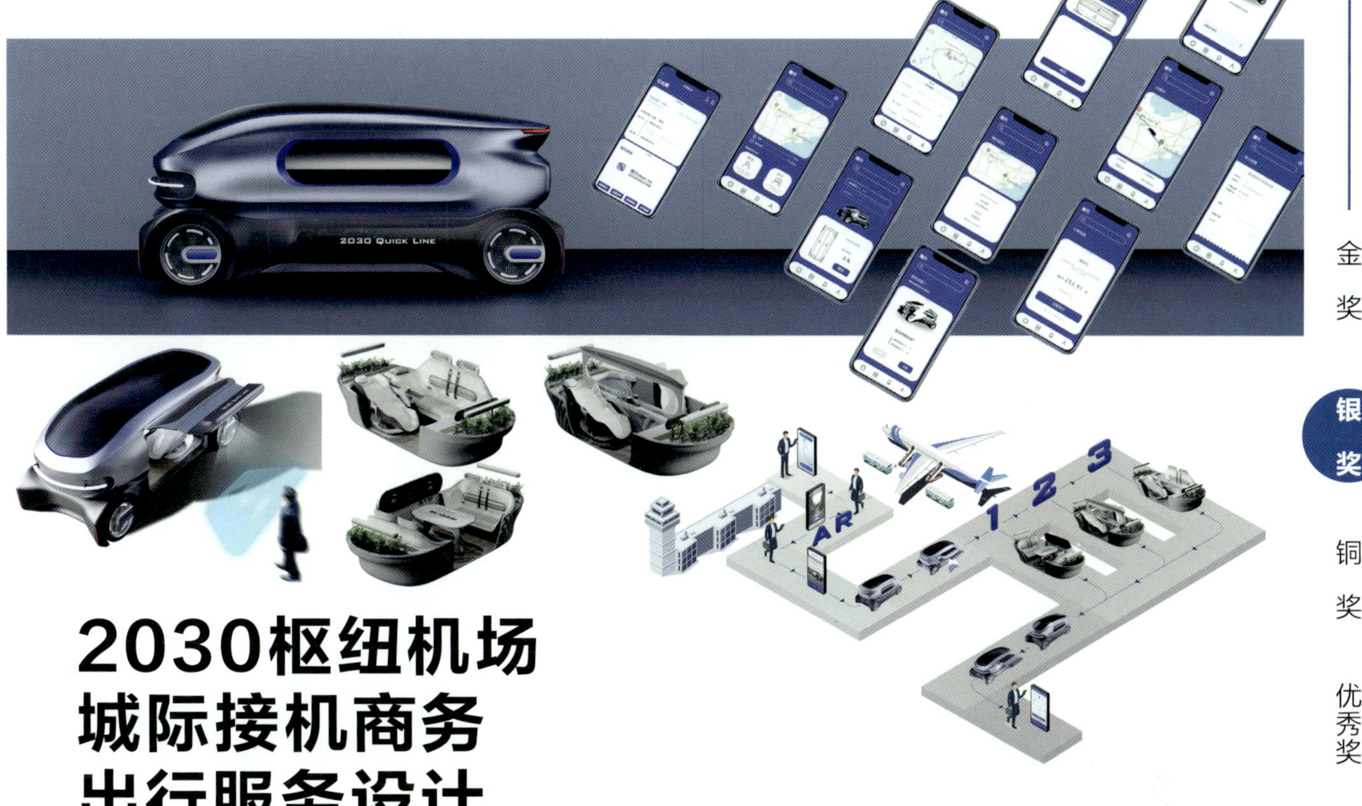


# 2030枢纽机场城际接机商务出行服务设计

## 2030 Hub Airport Intercity Pickup Business Travel Service Design

作　　者：王龙龙
指导老师：张　琳
所在院校：青岛科技大学

### 设计说明

该设计旨在解决城际接机专车出行中商务型旅客个性化需求未被满足的问题，并探索在2030年无人驾驶时代城际接机专车的服务模式。运用服务设计思维洞察“中高端”和“低端”商务型旅客两类用户的差异化需求，定义3种出行场景，通过“陪同出行”“多人同行”和“拼车出行”3种座舱方案予以满足，设计“捷行”品牌形象，优化移动端APP的预约与支付程序，在APP中引入AR实景导航以补充接机程序。无人驾驶汽车利用投影灯光、可视化屏幕和AI出行助手等创造多感官交互，利用AI出行助手和交互屏幕控制车内多功能座椅等车内设施，交付服务的使用、响应和关怀价值，打造无人驾驶场景下的城际接机商务出行服务模式。

### Design notes

The design aims to solve the problem of unmet individual needs of business passengers in intercity pick-up car travel, and to explore the service model of intercity pick-up car in the driverless era in 2030. Using the service design thinking to see the differential needs of "High-end" and "Low-end" business travelers, three travel scenarios are defined, to meet the requirements of the three cockpit schemes of "Escort travel" "Multi-person travel" and "Carpool Travel", to design the "Shortcut" brand image, to optimize the booking and payment procedures of the mobile APP, add AR Navigation to the APP to complement pick-up programs, and driverless cars use projection lights, visual screens, and AI travel assistants to create multi-sensory interactions, the use of AI travel assistant and interactive screen to control multi-functional seats and other facilities in the car, to deliver the use of services, response and care value, to create an intercity airport pick-up business travel service model under the driverless scene.

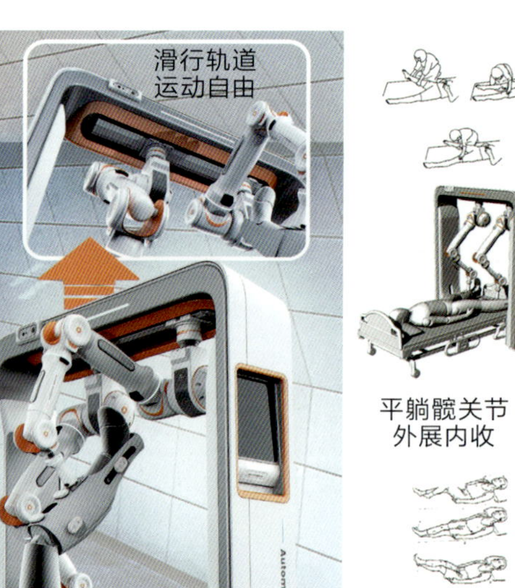
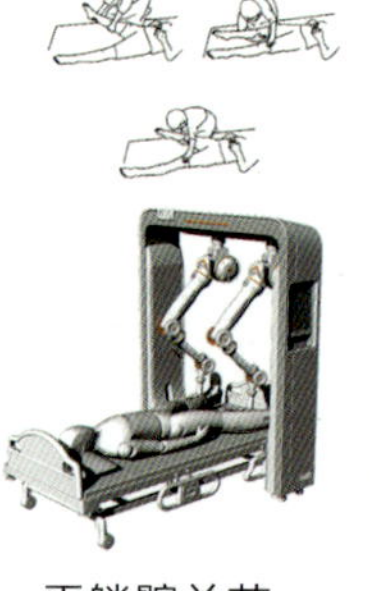

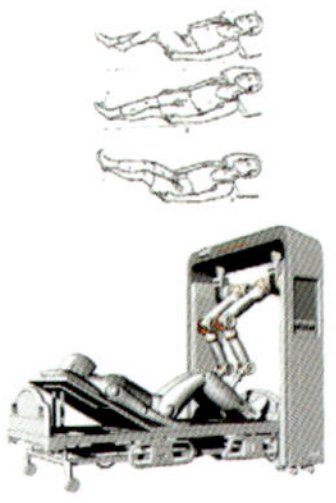
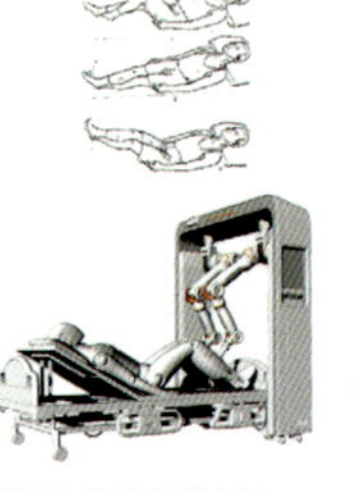
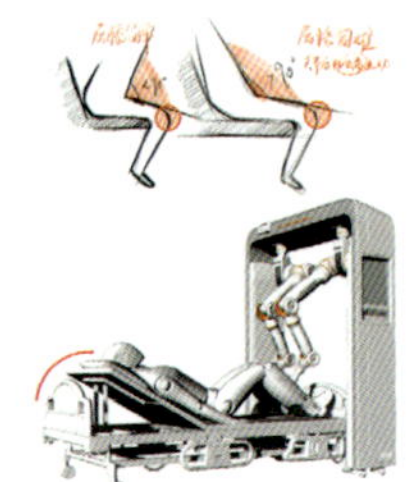


# 下肢游戏化自动音乐电脉冲康复训练器

## Gamified Automatic Music Electric Pulse Rehabilitation Trainer for Lower Limbs

作　　者：任紫涵　王龙龙
指导老师：付晓莉
所在院校：中原工学院

### 设计说明

这款专为下肢障碍的中老年脑瘫患者设计的下肢游戏化自动音乐电脉冲康复训练器，解决了传统人工辅助康复损耗精力的痛点，采取记忆性的康复系统，对于重复的康复动作，康复师仅需要示范一次让机器读取运动姿势，填补了早期床上康复的市场空缺。设备的使用贯穿患者康复的前5个周期，更好地衔接下床行走的后期康复。此外，设备可将患者的康复记录和个性化计划录入身份证中，将身份证插入操作屏幕下方的读卡槽即可导入患者信息。

为了缓解患者的心理压力，这款设备还配有音乐电脉冲腿套，医生可根据患者的具体情况制定音乐处方。随着下肢动作的改变给予患者听觉和触觉上的双重反馈效果。

### Design notes

This lower limb Gamification automatic music electric pulse training rehabilitation device designed for middle-aged and elderly cerebral palsy patients with lower limb disorders solves the pain points of energy loss in traditional artificial assisted rehabilitation, and adopts a memory rehabilitation system. For repeated rehabilitation actions, the rehabilitation teacher only needs to demonstrate once to let the machine read the movement posture. It fills the market vacancy of early bed rehabilitation. The use of equipment runs through the first five cycles of patient rehabilitation, better connecting the later rehabilitation of walking out of bed. In addition, the equipment can input the patient's rehabilitation record and personalized plan into the ID card, and insert the ID card into the card reading slot at the bottom of the operation screen to import the patient information.

In order to relieve the psychological pressure of patients, this device is also equipped with music electric pulse leg cover. Doctors can formulate music prescriptions according to the specific conditions of patients. With the change of lower limb movement, the patient will get double feedback effect on hearing and touch.

# MESM——模块化矿下CO智能检测处理装备设计

## MESM —— Design of Modular Underground CO Intelligent Detection and Processing Equipment

作　　者：张弼玥
指导老师：黄　晶
所在院校：河南理工大学

### 设计说明

目前，我国绝大多数煤矿工作面采用的是"U"型通风模式，这种模式往往导致上隅角区域通风不畅，风流容易形成涡流，从而引发一氧化碳（CO）的积聚问题，使得该区域的CO浓度频繁超标。针对这一问题，本设计聚焦于上隅角CO的消除方法和流程，通过结合具体的设计情境分析，并从系统的角度审视安全隐患的处理，提出了"消除＋抑制"CO抑消系统。该系统基于"三化"（即简化、优化和安全化）的设计原则，并融入了模块化设计理念，使得装备能够适应多种复杂的井下环境，有效减少安全事故隐患。

### Design notes

The vast majority of coal mine working faces in China are using the "U" ventilation mode, which often leads to poor ventilation in the upper corner area, and the airflow is easy to form eddy currents, which leads to the accumulation of carbon monoxide (CO), making the CO concentration in the area frequently exceed the standard. In response to this problem, this design focuses on the elimination method and process of CO in the upper corner, through the combination of specific design situation analysis, and from the perspective of the system to examine the treatment of security risks, put forward the "elimination + suppression" CO suppression system. The system is based on the design principles of "three" (i.e. simplification, optimization and safety), and incorporates a modular design concept, so that the equipment can adapt to a variety of complex downhole environments, effectively reducing safety risks.

金
奖

银
奖

铜
奖

优秀奖

# SHIFTER——多功能移位轮椅

## SHIFTER —— Multifunctional Wheelchair

作　　者：杨　怡　何润丰　龚雯璟
刘晓妍　王兆勇
指导老师：李晓英
所在院校：湖北工业大学

### 设计说明

随着社会人口老龄化问题的不断加剧，失能老年人的数量日益增加。这些老人由于身体行动障碍，往往需要照护者的帮助才能完成行动。长期依赖他人协助不仅增加了护理工作的压力，降低了照护效率，还可能因为操作不当带来二次伤害等安全风险。为此，SHIFTER应运而生，它能够辅助失能老人独立完成上下床移动、如厕、站立等日常生活动作，同时也可作为代步工具使用。通过提高失能老人的生活自理能力，SHIFTER不仅有效解决了他们的实际生活难题，提升了生活质量，也显著减轻了照护者的负担，提高了护理服务的效率。

### Design notes

With the increasing aging of the social population, the number of disabled elderly people is increasing. These elderly people often need the help of caregivers to complete the movement due to physical disabilities. Long-term dependence on the assistance of others not only increases the pressure of nursing work, reduces the efficiency of nursing care, but also may bring safety risks such as secondary injuries due to improper operation. SHIFTER is designed to help disabled people move up and down the bed, go to the toilet and stand on their own, and can also be used as a means of transport. By improving the self-care ability of disabled elderly people, SHIFTER not only effectively solves their practical problems in life and improves the quality of life, but also significantly reduces the burden of caregivers and improves the efficiency of nursing services.

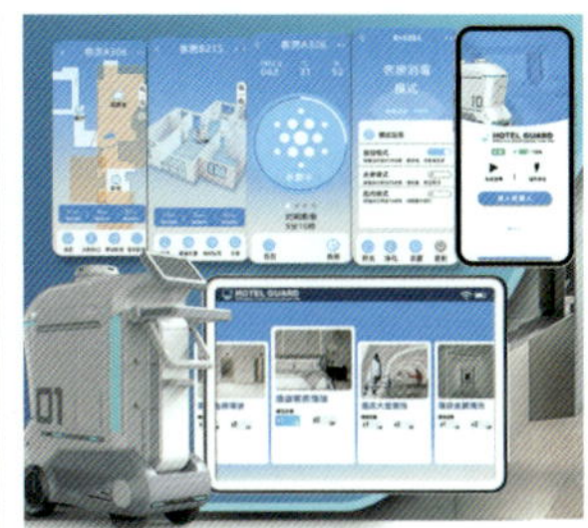

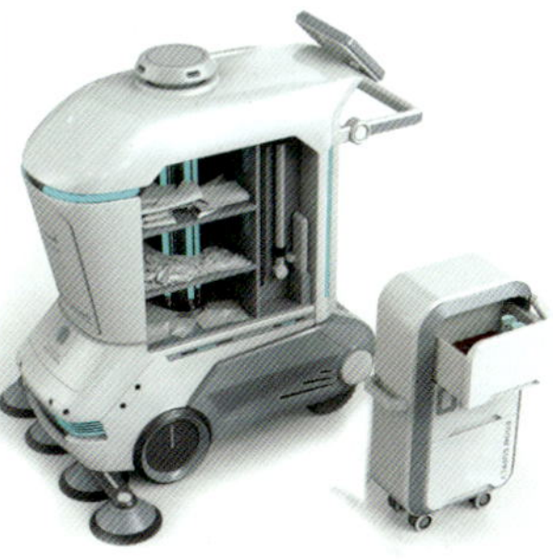

# HOTEL GUARD——酒店智能消毒清洁车设计

## HOTEL GUARD —— Design of Hotel Interlligence Disinfection and Cleaning Vehicle

作　　者：童　威　王　蕊
指导老师：石元伍
所在院校：湖北工业大学

### 设计说明

这款创新的酒店智能消毒清洁车集消毒机器人与手动清洁推车于一体。该产品依托机器人的先进计算能力和雷达传感器，实现了自动化与人工消毒模式的完美结合，极大地提升了清洁工作人员的效率和消毒效果。该消毒清洁车配备了可升降的紫外线装置，与空气净化器、消毒液和扫地设备相结合，能够对酒店环境进行高效的自动消杀。同时，高温高压蒸汽拖把和紫外线手持器为保洁人员提供了深度消毒清洁客房环境的有效工具。该产品采用了模块化设计，前端是一号模块，负责消毒清洁酒店公共区域；后端是二号模块，用于为客房提供日常补给。这一设计使得产品能同时满足酒店的消毒和服务需求。

### Design notes

This innovative hotel intelligent disinfection and cleaning vehicle is an integrated design that integrates disinfection robots and manual cleaning carts. The product relies on the robot's advanced computing power and radar sensors to achieve the perfect combination of automation and manual disinfection mode, which greatly improves the efficiency and disinfection effect of cleaning workers. The vehicle is equipped with an elevated UV device, which combined with an air purifier, disinfectant and sweeping equipment, can effectively and automatically eliminate the hotel environment. At the same time, high temperature and high pressure steam mops and ultraviolet handheld devices provide effective tools for cleaning staff to deeply disinfect and clean the guest room environment. The product has a modular design, with the front end being module 1, which is responsible for disinfecting and cleaning the public areas of the hotel, and the back end being module 2, which is used to provide daily supplies to the guest rooms. This design allows the product to meet both the disinfection and service needs of the hotel.

类叉车装置

车头底部装备类叉车装置，衔接负压舱底部空间，适配负压舱重量，牵引、放置负压舱

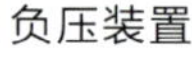

负压装置

位于负压舱顶端，经过滤系统过滤车内空气，再由排气系统排出

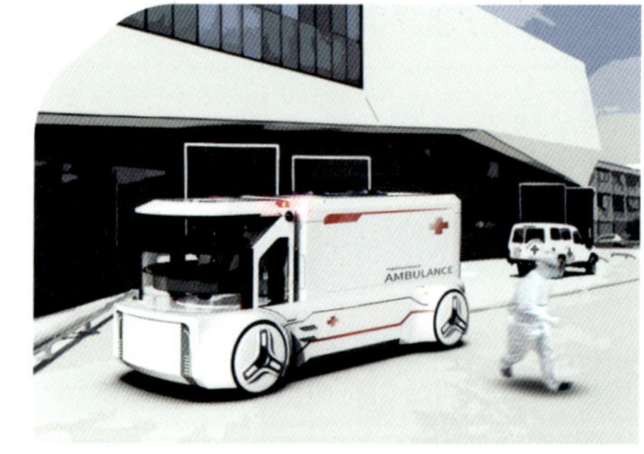

# 负压救护车概念设计

## Conceptual Design of A Negative Pressure Ambulance Based on COVID-19

作　　者：廖佳威　严祥玉
指导老师：杨　艺　郑素霖
所在院校：湖北美术学院

### 设计说明

为了提升对患者的服务质量，使医疗产品更好地实现医疗援助，并确保医护人员的安全，先对国内外先进的救护车技术进行了详尽的调研。通过分析配置布局、工艺材料和设计理念，同时结合方舱医院的特点和结构，设计了一款创新的模块化组装式负压救护车。

### Design notes

In order to improve the quality of service to patients, make medical products achieve better medical assistance, and ensure the safety of medical personnel, the advanced ambulance technology at home and abroad was investigated in detail. Through the analysis of configuration layout, process materials and design concept, combined with the characteristics and structure of the shelter hospital, an innovative modular assembled negative pressure ambulance was designed.

# 零碳集合体

## Zero Carbon Aggregation

作　　者：周奕博　颜潇晓　唐嘉鹏
指导老师：陈　丹
所在院校：湖南师范大学

### 设计说明

零碳集合体是一个碳中和主题展览馆，让建筑、人与碳中和产生联系，在此基础上向人们普及碳中和的概念与知识，以此服务社会。

我们将展览馆展区分为“过去”“现在”“未来”3部分，结合配套设计的互动游戏，通过空间和展陈营造不同时期环境氛围，并设计了“高－低－高”的情绪线。通过过去美好环境与当代环境现状形成对比，引发游客对环境问题的思考。最后通过“未来”展厅介绍改善环境的可行性，呼吁人们从改善身边环境做起！

### Design notes

Carbon neutrality, architecture, and people. We are thinking about the relationship and connection of the three.

Therefore, we want to build a carbon-neutral theme exhibition hall to connect buildings and people with carbon-neutral, and on this basis, popularize the concept and knowledge of carbon-neutral to people, so as to serve the society and people.

We divided the exhibition area of the exhibition hall into three parts: "the past", “the present" and "the future". We combined the interactive games designed to create the environment atmosphere of different periods through the space and exhibition display, and designed the "high-low-high" emotional line. Through the comparison of the past and the present situation of the beautiful environment, we triggered the tourists to think about the environmental problems. Finally, the "Future" exhibition hall introduces the feasibility of improving the environment and relevant means to look forward to the hope of improving the environment, and calls on people to act from around!

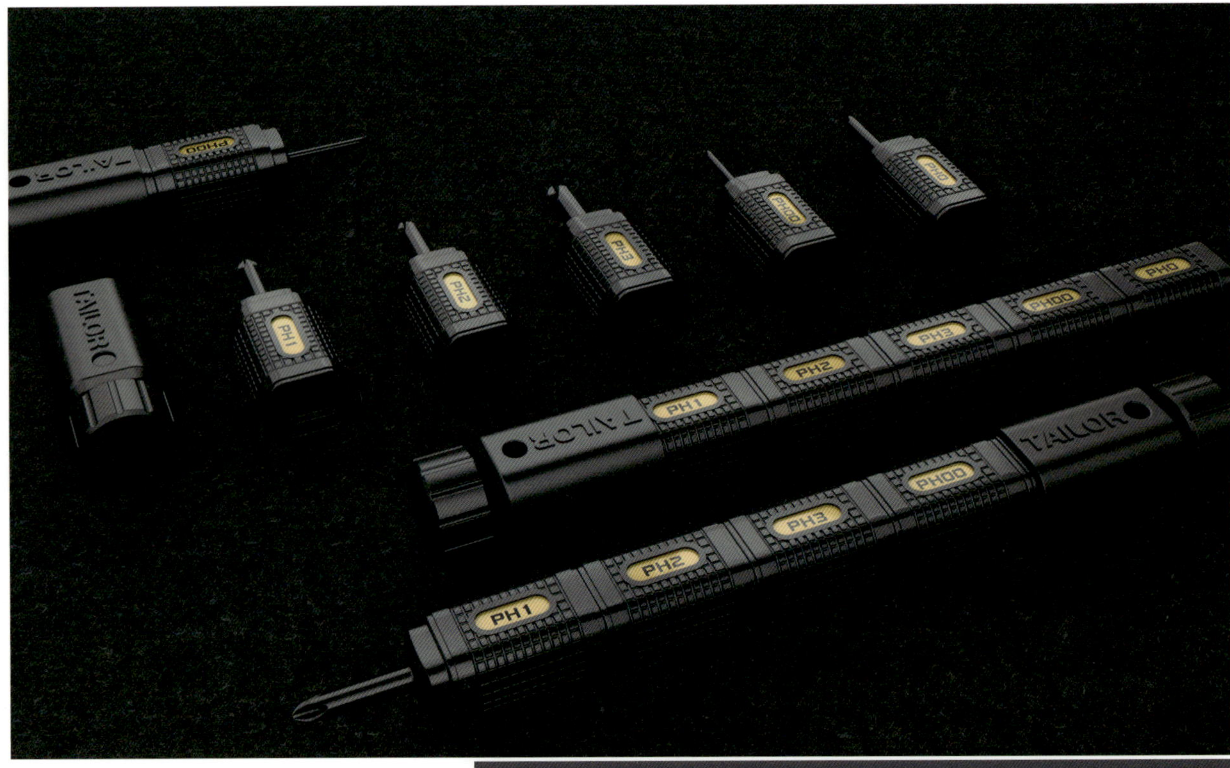

# 扣合型十字便携螺丝刀

# Snap-In Cross Portable Screwdriver

作　　者：吴思凯　向心仪
指导老师：戴　端　李志榕
所在院校：中南大学

## 设计说明

将传统十字螺丝刀的多种规格融合为一体，集成了PH00—PH3五种常用型号的十字螺丝刀。采用积木式的叠加设计，通过独特的异型几何卡扣，实现了前后模块的快速拼合。使用者只需简单地将所需的刀头置于最前端即可使用。这一产品设计避免了复杂的机械结构，便于组装和生产，同时具有成本低和便携的优点。

## Design notes

It integrates various specifications of traditional Phillips screwdrivers into one, and integrates PH00—PH3 five commonly used Phillips screwdrivers. The modular superposition design is adopted, and the front and back modules are quickly integrated through the unique special-shaped geometric buckle. The user simply places the desired batch header at the forefront to use it. This product design avoids complex mechanical structures, is easy to assemble and produce, and has the advantages of low cost and portability.

# RESCUERS——水下智能搜救装置

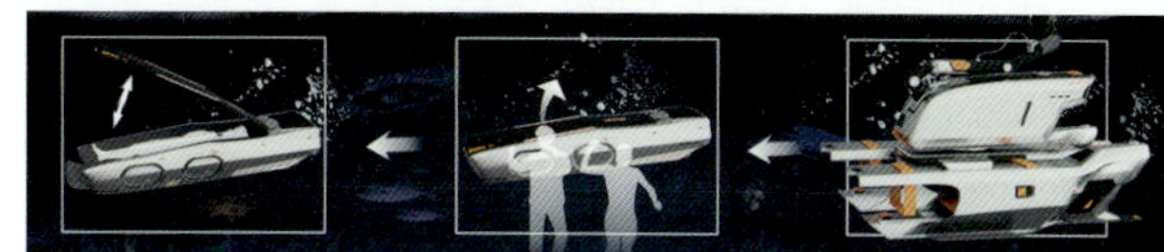

## RESCUERS —— Design of Underwater Intelligent Search and Rescue


作　　者：严　冲　阴　玲　龙润彬

劳杰成　龙启一帆

指导老师：张明星　柏小剑

所在院校：湖南科技学院


### 设计说明

大型水域水上安全事故一直是危及人类生命财产安全的重大隐患。因水上事故特别是大型水域环境的特殊性，救援工作往往难以有效开展，经常造成较大的生命财产损失，这其中包括相当比例的救援人员。本设计从江、湖、海等大型水域水上安全事故的紧急救援角度出发，通过智能无人搜查救援装置设计，以不同的功能部件应对不同的复杂环境，提高紧急救援的效率和成功率。同时采用无人化、远程化、智能化操作模式，在保证救援效率的同时确保救援人员的安全。

### Design notes

Large-scale water safety accidents such as shipwrecks and river disasters have always been a major hidden danger to the safety of human life and property. Water safety accidents have occurred frequently. Due to the special nature of water accidents, especially large-scale water environments, rescue work is often difficult to carry out effectively, often resulting in greater loss of life and property, including a considerable proportion of rescuers. This design starts from the perspective of emergency rescue of water safety accidents in large waters such as rivers, lakes, and seas. Through the design of intelligent unmanned search and rescue devices, different functional components are used to deal with different complex environments and improve the efficiency and success rate of emergency rescue. At the same time, the unmanned, remote and intelligent operation mode is adopted to ensure the safety of rescuers while ensuring the rescue efficiency.

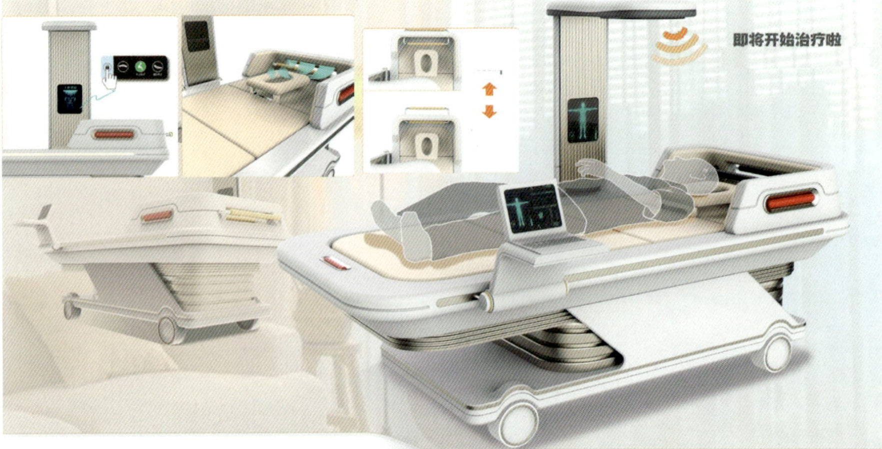


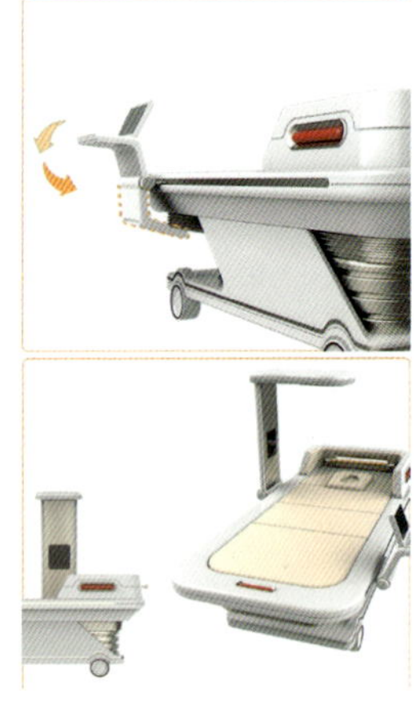

金奖

银奖

铜奖

优秀奖

# 移情视角下的社区老人康复理疗床设计

# Design of Rehabilitation Physiotherapy Bed for The Elderly in Community From The Perspective of Empathy

作　　者：杨　娟　郑　晨　王　巧
指导老师：那成爱
所在院校：湖南科技大学



## 设计说明

移情设计理论指的是站在用户需求的角度进行设计。众多研究表明，积极的心理状态可以改善治疗效果。现有的理疗床设计往往忽略了用户体验，这可能导致患者产生对治疗的抵触心理，故而在移情设计的理疗床的侧面配备了人脸识别和简易模式选择功能，增强了用户的参与度，简单需求不必实时呼求医护人员。此外，床顶装有视听反馈装置，能够在治疗过程中实时提醒患者治疗的开始、进程和预期效果，使患者对整个治疗过程有一个直观的了解，从而极大降低了对治疗的恐惧感。该理疗床还具备红外定位按摩、局部冷疗和热疗等常规理疗康复功能，医护人员可以通过操控屏监测数据、记录治疗信息，并控制紧急停止按钮，以进一步确保患者的安全。

## Design notes

Empathic design theory refers to the design from the perspective of user needs. Numerous studies have shown that positive mental states can significantly improve the therapeutic effect. Existing therapy bed designs often neglect the user experience, which can lead to patient resistance to treatment. Therefore, the side of the physiotherapy bed designed with empathy is equipped with face recognition and simple mode selection functions, which enhances the user's participation, and simple needs do not need to call the medical staff in real time. In addition, the top of the bed is equipped with audio-visual feedback devices, which can remind patients of the start, progress and expected effects of treatment in real time during the treatment process, so that patients have an intuitive understanding of the entire treatment process, thus greatly reducing the fear of treatment. The physiotherapy bed also has conventional physiotherapy rehabilitation functions such as infrared positioning massage, local cold therapy and heat therapy. Medical staff can monitor data, record treatment information, and control the emergency stop button through the control screen to further ensure patient safety.

吸引蝗虫

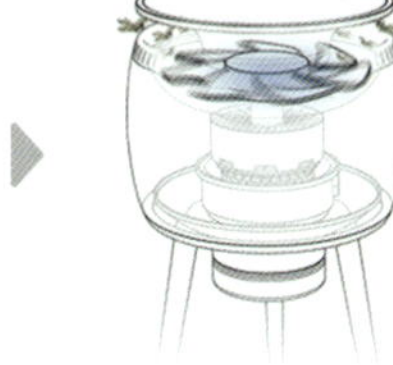
风扇吸入

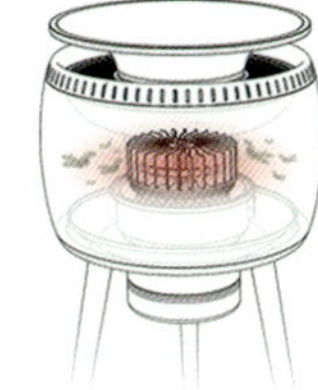
脱水干燥

搅拌粉碎

昆虫蛋白粉



# 蝗虫食物转化机

## From Locust to Nutrition

作　　者：汪启韬　郑小涵　陈文静　谢晓文
指导老师：陈朝杰
所在院校：广东工业大学

### 设计说明

蝗灾可能导致粮食产量下降30% 至50%，从而使1900万非洲人面临严峻的粮食安全威胁，为此我们设计了一款将蝗虫转化为食物资源的蝗虫食物转化机。该机器使用化学物质4VA 吸引并诱捕蝗虫，这种物质能够引诱蝗虫群聚。随后，通过风扇产生的气流，捕获的蝗虫被吸入机器中并迅速干燥。干燥后的蝗虫接着被磨成昆虫蛋白粉，这种蛋白粉可作为食物来源，为非洲食物短缺地区的居民提供必需的营养。

### Design notes

Because locust infestation can lead to a 30 to 50 percent drop in food production, putting 19 million Africans at serious risk of food security, so we propose an innovative solution to turn locusts into a food resource. The scheme involves attracting and trapping locusts using the chemical 4VA, which can lure them into swarms. Then, through the airflow generated by the fan, the captured locusts are sucked into the machine and quickly dried. The dried locusts are then ground into insect protein powder, which can be used as a food source to provide essential nutrients to people living in food-shortage areas in Africa.

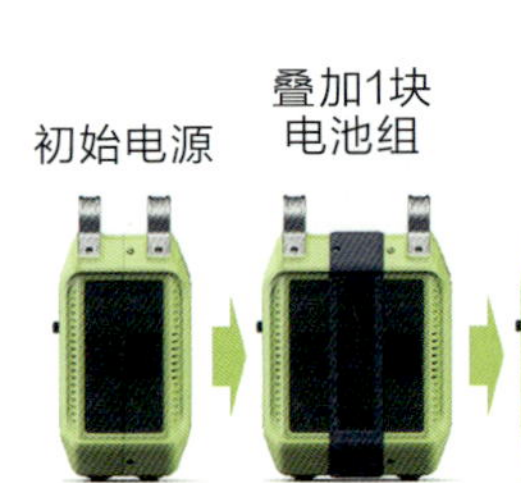

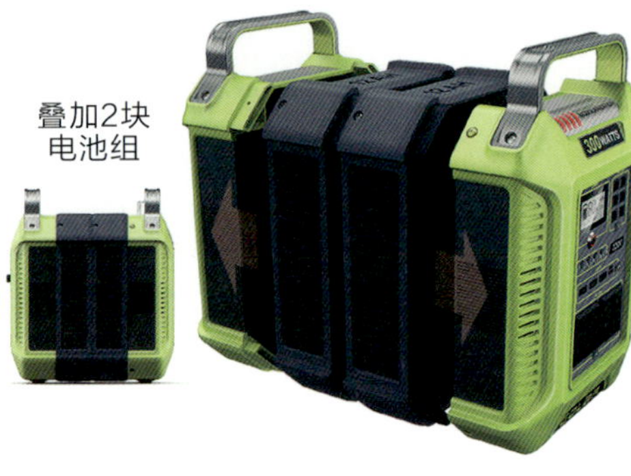

# 非洲家用太阳能供电设备



## Solar Generator for Africa Home Use

作　　者：林宗建
指导老师：杨　帆
所在院校：广州美术学院

### 设计说明

该作品是基于无电网地区的特殊环境下设计的家用太阳能供电设备，它采用了模块化设计，将电池组模块与主机模块分开，使产品能够在不淘汰旧设备的前提下，根据用户用电需求的增长灵活增加电池容量。这样既延长了设备的使用寿命，又降低了用户的使用成本。同时产品采取分期付款的商业模式，帮助那些因设备价格昂贵而难以一次性支付的用户解决了购买难题。此外，设备还特别设计了收音机学习播报功能，使因故无法正常上学的学生可以在无电网地区通过设备接收学校播报的课程内容。

### Design notes

The work is based on the BOP off-grid area in the special background of the design of the home solar power supply equipment, it uses a modular design, with the battery pack module and the host module separating, so that the product can not eliminate the old equipment, the premise of flexible increase in battery capacity according to the growth of the user's electricity demand. This not only extends the service life of the equipment, but also reduces the cost of the user. At the same time, the product adopts the business model of installment payment, which helps those users who are difficult to pay once because of the high price of equipment to solve the purchase problem. In addition, the equipment has been specially designed with a radio learning broadcast function, so that students who are unable to go to school normally under the impact of the epidemic can receive school and government broadcast curriculum content through the equipment in BOP off-grid areas.

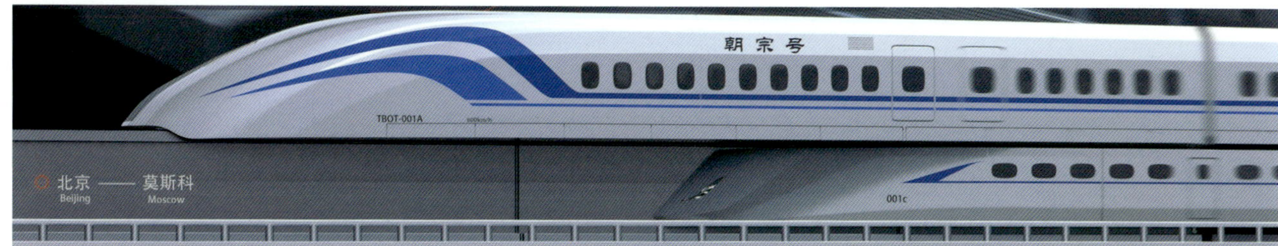


# 三体一轨不停换乘磁浮交通系统

## Three Body One Track Non-Stop Transfer Maglev Transportation System

作　　者：曹日龙　邹　瑞　易歆普
常真毓　王家统
指导老师：向泽锐
所在院校：西南交通大学

### 设计说明

该设计方案为面向世界的洲际磁悬浮列车的优化设计，以解决换乘等待时间长、不方便的问题。采用常导电磁磁浮技术，最高时速达600km/h，整体分为ABC3部分（三体）共用一个异形轨道，通过控制运行速度，使B/C脱离A后加速追赶至与目标换乘的车体A^达到相同速度保持相对静止后对接，通过开放换乘通道，完成不停换乘的功能。

### Design notes

This design scheme is an optimized design for intercontinental maglev around the world, to solve the problems of long transfer waiting time and inconvenient transfer. Using constant conduction electromagnetic levitation technology, with a maximum speed of 600km/h, it is divided into three parts (three bodies) and shares a shaped track. By controlling the running speed, B/C separates from A and accelerates to catch up with the target transfer vehicle A^. After reaching the same speed and maintaining relative stillness, it docks and opens the transfer channel to complete the function of non-stop transfer.

# 草方格沙障固沙车

## Grass Square Grid Sand Barrier Sand Fixation Vehicle

作　　者：黄瑞淇
指导老师：徐　刚
所在院校：四川轻化工大学

## 设计说明

治沙先固沙，设置草方格沙障是固沙的有效措施。但现阶段多为人工铺设，为减轻人工劳动强度并提高草方格沙障铺设的效率和质量，本设计是一款以自动压草播种为核心的草方格沙障铺设车，采用低矮宽大的车身，提高压草精度，加强转向能力和爬坡能力，可以在坡度更陡更高的沙丘上工作，提升铺设范围。通过机械和人工的通力合作，进一步加快土地沙漠化治理，为大自然做出一份贡献。

## Design notes

It is an effective measure of fixation by setting up the straw checkerboard . However, at the present stage, it is mostly manual operation. In order to reduce the manual labor intensity and improve the efficiency and quality of straw checkerboard sand barrier laying, I designed a straw checkerboard sand barrier laying vehicle with automatic grass pressing and sowing as the core. The low and wide body can improve the accuracy of the grass pressing, and strengthen the steering ability and climbing ability. It can work on the steeper and higher sand dunes and improve the laying range. Through the mechanical and artificial cooperation, to further accelerate the management of land desertification and make a contribution to the nature.

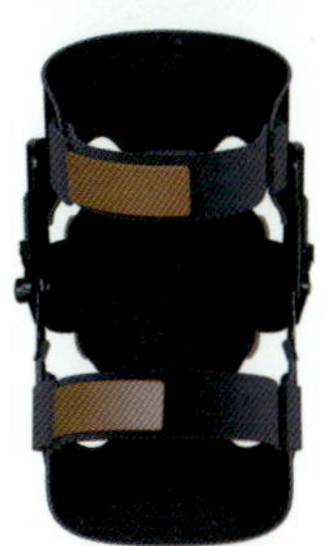

# 无源助力护膝

## Passive Assist Knee Pads

作　　者：靳创源

指导老师：闫　雯

所在院校：西安理工大学

### 设计说明

这是一种无动力源的辅助护膝。本产品专为膝关节损伤或退行性关节病变的人群设计，它能够帮助人们减轻运动过程中的疲损。当膝盖弯曲时，其内部机构能够将动能转换为势能，并在膝盖伸直时释放这些能量，从而起到预防或改善关节损伤的作用。

### Design notes

This product is designed for people with knee injury or degenerative joint disease. It is an auxiliary knee support without power source. It can help reduce fatigue during exercise, and when the knee is bent, its internal mechanism is able to convert kinetic energy into potential energy and release this energy when the knee is straightened, thus playing a role in preventing or improving joint injury.

# 基于光储直柔的办公建筑能流可视化设计

## Visualization Design of Energy Flow in Office Buildings Based on Light Storage, Direct Flexibility

作　　者：任婧轩
指导老师：张　军　俞　准
所在院校：湖南大学

### 设计说明

本设计是在当前“双碳”目标和能源结构调整大背景下，针对光伏建筑一体化办公场景能源管理控制系统的能流可视化设计。通过对管理系统与服务系统的梳理，协助区域负责人员理解园区能流状况，以更好的组织和可视化形式，方便其对能源的集中监控、管理以及分散控制。通过打通管理与员工信息渠道，满足柔性调整需求和综合能源的切换与利用，助力运维人员进行高效精细化管理。

### Design notes

Under the background of dual-carbon strategy and energy structure adjustment, the design is aimed at the energy management and control system of an enterprise's building-integrated photovoltaics office environment. Through the management system and the service system combing, the design assists the area responsible personnel to understand the park energy flow condition, to facilitate its centralized monitoring, management and decentralized control of energy resources in a better organizational and visual form, and to meet flexible adjustment needs and the switching and utilization of integrated energy resources by opening up channels for management and employee information, assist operation and maintenance personnel to carry out efficient and fine management.

# 冰种计划
# 北极熊海冰平台

## Self-Growing Ice

作　　者：胡妙笛　黄钰准　林科延
指导老师：张　军　王　魏　卢继武　俞　准
所在院校：湖南大学

### 设计说明

全球变暖导致北极浮冰融化，北极熊栖身地逐渐减少。“冰种计划”是一款能在水下制冰的漂浮平台。以极地富裕的阳光作为能源，重力球释放使富集绳垂落，海水中的杂质聚集促进海水结冰，冰体沿着富集绳逐渐凝结成提供更大浮力的海下冰川，为北极熊提供了赖以生存的冰面栖息环境。

动力模块与漂浮模块共同运作，快速部署，两种模块可拼接组合成不同形式以适应北极海冰环境。吸引灯利用海洋生物趋光性为北极熊提供食物。摄像头供工作人员远程监控周围环境、控制移动路径，并可为北极熊拍摄生活纪录片。“冰种计划”海冰平台为北极熊提供了更好的生存环境，以应对气候变暖和冰川融化带来的威胁。

### Design notes

Global warming has caused the Arctic ice floes to melt, and the platforms on which polar bears shelter and feed have almost completely disappeared, increasing the risk of drowning by swimming long distances. "Self - Growing Ice" is a floating platform that can make ice underwater. Using the rich polar sunlight as a clean energy source, the gravity sphere is released to make the enrichment ropes drop, impurities in the seawater collect to promote seawater freezing, and the ice body gradually condenses along the enrichment ropes to provide more buoyancy to the underwater glacier, providing the polar bears with the ice habitat to survive the threat of climate warming and glacial melting.

The power module and the floating module operate together for rapid deployment, and the two modules can be spliced and combined into different forms to suit the Arctic sea ice environment. Attractor lights provide food for polar bears by taking advantage of the phototropism of marine life. Cameras allow staff to remotely monitor their surroundings, control movement paths, and take documentary footage of polar bears' lives. "Self - Growing Ice" provides a better environment for polar bears to survive.

# COMMA——模块化居家健身产品设计

## COMMA — Modular Home Fitness Product Design

作　　者：叶琦钧　张德寅　芮绮苒
指导老师：霍春晓
所在院校：南京艺术学院

### 设计说明

作为一款模块化居家健身设备，COMMA打破了传统健身器材握把与配重的固定关系（将配重置于握把之间），模块化的拼接设计，成功地提高了新手用户使用健身器材时的舒适度、安全性，多个模块也使得新手用户可训练到多个部位。采用可扭动的把手设计，使得用户可以轻松改变抓握角度，提高手腕舒适度并减少受伤概率。COMMA采用柔和的造型语言，在使用钢材与透明硅胶材料的基础上，完美体现了健身器械的柔和感，力图做到与居家环境的融合。

### Design notes

As a modular home fitness equipment, COMMA breaks the fixed relationship between the grip and counterweight of traditional fitness equipment (reset the configuration between the grips). The modular splicing design successfully improves the comfort and safety of novice users when using fitness. Multiple modules also enable novice users to train any part of body. With a twritable handle design, users can easily change the grip angle, improve wrist comfort and reduce the probability of injury for novices. COMMA adopts a soft modeling language. Based on the use of steel and transparent silicone materials, it perfectly reflects the softness of fitness equipment, and strives to achieve the integration of the living home environment.

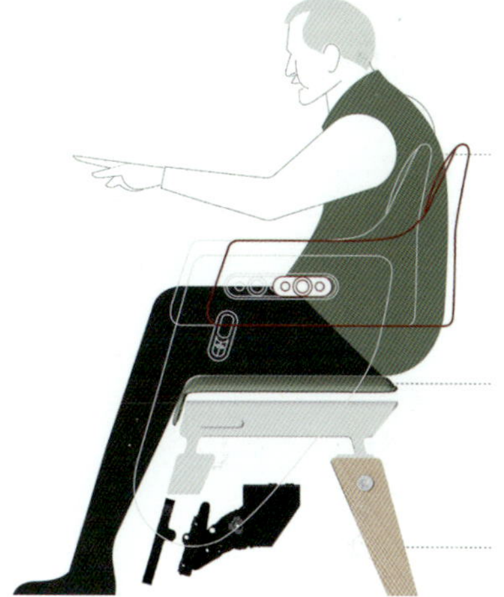

# 市井青雀——棋牌室健康座椅

# CITY GREEN SPARROW — Chess Room Seat for Health

作　　者：孙嘉霖
指导老师：霍春晓
所在院校：南京艺术学院

## 设计说明

市井青雀专注于改善老年社区娱乐设施，从健康角度出发重新设计棋牌室座椅，以减少老年人在娱乐活动中易患的下肢深静脉血栓风险，助力打造一个健康安全的老年社区娱乐场景。

## Design notes

City Green Sparrow focuses on improving the entertainment environment of the elderly community, especially the redesign of the chess and card room seats from the perspective of health, in order to reduce the risk of lower extremity deep vein thrombosis that the elderly are prone to in recreational activities, reduce the possibility of aggravating the risk, and help create a health management friendly entertainment scene for the elderly community.

# 全国大学生工业设计大赛
# 优秀作品集·2022

Collection of Award-winning Works of
China Universities Industrial Design Competition
2022

# 铜奖

# Bronze Award

# 胰岛素注射器套装设计

## Insulin Syringe Set Design

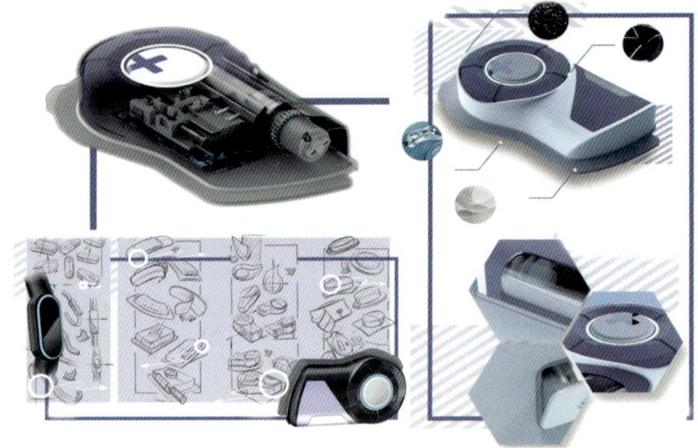

作　　者：马开鑫　贺湟溶　史倩文
高　天
指导老师：姜　蕊
所在院校：北方工业大学

### 设计说明

本设计是针对老年人群体的胰岛素注射套装，它充分考虑老年人身体机能衰退等各类特点，结合材料学、人机工程学、人机交互等设计原则，方便年迈的糖尿病患者注射胰岛素。

### Design notes

Diabetes is a common chronic disease that affects about one in ten adults. This project is designed for an insulin injection kit for the elderly. With full consideration of various characteristics such as the decline of the physical function of the elderly group, and in combination with the design principles of material science, ergonomics and human-computer interaction, this suit is designed to facilitate the insulin injection of elderly diabetic patients.

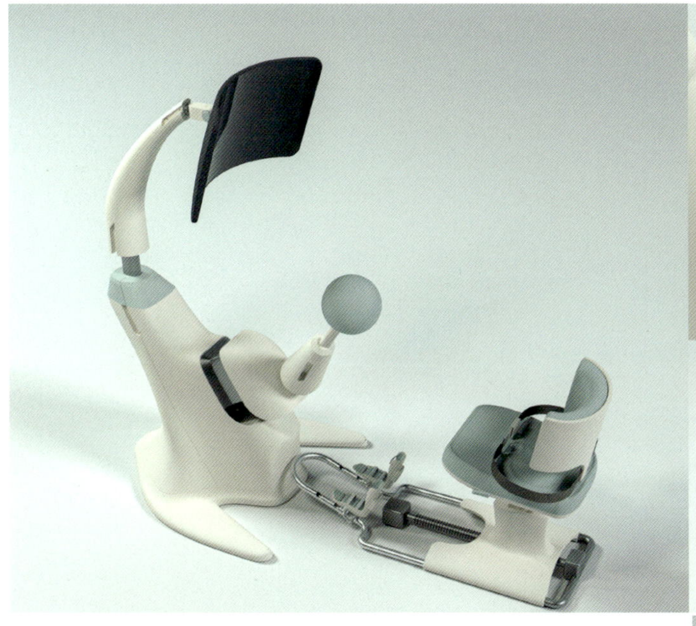

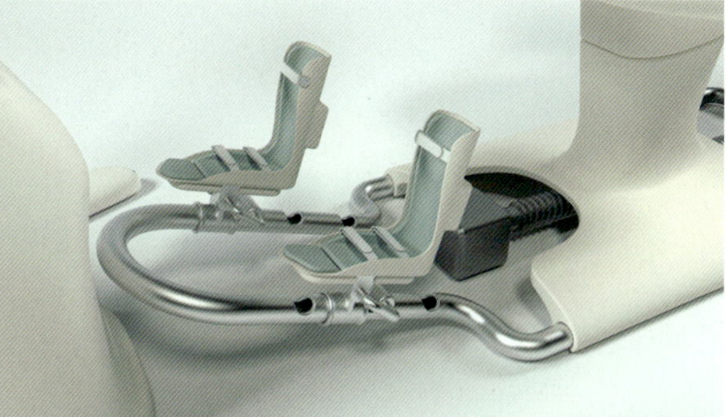

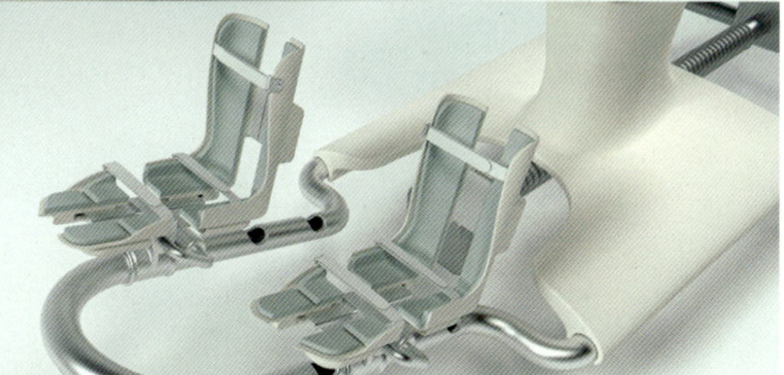

# 杜氏肌营养不良患者辅助康复训练产品设计

## Design of Auxiliary Rehabilitation Training Products for Patients with Duchenne Muscular Dystrophy

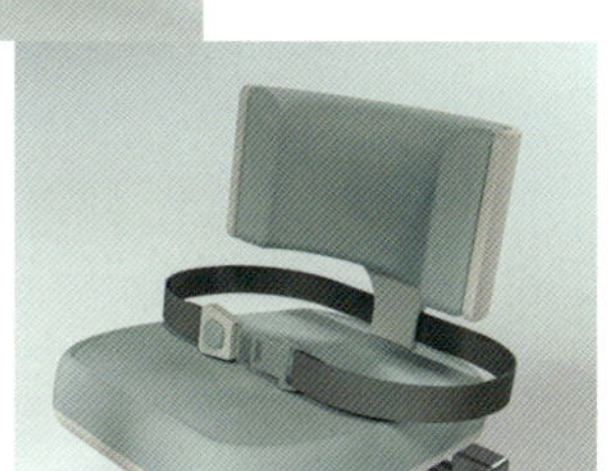

作　　者：卢　灿　秦　川　贾佳怡
指导老师：金　鑫
所在院校：北京工业大学

### 设计说明

杜氏肌营养不良（DMD）是一种遗传性肌肉萎缩病，人工复健是DMD康复训练中最常见的训练手段。为解决目前DMD患者缺少辅助康复训练器材、康复训练方式单一以及患者无法体会运动趣味性的问题，对产品进行外观、功能、结构以及交互流程的设计，保证了患者上肢在三维空间范围内的锻炼，使患者的关节保持灵活、肌肉保持力量。结合交互游戏进一步增加DMD患者训练的趣味性，满足了趣味性训练的设计要求。

### Design notes

Manual rehabilitation is the most common means of training in DMD rehabilitation. In order to solve the problems of the lack of assisted rehabilitation training products for DMD patients and the single way of rehabilitation training makes the patients unable to experience the interest of sports, this product designs in terms of appearance, function, structure and interaction process. The final solution ensures the exercise of patients' upper limbs in a three-dimensional space, so that patients' joints remain flexible and muscles maintain strength. Through interactive games, this design fulfilled the further increase which meets with the requirements of interest training.

金
奖
银
奖
铜
奖
优秀奖



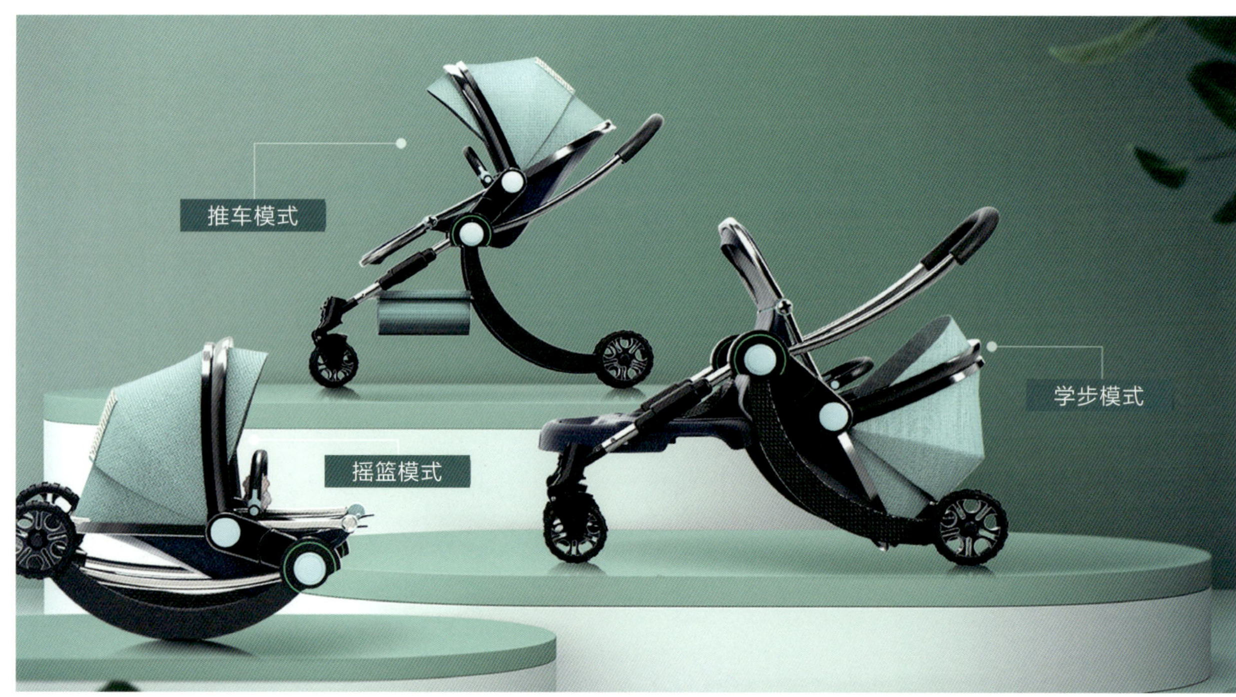


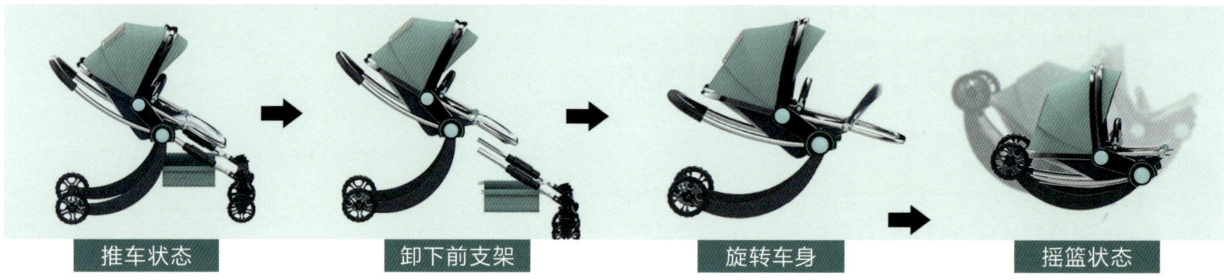


# 可拼装式多功能儿童车

## Assembled Multifunctional Children's Car

作　　者：吴成富
指导老师：朱天阳
所在院校：北京化工大学

### 设计说明

可拼装式多功能儿童车针对8岁以内儿童进行多场景的产品整体服务体验设计。这款产品可在婴儿推车的基础形态上，通过拼装拆卸转变为摇篮车形态、学步车形态、滑板车形态来满足儿童不同阶段的使用需求，既节省了频繁换车导致的家庭空间问题，也更好地满足了父母与儿童之间的情感交流。

### Design notes

The integrable multi-functional children's car is designed for the overall service experience of the product in multiple scenes for children under eight years old from the perspective of user experience. On the basis of the stroller, through assembling removal, this product can be transformed into a bassinet, baby walker, skateboards car to meet the demand of children at their different stages of use, and better save family space problem caused by the frequent transfer, and better meet the emotional communication between parents and children.

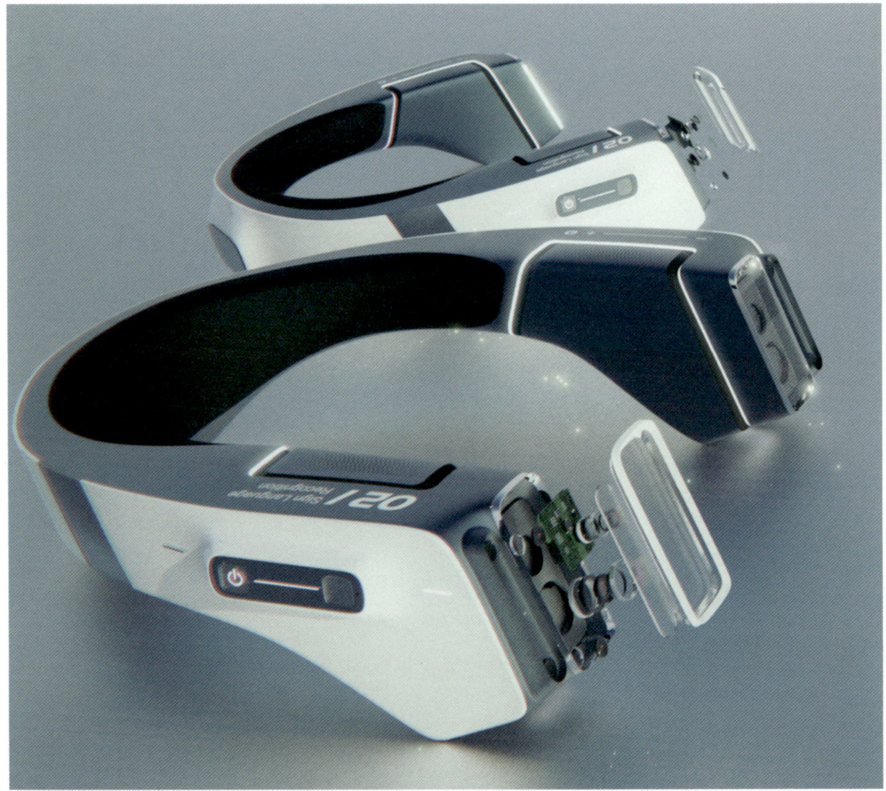

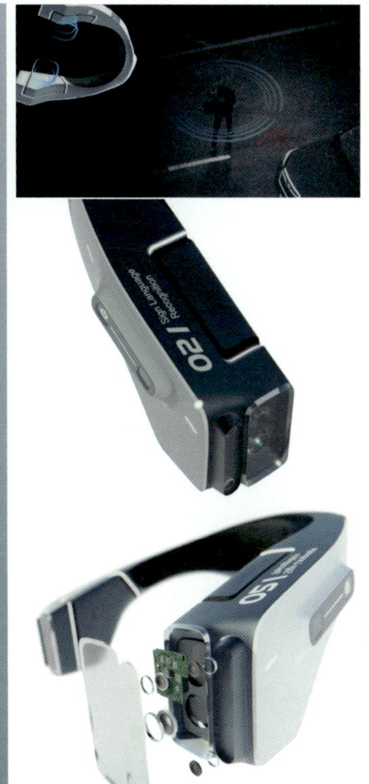



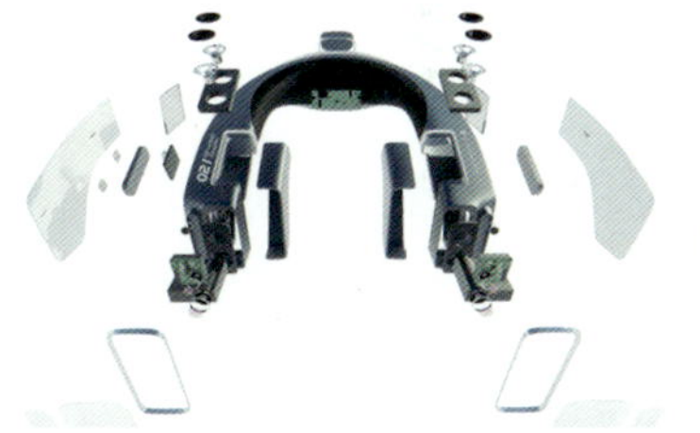



# 听障人士手语识别辅助交流设计

# Design of Sign Language Recognition Assisted Communication for Hearing-Impaired Individuals

作　　者：石　峋
指导老师：秦文婕
所在院校：天津美术学院

## 设计说明

本设计旨在保证听障人士与正常人之间的顺畅沟通，通过语音转换为文本并将其投影到手掌上，以及手语手势转换成语音的方式，实现交流双方的信息传递。这种设计解放了双手，更符合听障人士手语交流的习惯。此外，贴合颈部的触觉传感器能够为听障人士预警周围的车辆，保障出行安全。由此使听障人士更好地融入到社会生活中。

## Design notes

Hearing impairment is a barrier to interpersonal contact. In order to ensure smooth communication between the hearing-impaired and normal people, this design liberates both hands and is more in line with the personal habits of human-computer interaction, which is the greatest distinction between this gadget and conventional hearing aids. By turning voice into text and projecting it onto the palm, the distance between hearing-impaired and normal people can be shortened.In addition, tactile sensors that fit snugly around the neck can alert hearing-impaired individuals to nearby vehicles, ensuring safe travel. This enables hearing-impaired individuals to better integrate into social life.

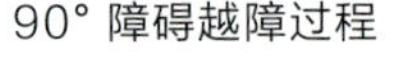

90° 障碍越障过程

# ASCENDER——地形自适应机器人

## ASCENDER —— Terrain Adaptive Robot

作　　者：于卓远
指导老师：解嘉慈
所在院校：河北工业大学

### 设计说明

ASCENDER 地形自适应机器人是针对灾后复杂而难以进入的救援环境设计的地面智能无人机械，整合了行星轮履带结构和双履带结构，负载不同的功能模块完成救援任务，功能模块同时接入机器人控制系统，实现复杂地形内的点对点智能救援。地形自适应机器人能够智能识别多种复杂地形并自适应翻越障碍，代替人类到难以进入的环境进行搜救任务，最大程度上降低灾害带来的损失。

### Design notes

ASCENDER terrain adaptive robot is a ground intelligent UAV designed for the complex and difficult to access rescue environment after the disaster. It integrates planetary wheel track structure and double track structure, and carries different functional modules to complete the rescue task. The functional modules are connected to the robot control system at the same time to achieve point-to-point intelligent rescue in complex terrain. Terrain adaptive robots can intelligently identify a variety of complex terrain and adaptively surmount obstacles, instead of human beings to search and rescue the environment that is difficult to access, and minimize the losses caused by disasters.

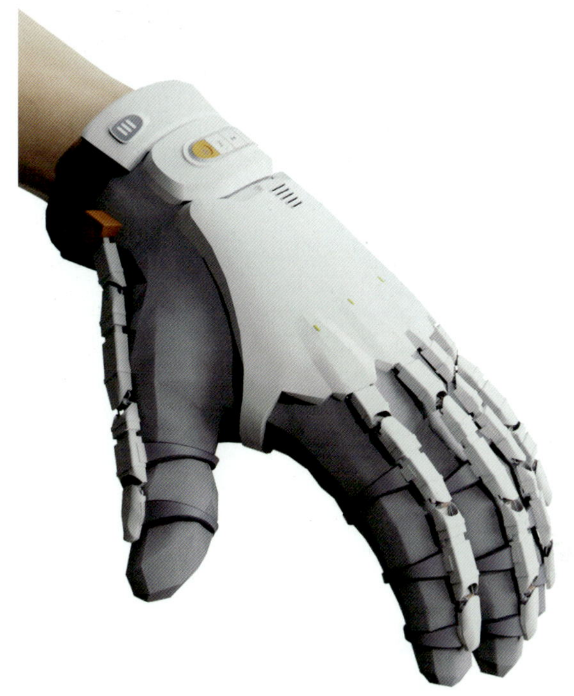

# 基于人机融合的掌指关节康复机械手

# A Palm Finger Joint Rehabilitation Robotic Arm Based on Human-Machine Fusion

作　　者：黄　涛　左子滢　薛梦姝
指导老师：王年文
所在院校：燕山大学

## 设计说明

本设计是基于人机融合的掌指关节康复机械手研究，在传统康复机械手的基础上进行功能改造和外观创新的一款产品。由于每个用户的骨骼特征不同，考虑用户使用产品的适配性，通过掌上小程序的功能设置，能够及时收到反馈信息，对康复过程的难度反馈及时作出调整。能有针对性地辅助患者完成手部康复锻炼，在脑卒中患者康复治疗中发挥着极大的作用。

## Design notes

This design is based on the research of human-machine fusion palm finger joint rehabilitation robotic arm. It is a product that has undergone functional transformation and appearance innovation on the basis of traditional rehabilitation robotic arm. Due to the different bone characteristics of each user, considering the adaptability of the user, feedback information can be received in a timely manner through the function settings of the handheld mini program, and adjustments can be made to the difficulty feedback of the rehabilitation process in a timely manner. Being able to assist patients in completing targeted hand rehabilitation exercises plays a significant role in the rehabilitation treatment of stroke patients.

# 多场景智慧公共服务设施设计

## Design of Multi-scenario Smart Public Service Facilities

作　　者：刘梦歌
指导老师：赵芳华
所在院校：河北工业大学

### 设计说明

多场景智慧公共服务设施设计是以用户体验为核心的产品设计。通过深度市场调研和用户行为分析，我们洞察到各场景下的真实需求和潜在痛点。结合物联网、大数据和人工智能等前沿技术，我们成功地将智能化、便捷化、舒适化等元素融入到设计之中。

在办公场景，我们设计了智慧办公设施，实现智能空间布局和办公环境自适应调节，极大提升了工作效率和用户体验。在母婴出行场景，智慧母婴设施的设计解决了携带婴幼儿出行时的诸多不便，为母婴群体提供了更加贴心和人性化的服务。而在社区康养场景，我们设计了智慧康养舱，为老年人提供了安全、舒适的康养环境，实现了健康管理与紧急救援的智能化。

### Design notes

The design of multi-scenario smart public service facilities is a product centered on user experience. Through in-depth market research and user behavior analysis, we have gained insights into the real needs and potential pain points in various scenarios. By combining cutting-edge technologies such as the Internet of Things, big data, and artificial intelligence, we have successfully integrated elements of intelligence, convenience, and comfort into our design.

In the office setting, we have created smart office facilities to achieve intelligent space layout and adaptive adjustment of the office environment, greatly improving work efficiency and user experience. In the scene of mother and baby travel, the design of smart mother and baby facilities solves many inconveniences when carrying infants and young children, providing more considerate and humanized services for the mother and baby group. In the community health care scene, we have launched a smart health care cabin, providing a safe and comfortable health care environment for the elderly, and achieving intelligent health management and emergency rescue.

# VANGUARD——轨道式隧道消防机器人

## VANGUARD——Orbital Tunnel Firefighting Robot

作　　者：周子豪　贾　墨　朱昊然
指导老师：姚小清
所在院校：华北电力大学（保定）

## 设计说明

轨道式隧道机器人是利用隧道上方轨道将本装置投入火场进行监控，起到现场指挥和扑灭火源的作用。隧道空间狭窄，遇到火情时如不及时采取措施极易发生扩散和爆炸，消防车体量大难以入内，使用消防栓也会对消防员生命造成不小的威胁，加上极低的可视度和混乱的人群，灭火工作更加困难。本装置检测到火源时可利用轨道在第一时间进入火场，并对火情进行控制和监控，为后续救援争取更多时间。

## Design notes

The track type tunnel robot uses the track above the tunnel to put this device into the fire scene for monitoring, playing the role of on-site command and extinguishing the fire source. When encountering a fire in a narrow tunnel space, if measures are not taken in a timely manner, it is prone to spread and explosion. The large size of the fire truck makes it difficult to enter, and the use of fire hydrants can also pose a significant threat to the lives of firefighters. Coupled with extremely low visibility and chaotic crowds, firefighting work is even more difficult. When this device detects a fire source, it can use the track to enter the fire scene in the first time to control and monitor the fire situation, which will buy more time for subsequent rescue.

# 沙漠愚公——基于沙漠植树的智能植树车设计

## DESERT FOOLISH OLD MAN——Design of Intelligent Desert Tree Planting Vehicle

作　　者：白沐凡
指导老师：赵芳华
所在院校：河北工业大学

### 设计说明

沙漠化减少了耕地和草原面积，带来了肆虐的沙尘暴、水土流失、洪涝灾害等。最好的处理方式是对沙漠进行植树造林，但传统的人工植树不仅成活率不高而且种植十分辛苦。此款沙漠智能植树车能实现24小时挖孔植树，面对特殊的治沙环境，机器人大大提高了造林效率。

### Design notes

Desertification has reduced the area of arable land, grassland and brought about rampant sandstorms, soil erosion, and flooding. The best way to deal with this is to plant trees in the desert, but traditional artificial tree planting not only has a low survival rate but also is very hard to plant. The intelligent tree planting vehicle design is based on desert tree planting, digging holes and planting trees for 24 hours a day, in the face of the special sand control environment, the robot greatly improves the efficiency of sand control and puts on an embroidered suit to protect the desert from wind and rain.

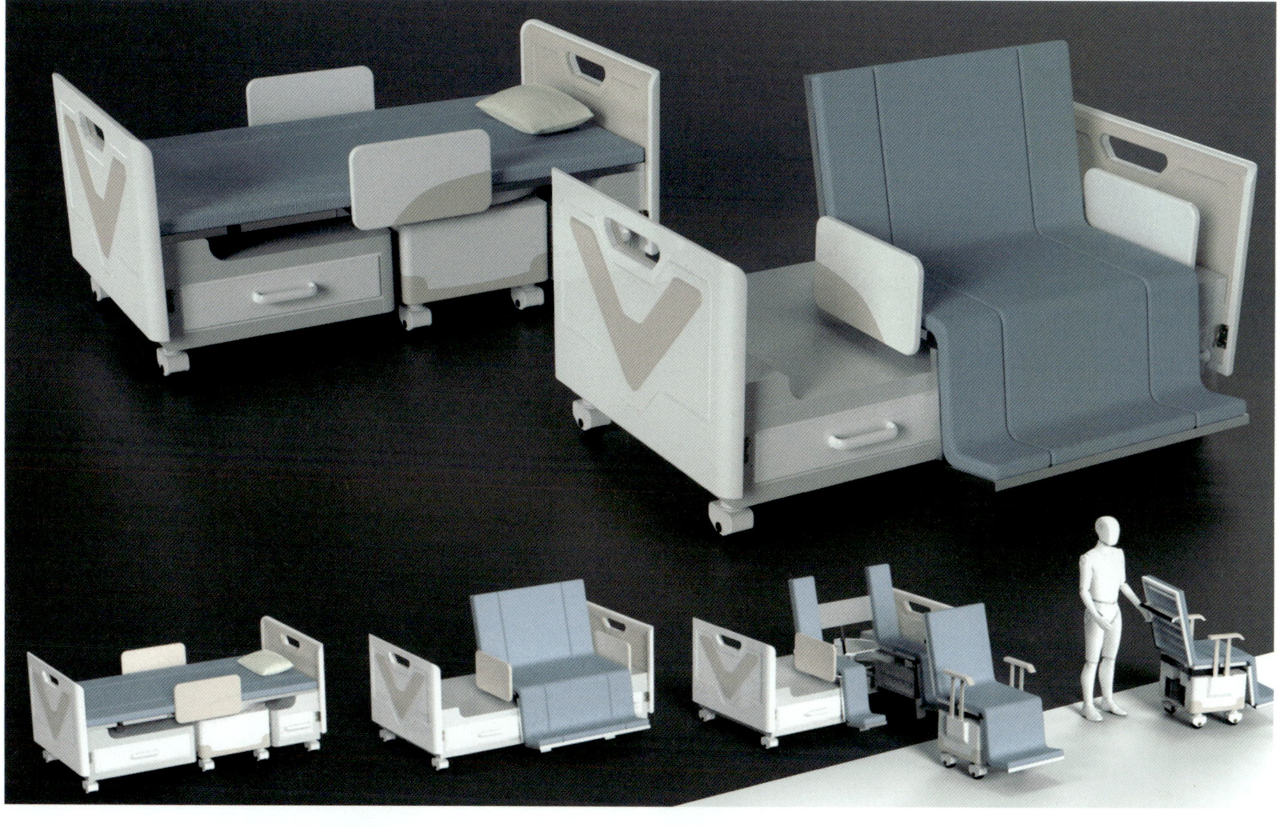



# 分体式居家轮椅护理床设计

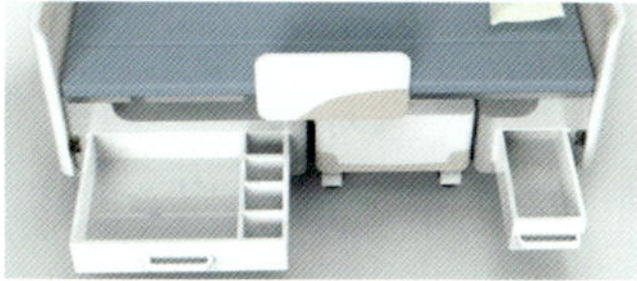
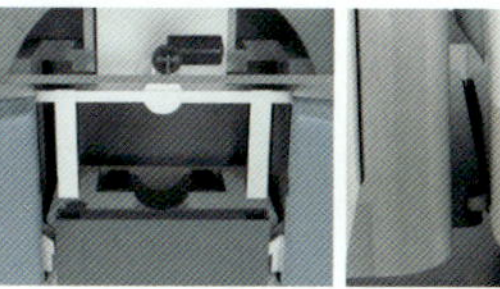
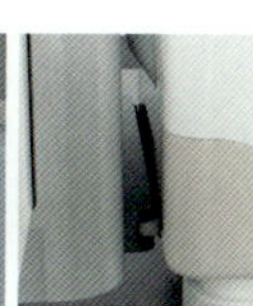



# Design of Split Type Home Wheelchair Nursing Bed

作　　者：王　莉　蒋　瑞　王鹏菲　崔智綮
指导老师：杨冬梅
所在院校：河北工业大学

## 设计说明

分体式居家轮椅护理床的设计，既可以实现病患抬背、曲腿等基础姿态调节功能，又可将床架旋转成坐姿状态，方便辅助病患下床以及活动。旋转后的床架一部分构成轮椅模式，可实现轮椅与护理床床体的分离与对接功能，扩大病患的活动空间，从而减轻护理者的负担和病患因位置转移带来的身体不适，流程简单便捷，适于推广学习使用。

## Design notes

The design of the split home wheelchair nursing bed can not only realize the basic posture adjustment function such as the patient's back and leg bending, but also rotate the bed frame into a sitting position, which is convenient for assisting the patient to get out of bed and chat with others for entertainment. Part of the rotating bed frame constitutes a wheelchair mode, which can realize the separation and docking function of wheelchair and nursing bed body, and can expand the patient's activity space, thereby reducing the burden of caregivers and the physical discomfort caused by the patient's position transfer, and the process is simple and convenient, suitable for promoting learning and using.

# 海洋塑料收集系统设计

## Design of Marine Plastic Collection System

作　　者：周　悦
指导老师：王　鹏
所在院校：河北师范大学

### 设计说明

海洋塑料收集系统包括海洋设备回收基站和无人海洋塑料收集器。收集器采用双体船结构设计，含有收集口和收集仓，收集器上的热解系统可以将一部分收集的废料转化成燃料，不能转化的废料则通过回收基站转运到陆地上进行再加工处理。

### Design notes

The marine plastic collection system includes a marine equipment recycling base station and an unmanned marine plastic collector. The collector adopts a catamaran structure design, including a collection port and a collection bin. The pyrolysis system on the collector can convert a part of the collected waste into fuel, while the non-converted waste is transported to land through the recycling base station for further processing.

# 洪涝灾害救援车设计

# Design of Flood Disaster Rescue Vehicle

作　　者：丁小怡　卫瑞洁　鞠勇超
指导老师：卢　颖
所在院校：燕山大学

## 设计说明

本产品是针对洪涝灾害设计的一款水陆两栖救援车，可以应对不同救援状况，满足洪涝救援过程中的救援装备需求。洪涝救援车配备多种救援设备，如无人机、水面救生机器人、救生圈等，同时运用新技术，与驾驶室的智能操作系统形成体系，以提高救援效率。

## Design notes

This product is an amphibious rescue vehicle designed based on flood scenarios, which can cope with different rescue situations and meet the rescue equipment needs in the flood rescue process. The flood rescue vehicle is equipped with a variety of rescue equipment such as drones, surface lifesaving robots, lifebuoys, etc., and uses new technologies to form a system with the intelligent operating system of the cab to improve rescue efficiency.

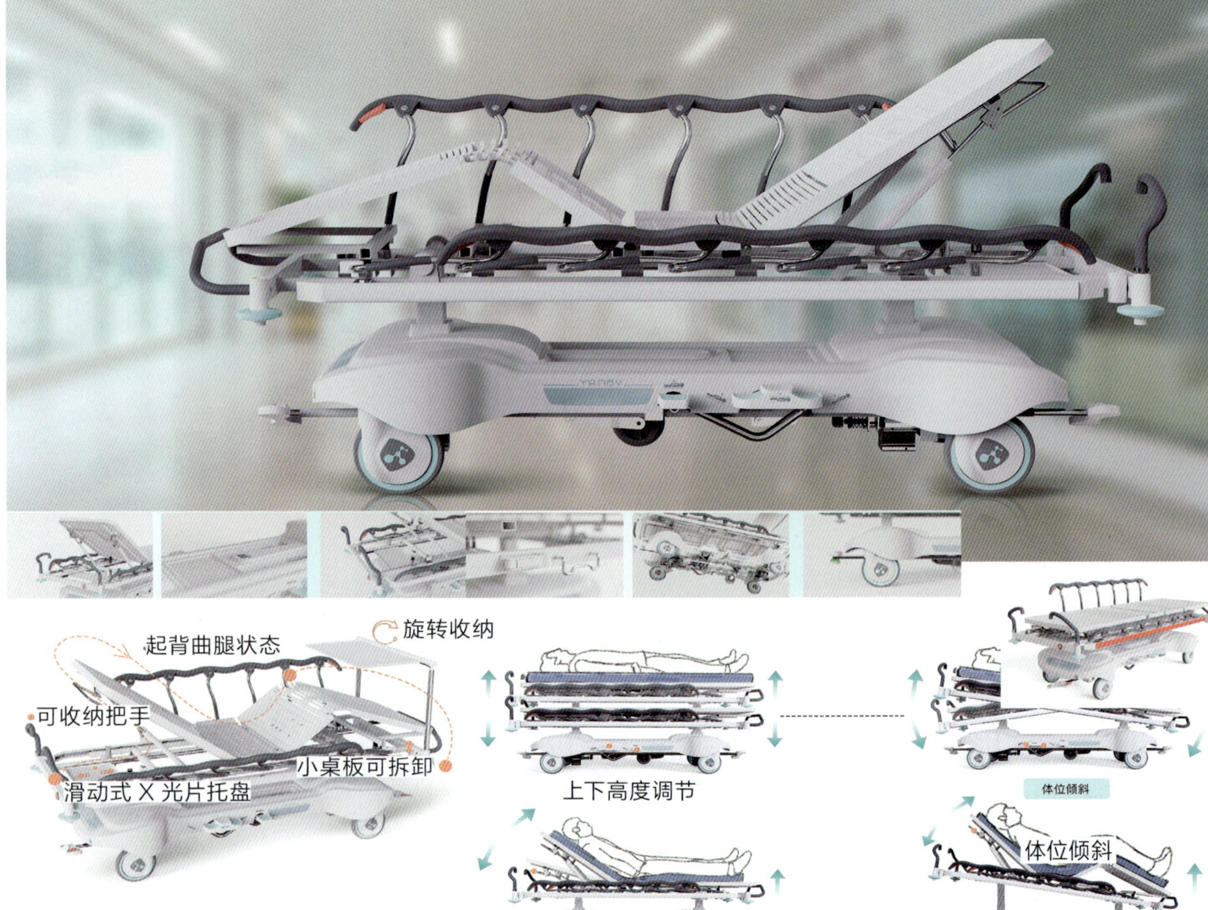


起背状态

起背曲腿倾斜状态

# EFFICIENT ——医疗转运车

## EFFICIENT——Medical Transport Vehicle


作　　者：李原慧　任思凡　秦文静
　　　　　武绍华　李　燕
指导老师：王年文
所在院校：燕山大学


### 设计说明

EFFICIENT——医疗转运车基于人体工程学，通过研究人体结构与操作状态的尺寸，使其更贴合人体结构和符合医患使用习惯而设计。主要解决如何提升患者的就医效率及医护人员的救治效率，为医疗救治提供更多的可能性。主要功能有预警功能、起背曲腿、高度调节、体位倾斜等，是一款追求速度、安全的高端医疗转运车。

### Design notes

EFFICIENT——the medical transfer vehicle is designed based on ergonomics. By studying the size of the human body structure and operating state, we make it more fit to the human body structure and conform to the use habits of doctors and patients. The product function design is mainly aimed at how to improve the efficiency of patients' medical treatment and how to efficiently assist medical staff in the treatment, so as to provide more possibilities for medical treatment. It is a high-end medical transfer vehicle designed for speed, safety and efficiency with the main functions of early warning, back and leg lifting, height adjustment, body position inclination, etc.

# 医值担保——室内救护担架包设计

## MEDICAL VALUE GUARANTEE——Indoor Rescue Stretcher Bag Design

作　　者：杨玉欣　孙雪晴
指导老师：王年文　王　鹏
所在院校：燕山大学

### 设计说明

医值担保——室内救护担架包利用折叠式设计将担架与背包结合，旨在解决现有救护担架体积大、不易搬运的问题，为高层、老旧小区、电梯间的营救提供便利性。它能够根据伤者不同的体位需求作出适宜的体位折叠方式。担架主体采用聚乙烯材料，控制在10千克，能够最大化减少重量，减少救援人员疲惫感；主骨架部分采用铝合金材料，做到坚固稳定，提高救援安全与救护效率。

### Design notes

Medical value guarantee——indoor first-aid stretcher speeder combines stretcher and backpack by folding design, aiming at solving the problem that the existing first-aid stretcher is large and difficult to handle, and providing convenience for high-rise, old residential areas, and elevator rescue. The speeder can make proper posture folding according to the different needs of the injured. The main body of stretcher is made of polyethylene material, controlled at 10Kg, which can minimize the weight and reduce the fatigue of rescuers; the main frame of the product is made of aluminum alloy, firm and stuble, which can improve the efficiency and safety of rescue.

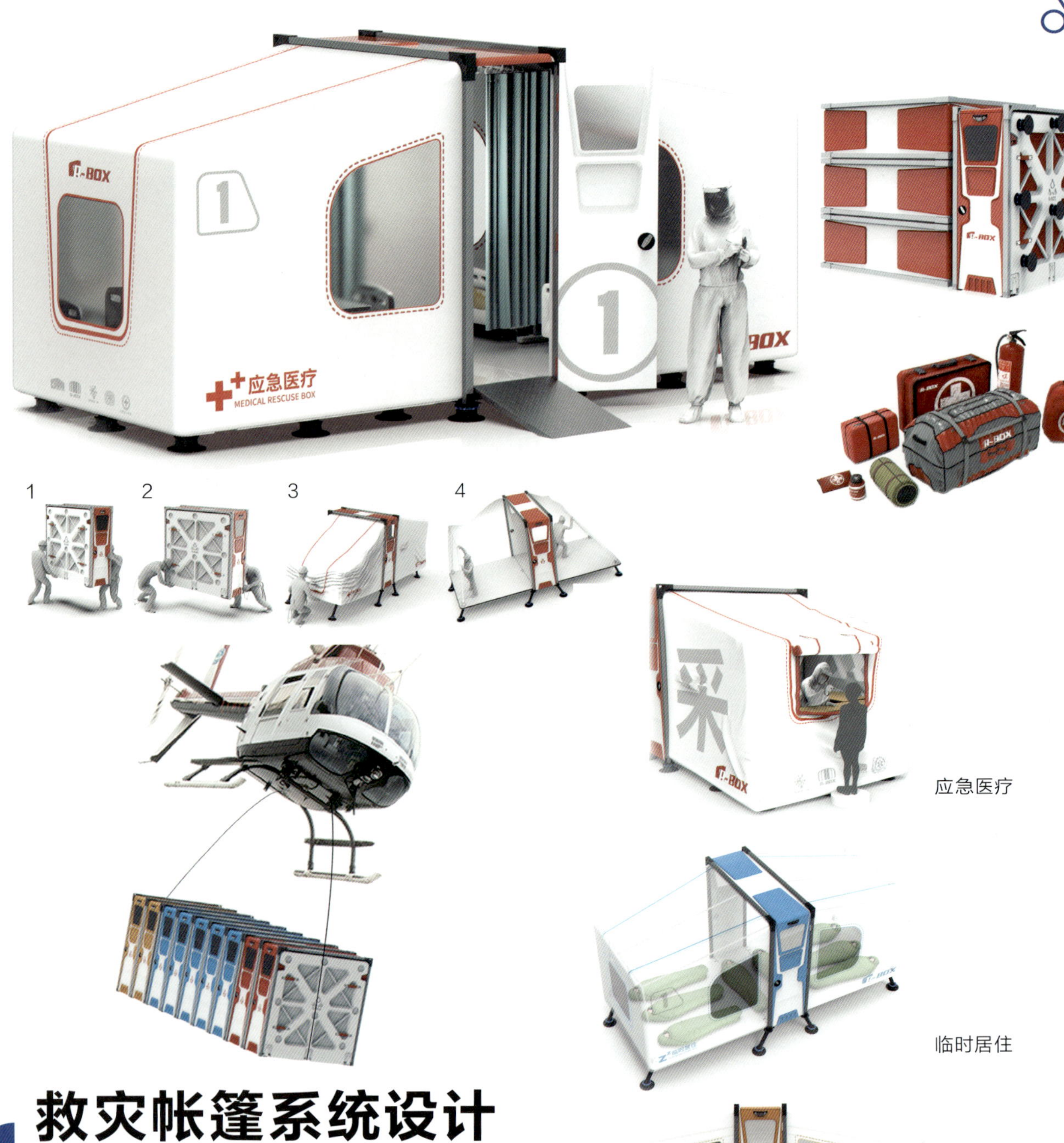


# 救灾帐篷系统设计

## Design of Disaster Relief Tent System

作　　者：桑鑫浩
指导老师：王　鹏
所在院校：河北师范大学

### 设计说明

此产品为一种便携式可折叠的救灾帐篷，能为灾民快速提供医疗急救、临时居住、物资储备的场所，满足受灾群众的需要，保障灾后人民群众的生命安全。

### Design notes

This product is a kind of portable, foldable disaster relief tent design, quickly provideing refugees medical first aid,temporary accommodation, material reserves, etc. , to meet the needs of disaster victims, and to ensure the safety of the people after the disaster.

# 单人一体化环卫装备设计

## Design of Single Person Integrated Sanitation Equipment

作　　者：温巧灵
指导老师：赵　静
所在院校：太原理工大学

### 设计说明

环卫装备的设计，能满足环卫工人代步和工具收纳的需求，收纳箱内部设置隔间收纳不同类别的清洁工具，上方有坐垫，环卫工人可以坐姿驾驶或站姿驾驶；垃圾捡拾箱固定在车身上，可将垃圾直接投放在前方的捡拾箱里，减轻了行走和弯腰负担，提高工作效率。充电站的设计，方便环卫装备集中收纳和规范管理；与环卫装备充电口相连接直接进行充电。构建智慧环卫系统：规范管理（环卫装备 + 充电站）+ 智能交互（环卫工人信息匹配+工作行程统计）。

### Design notes

The design of the sanitation equipment meets the needs of sanitation workers for mobility and tool storage at the same time. With compartments inside the storage box to store different types of cleaning tools and a cushioned seat above, sanitation workers can drive in a seated position or in a standing position; the waste pick-up box is fixed to the body of the vehicle, so that sanitation workers can sit on the tool storage box or stand on the pedals of the counterbalance vehicle during long-time waste pick-up, and put the waste directly. This reduces the burden of walking and bending and improves work efficiency.

The design of the charging station: as a fixed centralised point, it is convenient for the centralised storage and standardised management of sanitation equipment; it is connected to the charging port of sanitation equipment for direct charging.

Build a smart sanitation system: standardised management (sanitation equipment + charging station) + intelligent interaction (sanitation worker information matching + work trip statistics).

金奖

银奖

铜奖

优秀奖

# 智能乡村无人果蔬配送车

# Intelligent Rural Unmanned Fruit and Vegetable Delivery Vehicle

作　　者：武海川　李佳乐　田佳杰　章胡喆
指导老师：于学斌
所在院校：沈阳工业大学

## 设计说明

智能乡村无人果蔬配送车设计宗旨为“助力三农，振兴乡村”，以物联网云端数字平台为依托，构建“乡村&城市”物流运输一体化架构。它可根据调度平台发出的指令，规划合理路线，到达指定社区进行配送；用户到配送车前进行人脸识别提取货物。同时，该果蔬配送车的功能还有自动避障、智能识别、调节温度、智能称重等。

## Design notes

The design purpose of this vehicle is to "help agriculture, rural areas and farmers and revitalize the countryside" , relying on the cloud digital platform of the Internet of things. Build a "rural & Urban" logistics and transportation integration framework, and the intelligent rural unmanned fruit and vegetable distribution vehicle will be based on the instructions issued by the dispatching platform, plan a reasonable route to the designated community for distribution; notify the user to receive the goods through the mobile app, and the user will go to the driverless vehicle to carry out the inspection face recognition to pick up the goods. At the same time, the functions of this vehicle include automatic obstacle avoidance, intelligent identification, temperature adjustment, intelligent weighing, etc.

# JAMPING JACK——便携式儿童安全座椅

## JAMPING JACK——Portable Child Safety Seat

作　　者：李唯一
指导老师：曹伟智　刘勃峥
所在院校：鲁迅美术学院

### 设计说明

JAMPING JACK——便携式儿童安全座椅设计旨在有效解决现有产品普遍存在的占用车内空间过大、移动困难以及存放不便等问题。我们从便携式设计的角度出发，通过运用先进的折叠伸缩技术，显著缩减了产品闲置时的占用空间，极大地提升了产品的便携性和空间利用效率。

### Design notes

The design of the JAMPING JACK——portable child safety seat is aimed at effectively solving the common problems of occupying too much space in the car, making it difficult to move, and inconvenient to store existing products. From the perspective of portable design, we have significantly reduced the space occupied by the product when it is idle by using advanced folding and stretching technology, greatly improving the portability and space utilization efficiency of the product.

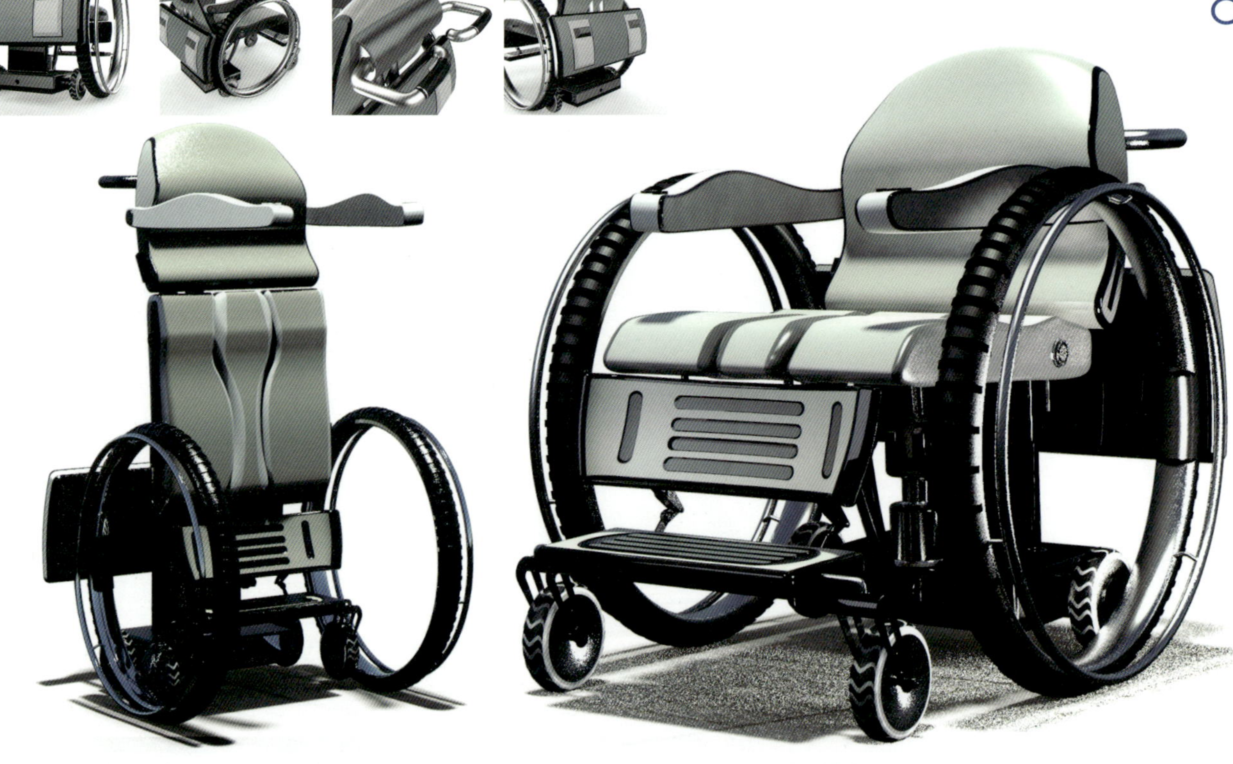

# 轮椅平衡车

## Wheelchair Balance Vehicle

作　　者：王建明　赵越久　王新荣
王雨馨　夏伟然
指导老师：刘华伟
所在院校：北华大学

### 设计说明

轮椅平衡车是一款创新的交通工具，它巧妙地结合了轮椅与平衡车的特性。这款产品利用先进的机械结构，可以将轮椅迅速变形成为一辆具备站立功能的平衡车。这款轮椅平衡车旨在为腿部残疾或行走有障碍的人士提供极大的便利，让他们能够重新体验站立的喜悦。它不仅能够辅助他们进行移动，还极大地提升了他们的生活质量。通过这款工具，他们不仅可以轻松地移动，更能享受到与常人无异、自由行走的乐趣。

### Design notes

Wheelchair balance vehicle is an innovative means of transportation that cleverly combines the characteristics of wheelchairs and balance vehicles. This product utilizes advanced mechanical structure to quickly transform a wheelchair into a standing balance vehicle. This wheelchair balance bike is designed to provide great convenience for people with leg disabilities or walking disabilities, allowing them to experience the joy of standing again. It not only assists them in movement, but also greatly improves their quality of life. Through this tool, they can not only easily complete transportation tasks, but also enjoy the pleasure of walking freely and being no different from ordinary people.

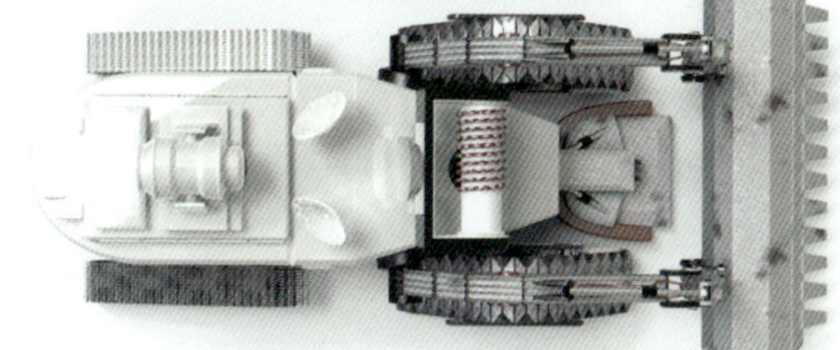





# 除雪机

## Snow Removal Machine

作　　者：何雨昂
指导老师：端文新
所在院校：长春工程学院

### 设计说明

本设计是一款大型除雪机产品，主要用于农村道路和城市街道除雪。在大雪封路时，帮助清扫出一条拥有足够宽度的道路供人们出行使用，以解决大雪封堵给生产生活带来的诸多不便。

### Design notes

This design is a kind of large-scale snow removal machinery for snow removal on rural roads and urban residential streets. When heavy snow closes the road, it can help clear a road with enough width for people's daily travel to solve the inconveniences caused by heavy snow blockage to work and life.

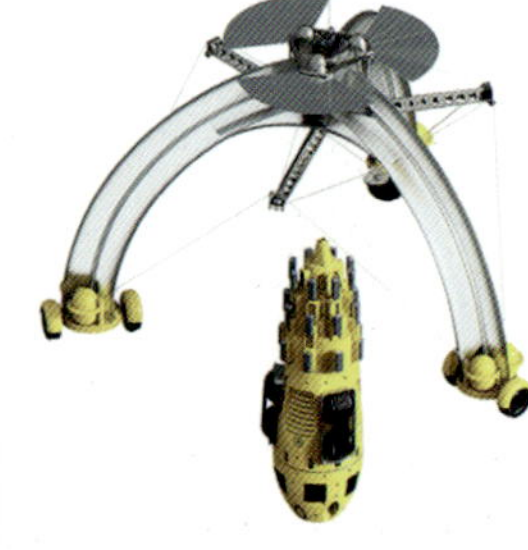

# 深井救援概念机器人设计

## Design of Deep Well Rescue Concept Robot

作　　者：邵志伟　闫宇廷
指导老师：安　猛
所在院校：吉林动画学院

### 设计说明

一些小口径深井，对于安全防范意识较差的儿童、青少年来说存在很大的安全隐患，由于空间狭窄、空气不循环，造成救援难度大，生存率降低。针对以上问题，结合常见的救援方法，通过利用5G、柔性臂和柔性包裹等先进技术，设计出一套深井救援概念机器人。主要设备包含固定装备、地面操作设备、运输设备、机械手爪、机器人柔性臂和充气救援装置。装备具有红外生命精确检测、语音、照明、摄像记录等功能，能对井下状况实现实时探测并展开救援。运输装备也能在收到救援信号后快速抵达。

### Design notes

With the development of society, there are still some small-caliber deep well risk factors, and this is a great hidden danger for children and adolescents with poor awareness of safety precautions. Due to the existence of narrow space, air circulation and other issues, resulting in rescue difficulties, survival rate is reduced. For solving this problem, combined with common rescue methods, through the use of 5G, flexible arms and flexible packages and other advanced technologies, we designed a set of deep well rescue concept robot. The main equipment consists of stationary equipment, ground operating equipment, transport equipment, robotic claws, robotic flexible arms and inflatable rescue devices. Equipped with infrared life accurate detection, voice, lighting, camera recording and other functions, it can achieve real-time detection and rescue of underground conditions. Transport equipment can also arrive quickly after a rescue signal is issued.

# DOREMI——高端租车场景下的多感官车载交互系统

## DOREWMI——A Multi Sensory Vehicle Interaction System In High End Car Rental Scenarios

作　　者：张峻玮　冯　捷　孙雪雯　张兴国
指导老师：胡　洁　董占勋　张立群　顾振宇
所在院校：上海交通大学

### 设计说明

本项目聚焦于L4—L5自动驾驶阶段，针对旅游、出差等需要异地租车的场景，以AR-HUD、情绪识别、眼动识别、手势识别等多感官交互为核心，为用户提供动态、共情、细致入微的全新车载交互体验。针对高端租车场景，我们提供了一键备车服务，通过区块链技术识别用户的数字身份，根据用户的信用等级和个人偏好快速备车，个性化调节车辆设置。根据不同的情绪状态和驾驶场景进行小精灵的主动交互和情景模式的调节，从而让车从一个机器，变成一个富有情感的驾车伙伴。

### Design notes

This project focuses on the L4—L5 autonomous driving stage, targeting scenarios such as tourism and business trips that require remote car rental. With AR-HUD, emotion recognition, eye movement recognition, gesture recognition, and other multisensory interactions as the core, it provides users with a dynamic, empathetic, and imperceptible new experience in car interaction . For high-end car rental scenarios, we provide a one-click backup service that uses blockchain technology to identify the user's digital identity, to quickly backup based on the user's credit rating and personal preferences, and personalize vehicle settings. According to different emotional states and driving scenarios, the elf actively interacts and adjusts the scene mode, thereby transforming the car from a machine to an emotional driving companion.

# 城市下水道清淤机器车设计

## Design of Machine Car for Desilting Urban Sewers

作　　者：姚奕奕
指导老师：王自强
所在院校：北京理工大学

### 设计说明

城市下水道作为保障居民生活的基本设施，在城市运转中发挥着至关重要的作用。为了确保城市的正常运作，防止内涝和污水反流的发生，定期的疏通和维护是必不可少的。下水道管理和疏通工作通常利用绞车、高压水枪等工具，并配合吸污车进行清洁。然而，在堵塞物过多或管道直径受限的情况下，就需要清淤工人亲自下井进行手工清理，不仅危险、劳动强度大，而且效率低。下水道清淤机器车采用更高效的科技手段替代传统的人工作业，不仅能显著改善工人的工作条件，还能更有效地维护城市基础设施的正常运作。它可以从管道排水口进入，使用吸污车管道将机器车从尺寸合适的井口处吊入下水道中。

### Design notes

As a basic facility to guarantee the life of residents, urban sewer plays a vital role in the operation of the city, and is a key link to determine the image and development of the city. In order to ensure the normal operation of the city and prevent phenomena such as waterlogging and sewage reflux, regular dredging and maintenance are essential. Sewer management and dredging work usually use winches, high-pressure water guns and other tools, and with the suction truck to clean. However, in the case of excessive blockage or limited pipeline diameter, it is necessary to depend on silt workers personally going down the well to manually clean. Urban pipeline desilting work is not only dangerous, labor intensity, but also low efficiency. If more efficient technological means can be used to replace traditional manual work, not only can significantly improve the working conditions of workers, but also more effectively maintain the normal operation of urban infrastructure. The sewage dredging machine truck can enter from the drainage outlet of the pipeline, and the suction truck can be used to lift the machine truck into the sewer from the appropriately sized wellhead.

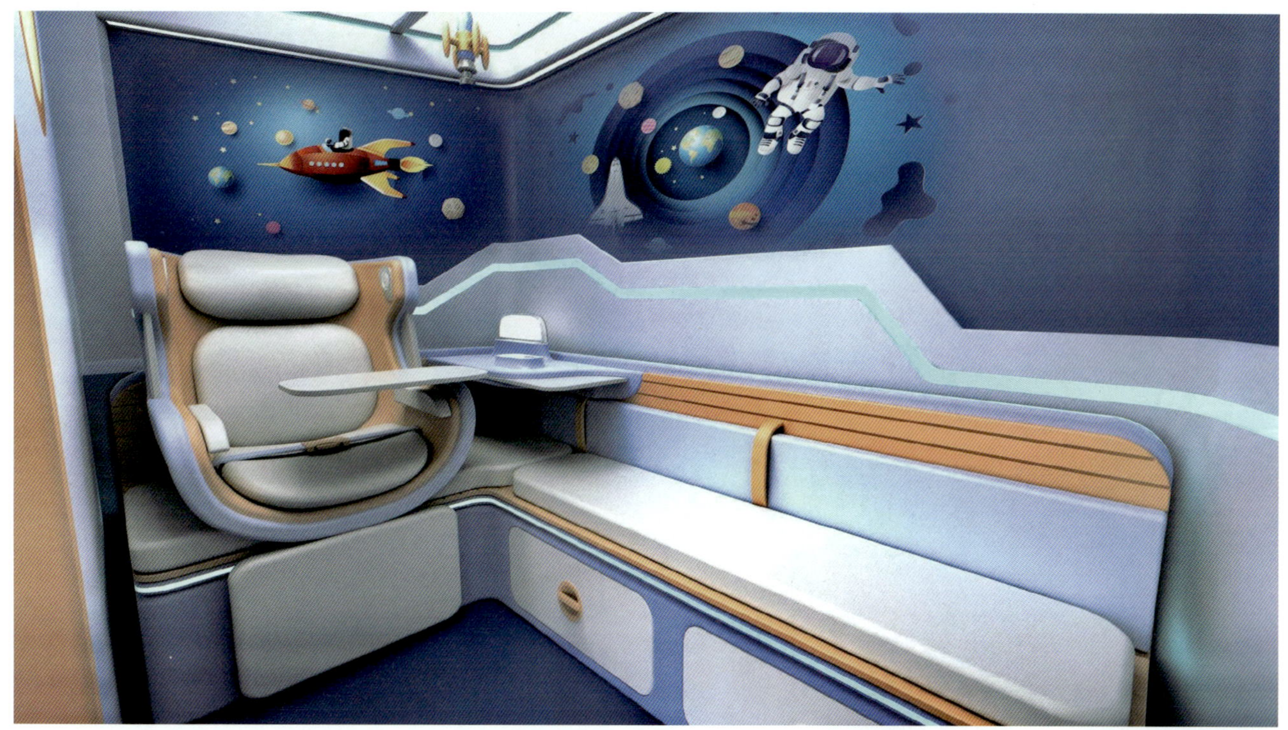

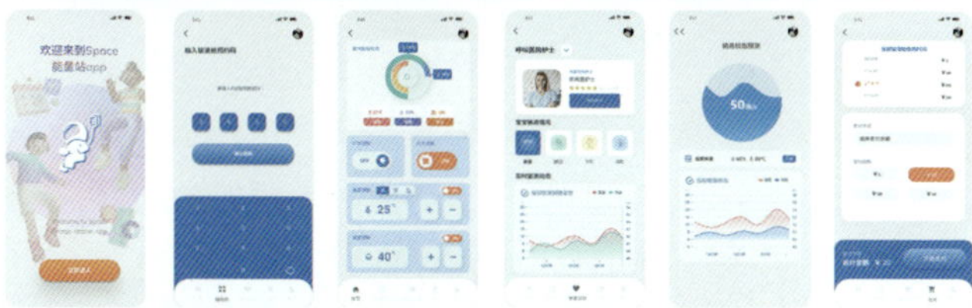

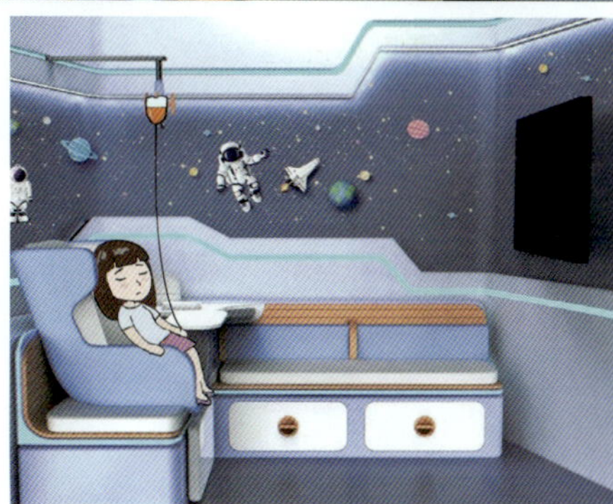



# 租赁式智能型儿童输液舱设计

## Lease Type Intelligent Children Infusion Cabin Design

作　　者：潘钰婕　吴宇辰
指导老师：倪　瀚
所在院校：上海理工大学

## 设计说明

本产品专为儿童医院输液室设计，以智能设备租赁为切入点进行输液设备的创新。结合智能型输液管理系统设计出一款“租赁式智能型儿童输液舱”。这种独立的输液空间概念不仅具有良好的密封性，还有助于减少人员之间的交叉感染。通过采用租赁的商业模式，既减轻了医院的财务负担，又为不同经济条件的病患提供了更多选择，从而改善了就医环境，并提升了患者的就诊体验。

## Design notes

This product is designed for the infusion room of children's hospital, and the innovation of infusion equipment is carried out with intelligent equipment rental as the starting point. Combined with the intelligent infusion management system, a "Lease Type intelligent child infusion cabin" is designed. This independent infusion space concept not only has good sealing, but also helps to reduce cross-infection between personnel. By adopting the leasing business model, it not only reduces the financial burden of the hospital, but also provides more choices for patients with different economic conditions, thus improving the medical environment and enhancing the patient's medical experience.

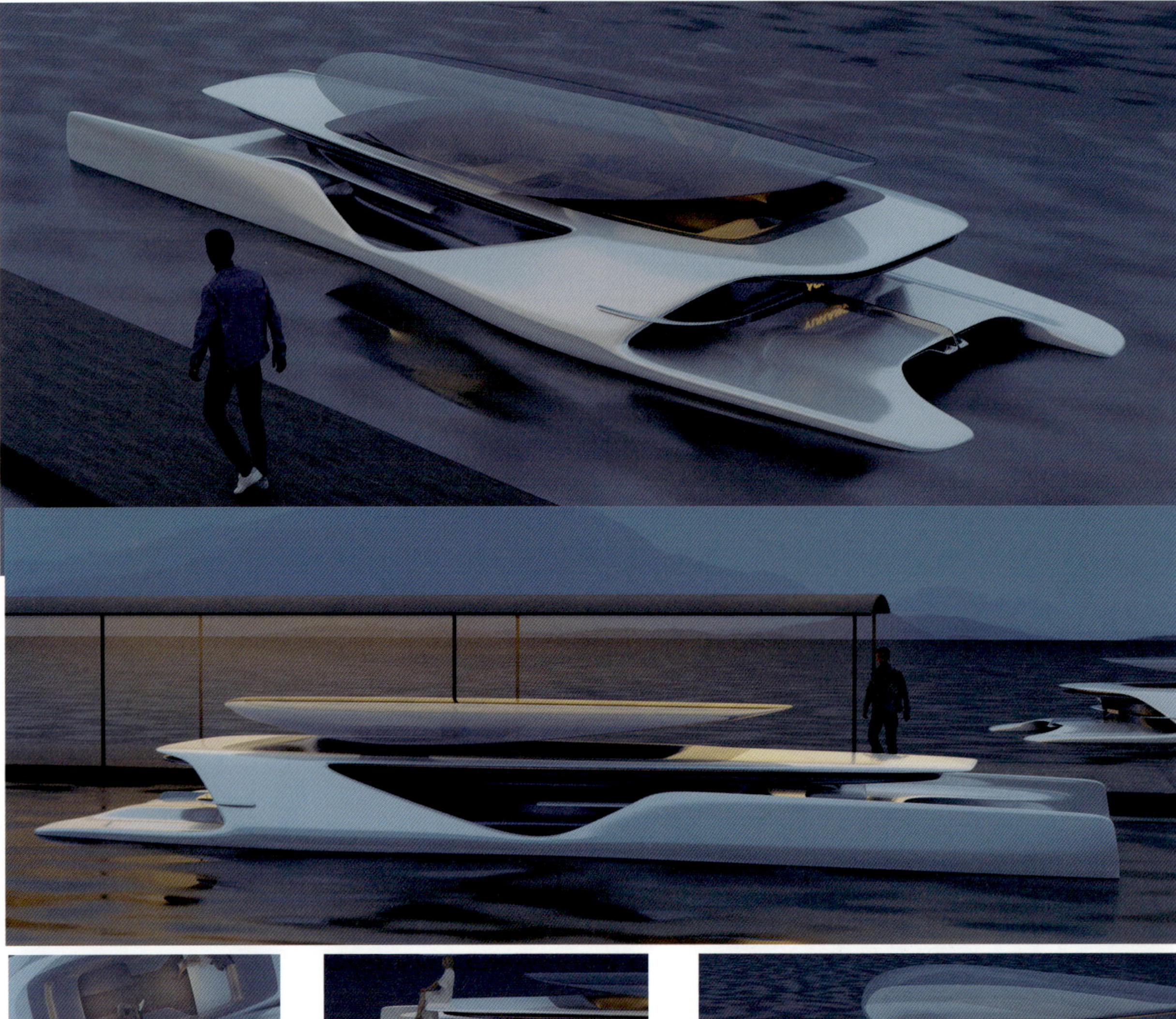

# 禹航

## YUHANG

作　　者：沈彦韬
指导老师：李雪楠
所在院校：华东理工大学

## 设计说明

这是一艘概念观光艇，为游客和市民提供租船自驾的服务，可以休闲娱乐，也可以跨水域畅游。作品名来源于中国杭州市的曾用名："禹航"。设计借鉴了汉字"禹"的造型，将船体线条与汉字字形相对应，具独特的功能和审美体验。

## Design notes

This is a concept sightseeing boat that provides rental and self driving services for tourists and citizens. It can be used for leisure and entertainment, as well as cross water tours. The title of the work, "Yuhang", comes from the former name of Hangzhou, China. The design draws inspiration from the shape of the Chinese character "禹", matching the ship's lines with the Chinese character shape, forming a unique unity of function and aesthetics.

# HUMAN X NATURE
# ——归野计划
# 整车概念设计

## HUMAN X NATURE —
## Return to the Wild Program
## Vehicle Conceptual Design

作　　者：魏怡文　田　源　孙琦瑞
　　　　　史书圆　徐芷婧
指导老师：韩　挺
所在院校：上海交通大学

### 设计说明

我们希望未来的车主都是地球守护者，人与自然能够诗意地邂逅。概念车身和车顶的设计灵感来自于蝴蝶造型，悬浮式的车身设计与可灵活组合的座椅为未来车主提供安全与舒适的自然体验。

### Design notes

We hope that future car owners will be the guardians of the Earth, where humans and nature can meet in a poetic way. The design inspiration for the conceptual body and roof comes from the butterfly shape, with a suspended body design and flexible combination of seats, providing future car owners with a safe and comfortable natural experience.

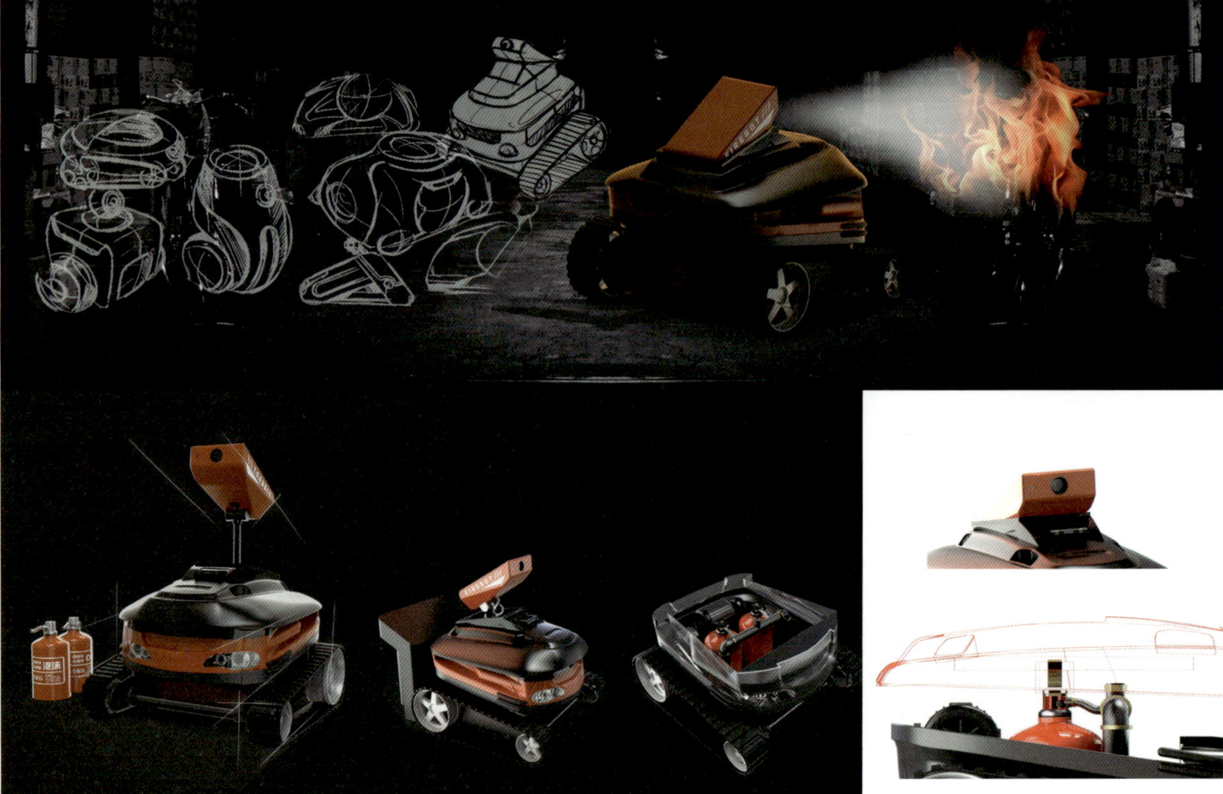



# FIRBOT——智能大型仓库灭火机器人

## FIREBOT——Intelligent Large-Scale Warehouse Fire Extinguishing Robot

作　　者：杨世楠　王文君
指导老师：刘　淼
所在院校：华东理工大学

### 设计说明

这是一款为大型仓库设计的智能灭火机器人，能快速、精准灭火。通过与烟雾探测器联网，能在检测到烟雾后快速定位火灾发生位置，利用自身携带的灭火罐快速扑灭，减少其他货物受到损坏，以防引发更大的火灾。

### Design notes

This is a fire-fighting robot designed for large warehouses. It is designed for the problem that traditional fire-fighting equipment cannot extinguish quickly and accurately. By networking with the smoke detector, it can quickly locate the location of the fire after detecting the smoke, and use its own fire extinguisher to quickly extinguish the fire, so as to reduce the damage of other goods due to fire extinguishing substances in the process of extinguishing the fire, and prevent more fires from being caused by not extinguishing the fire in time.

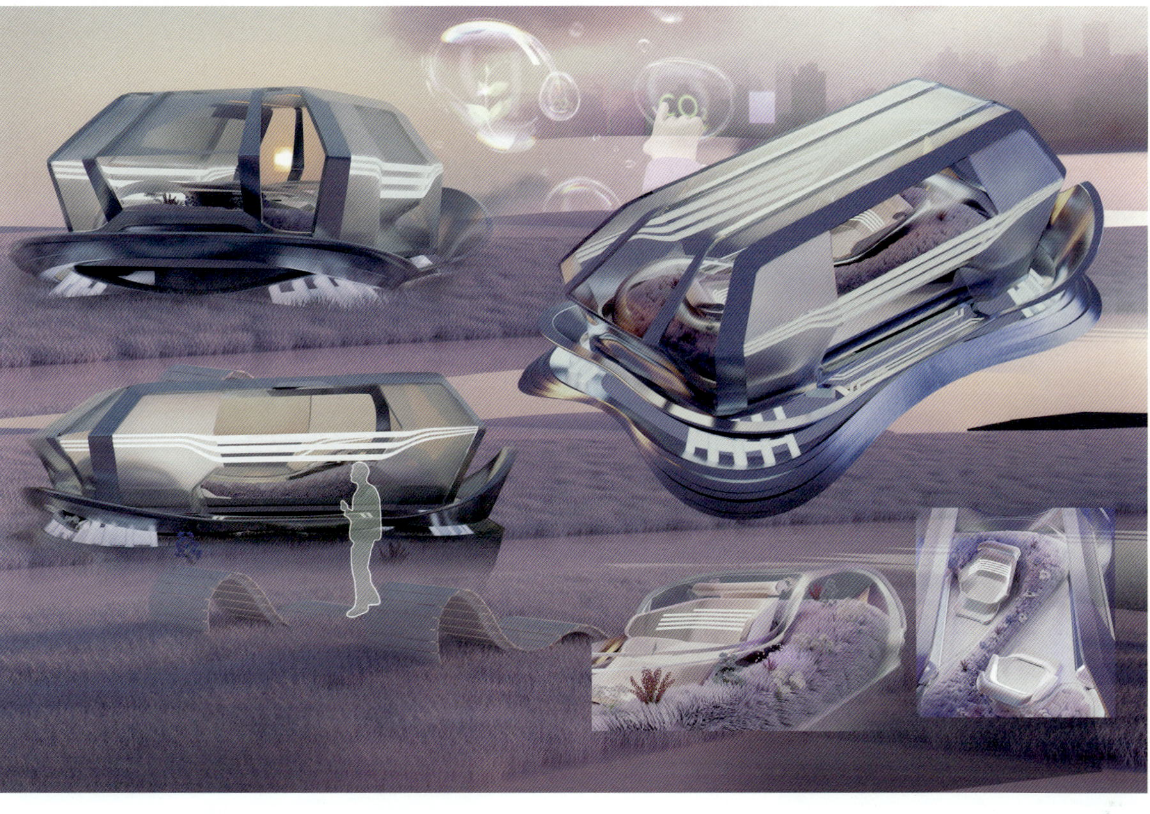



# 植物单元智慧出行工具概念设计

# Plant Unit Intelligent Transportation Tool Conceptual Design

作　　者：刘盛磊　陆轶凡　谭逸蓝
岳佳彤　徐剑韬
指导老师：韩　挺
所在院校：上海交通大学

## 设计说明

植物单元（Plant Unit）是2035年未来移动出行场景下的概念车设计。车身内置空气捕捉（DAC）装置，可收集空气中的二氧化碳。具有人工光合作用的功能模块将收集到的二氧化碳转变为氧气，供给发光藻模块，提供车内辅助氛围灯。

## Design notes

Plant Unit is a conceptual vehicle design for future mobile travel scenarios in 2035. The vehicle is equipped with an Air Capture Device, which can collect carbon dioxide from the air. The functional module with artificial photosynthesis converts the collected carbon dioxide into oxygen, supplies it to the luminescent algae module, and provides auxiliary ambient lighting in the car.

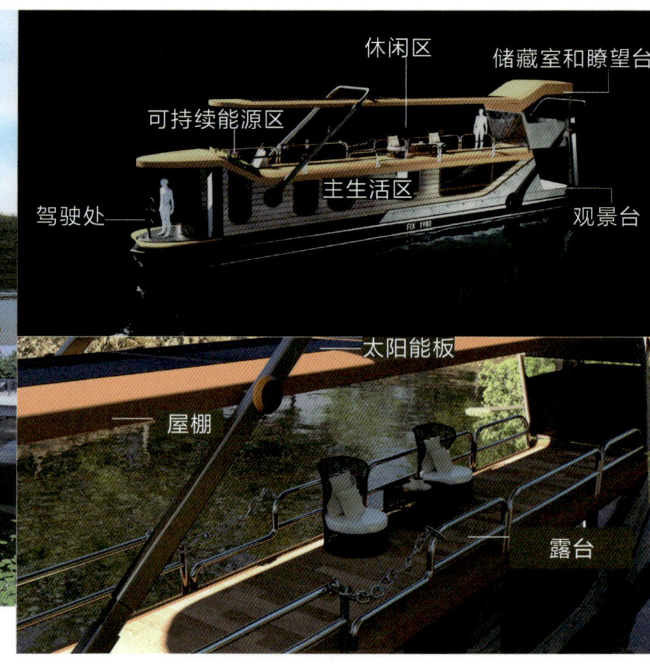


# 可升降窄船船屋

## Adjustable Narrow Boat House

作　　者：吴静怡　陈梦园
指导老师：William Volcoff　陈寿年
　　　　　田坤坤　Mattew Rhoades
所在院校：上海视觉艺术学院

### 设计说明

针对现有船屋老旧、屋内狭窄、屋顶积水、缺少娱乐空间等问题，我们设计出全新的可升降船屋。新设计的船屋整体外观舒适自然，升降平台最大限度利用了空间，开拓了二楼屋顶的功能，使之成为一个可供休闲的区域。屋顶的太阳能板让资源利用更合理，屋面花架的设计也让船上生活充满情趣。整体设计给人宜居舒适的感觉。

### Design notes

In response to the old age of the existing houseboats, the cramped interior, the problems with water on the roof and the lack of space for entertaining, we designed a new liftable houseboat and strictly limited the width of the houseboat to take into account the narrowness of the English rivers. The overall appearance of the newly designed boathouse is comfortable and natural, the lift platform maximises the use of space and opens up the function of the first floor roof as an area for relaxation. The solar energy converting board on the roof allows for a more rational use of resources and the design of the roof trellis makes life on board interesting. The overall design gives a pleasant and comfortable feeling of living.

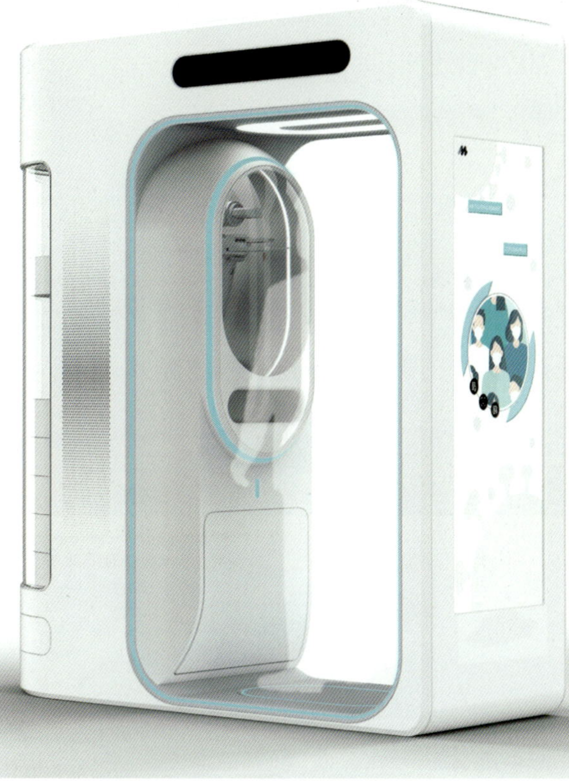

1. 机械臂准备，AI 摄像头自动识别登记用户打开舱门

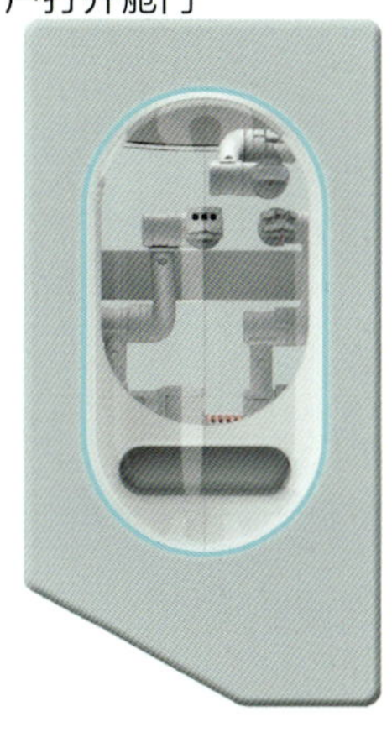

2. 咽拭子弹出，机械臂取用

4. 机械臂拧紧试管盖

6. 机械臂消毒废咽拭子放入回收口

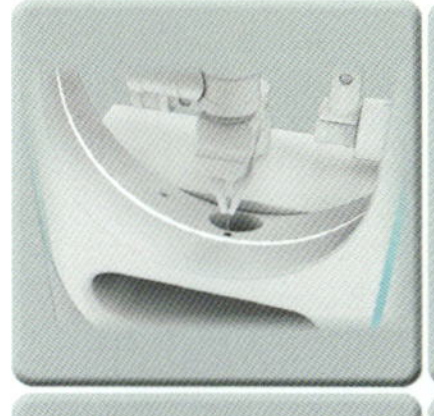

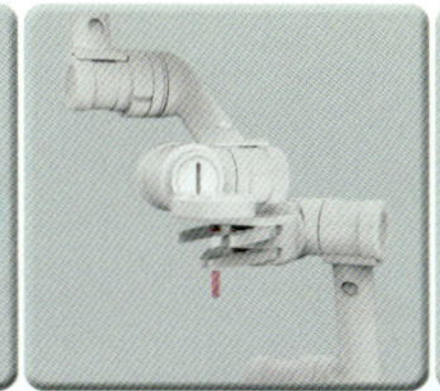

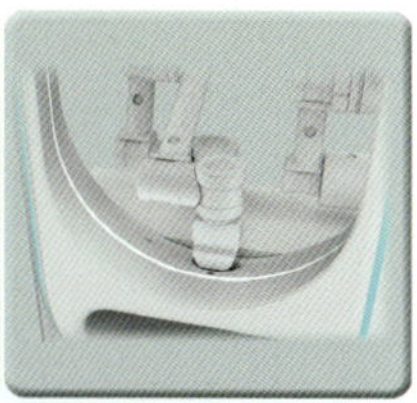

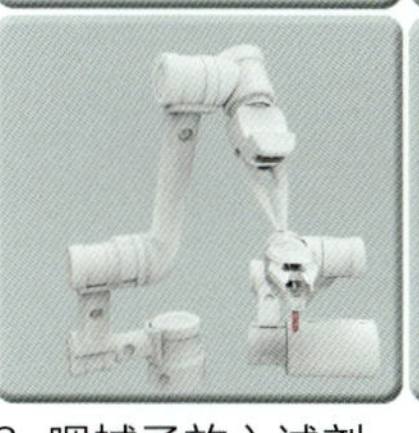

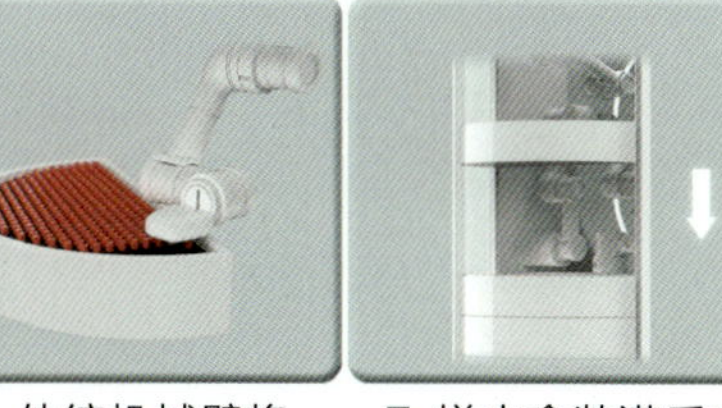

3. 咽拭子放入试剂管，机械臂剪断咽拭子，掉入试管

5. 伸缩机械臂将样本放入样本盒

7. 样本盒装满后向下移动

# 智能核酸检测系统设计

# Design of Intelligent Nucleic Acid Detection System


作　　者：崔　飞　钟　棋　陈亦颖　杨艳婷　钱　苑

指导老师：钟　辉

所在院校：上海师范大学


## 设计说明

智能核酸检测系统整合了人脸识别、口腔识别、数字哨兵和 AI 机械臂等技术，用户可以通过手机 APP 搜索、预约检测点并完成核酸检测，检测结果也将通过 APP 及时反馈给用户。这款智能核酸检测系统具有快速部署和便于移动的特点，适合在医院、机场、高铁站、地铁站和商场等人群密集场所使用。

## Design notes

In view of these problems, this design aims to develop an intelligent nucleic acid detection system under the background of epidemic prevention and control. The system integrates face recognition, oral recognition, digital sentry and AI robotic arm technologies. Users can search, reserve detection points and complete nucleic acid detection through the mobile APP, and the test results will also be timely fed back to users through the APP. This intelligent nucleic acid detection system has the characteristics of rapid deployment and easy to move, suitable for use in crowded places such as hospitals, airports, high-speed rail stations, subway stations and shopping malls.

# 帕加尼海神——基于帕加尼跑车家族化设计特征的游艇设计

## PAGANI SEA GOD——Yacht Design Based on the Family Design Features of Pagani Sports Cars

作　　者：刘西颜
指导老师：王　凯
所在院校：上海海事大学

### 设计说明

帕加尼海神将帕加尼超级跑车的家族化造型元素转化运用于游艇的外观与内饰造型，同时，使用了碳纤维等新型材料并搭载了喷水推进系统和螺旋桨推进系统，在保证可靠性的同时大幅提升了游艇性能，是一款高性能游艇。

### Design notes

While designing, I applied the family styling elements of the Pagani supercar to the exterior and interior styling of the yacht, used new materials such as carbon fiber, as well as a waterjet propulsion system and propeller propulsion system, which greatly improving the yacht's performance while ensuring reliability. It inherits its brand culture and design concept of "combining art and technology".

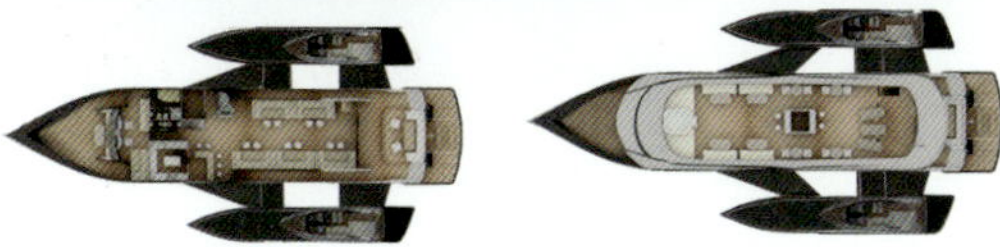

# MARI-WAVE ——外出游艇

## MARI-WAVE——Outgoing yacht

作　　者：高雅萱　王欣奕
指导老师：William Matthew　陈寿年　田坤坤
所在院校：上海视觉艺术学院

### 设计说明

这是一艘游艇的概念设计。采用混合动力装置，可以航行5000海里。船上有太阳能电池阵列，能快速为锂电池充电。游艇可以容纳12人，并带有一艘双人潜水艇，可以让乘客欣赏大海和海底的景色。

### Design notes

This is the conceptual design of a yacht. Using a hybrid system, it can navigate up to 5000 nautical miles. There is a solar cell array on board that can quickly charge lithium batteries. The yacht can accommodate 12 people and comes with a two seater submarine, allowing passengers to enjoy the sea and underwater scenery.

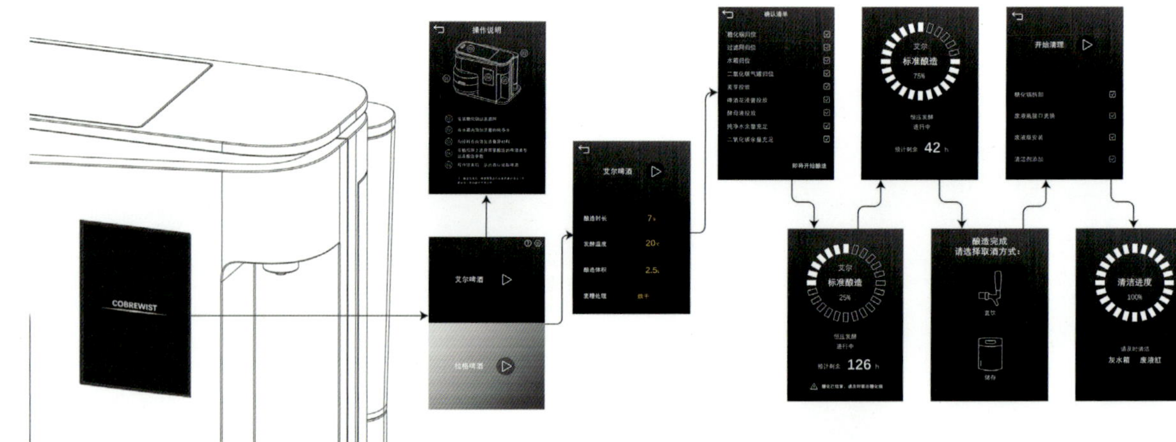

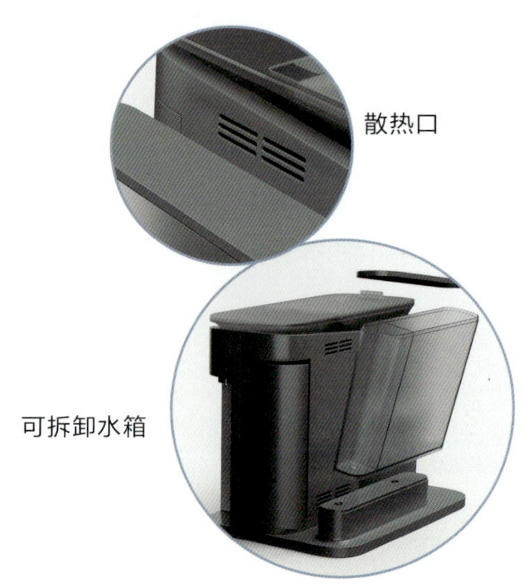




# COBREWIST ——家用啤酒酿造机

## COBREWIST——Domestic Beer Brewing Machine

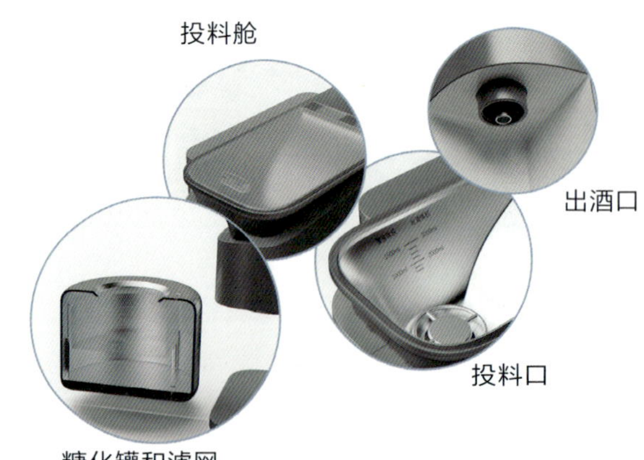


作　　者：彭子凯　江　苏　黄卓男　王瑞锋
指导老师：郭　琪
所在院校：华东理工大学

### 设计说明

COBREWIST是一台能够自动完成麦芽粉碎、糖化、发酵、起泡等啤酒酿造流程的家用啤酒酿造机。它通过超声波催化发酵过程，能够最快在48个小时内完成全部的酿造过程，为人们带来新鲜的精酿啤酒。

### Design notes

COBREWIST is a home brewing device that automates the process of malt crushing, saccharification, fermentation and bubbling. It can complete the entire brewing process within 48 hours through ultrasonic catalytic fermentation process, bringing the freshest craft beer to people.

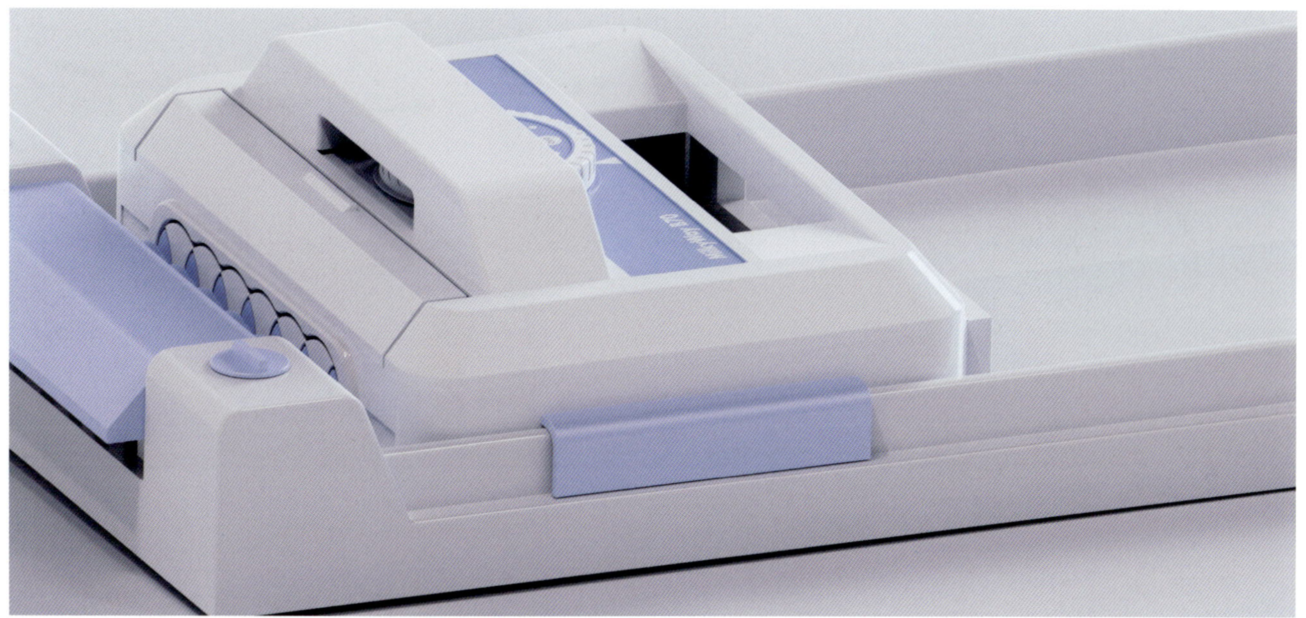

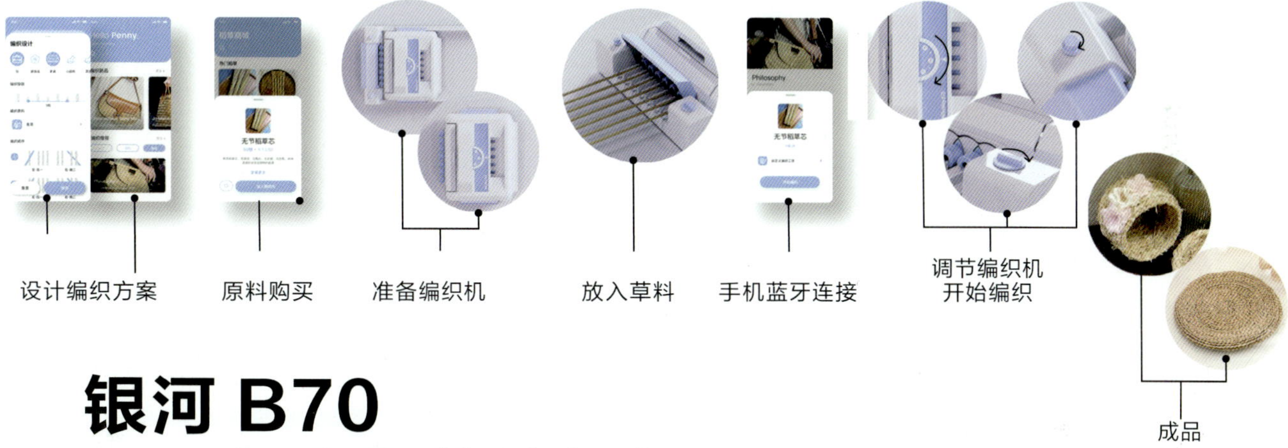




# 银河 B70
# ——七孔稻草编织机

## MILKY WAY B70
## ——Straw Knitter

作　　者：潘江鱼　李超群
指导老师：沈　琼　倪敏娜
所在院校：东华大学

### 设计说明

银河 B70七孔稻草编织机，探究了当代常见的稻草编织工艺，寻找合适的方法将现有的稻草编织工艺引入普通人生活中，从而更便捷地完成稻草编织，进而提高大众对稻草编织的热情和参与度，促进稻草编织文化的发展和传承。

### Design notes

This product explores the common contemporary straw weaving process and looks for suitable methods to introduce the existing straw weaving into the life of ordinary people, so that ordinary people can do straw weaving conveniently at home. This will improve the public's understanding of straw weaving, increase the public's enthusiasm for straw weaving, and help the development and transmission of straw weaving culture.

金奖 银奖 **铜奖** 优秀奖



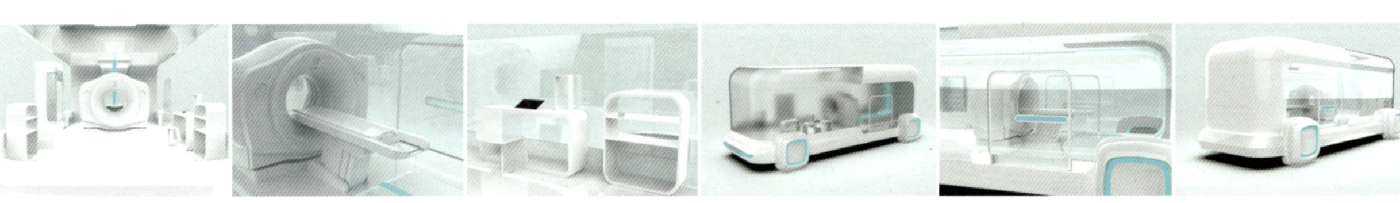

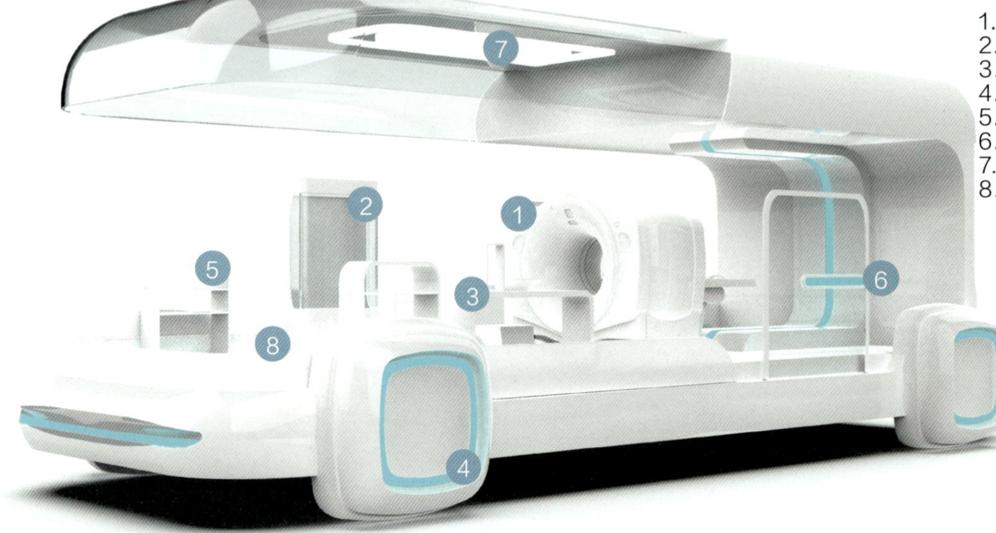


1. 电子计算机断层扫描
2. 医用物品、高精度消毒柜
3. 线上线下工作台
4. 外轮廓提示灯带
5. 药台、杂物台
6. 自动门
7. 多功能消毒灯、顶灯
8. 地面灯带

# 主动医疗服务移动 CT 车设计

## Design of A Mobile CT Vehicle for Active Medical Services

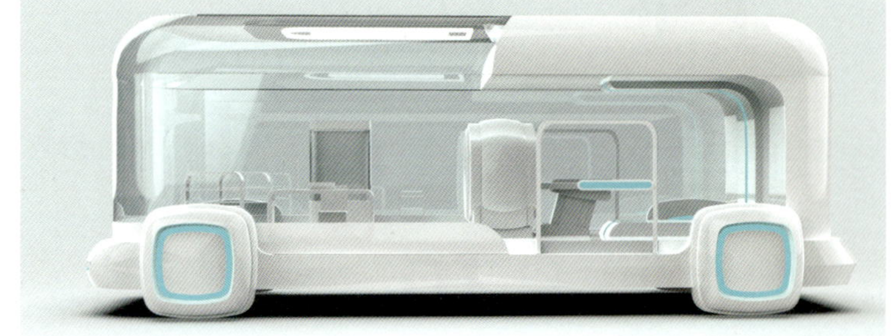

作　　者：刘吴鑫　丁　栋　屈高瑾　骆君言
指导老师：高　瞩
所在院校：上海工程技术大学

### 设计说明

这是一辆为了解决主动医疗服务需求而设计的社区医疗体检车辆。它采用模块化设计，具有全透明有雾化调光功能的钢化玻璃，模块化的新能源电池组底盘，高功率电机组模块，高效率的通风模块。车身内部具有 CT 设备，医用消毒柜，多功能照明系统。

### Design notes

In order to meet the demand for proactive medical services, this is a community medical examination vehicle. It adopts a modular design, with fully transparent tempered glass with atomization dimming function, modular new energy battery pack chassis, high-power electric unit module, and high-efficiency ventilation module. The interior of the vehicle is equipped with CT equipment, a medical disinfection cabinet, and a multifunctional lighting system.

# APSU——新型公园清洁垃圾车

## APSU —— New Park Cleaning Garbage Truck

作　　者：高雅萱　庄裕彦　吴静怡
指导老师：William Matthew　陈寿年　田坤坤
所在院校：上海视觉艺术学院

### 设计说明

APSU是一款为老年人设计的新型公园清洁垃圾车。这款垃圾车轻便快捷，能够存放清洁工具与生活用品，提高清洁效率与工作舒适度，而且造型美观，具现代感。产品考虑了用户感受，还考虑了游客的游览心情，在公园中不显突兀，同时兼顾实用性与创造性。

### Design notes

APSU is a new garbage truck designed for the elderly to clean the park. This garbage truck fully considers man-machine engineering, as well as the living habits and physical conditions of the elderly, without unnecessary design, light and fast, can store cleaning tools and daily necessities, improve cleaning efficiency and work comfort. It has beautiful appearance, with a modern sense, product CMF thoughtful, not only considers the user's feelings, but also considers the tourist mood. It doesn't stand out in the park.

轮胎可旋转，更好地适应各种地形

抓取货物后可以上翘，增加货物稳定性

前叉手可调节以适应不同货物

驾驶舱可左右旋转调节

座舱与底座可分别旋转

# 全地形智能调节叉车

## All Terrain Intelligent Adjustable Forklift

作　　者：刘　栋　钟佳伟　王　旻
　　　　　韦　城　张健瑞
指导老师：孙宁娜　张　凯
所在院校：江苏大学

## 设计说明

该设计是一款全地形智能调节叉车，座舱与底座可以360°旋转工作，叉手通过宽度调节可以叉取不同大小的货物，工作中叉取货物更加稳定，并且履带可以旋转调节，能轻松应对复杂地形。

## Design notes

The design is an all-terrain intelligent adjustment forklift, and its base can rotate 360°, fork hands can fork different sizes of goods through width adjustment, so that the work of cargo forking becomes more stable, and the track can be rotated and adjusted to easily cope with complex terrain.

# 绿化带模块化修剪机

## Modular Trimming Machine for Green Belt

作　　者：刘　栋　钟佳伟　王　旻
韦　城　张健瑞
指导老师：孙宁娜　张　凯
所在院校：江苏大学

### 设计说明

该产品是一款模块化绿化带修剪机，将前端割头进行替换，可修剪出方形、圆柱、圆台、圆球等形状的绿化带。采用遥控手柄控制方式，麦克纳姆轮使工作更加灵活，满足了城市绿化带的多样性与美观性，并且解决了户外工作者工作艰辛、人工成本高的问题。

### Design notes

This product is a modular green belt trimmer. It trims square, cylindrical, round table, round ball and other shapes of green belt by changing front cutter heads. It adopts remote control handle and McNamee wheel, which makes the work more flexible, satisfies the diversity and beauty of urban green belt, and solves the problems of hard work and high labor cost of outdoor workers.

# 智慧双辅助叉车设计

## Design of Smart Dual Assist Forklift

作　　者：张　滢　宫　宇
指导老师：邵　将　刘　勇
所在院校：中国矿业大学

### 设计说明

叉车作为传统机械设备广泛运用于工业，本设计不仅拥有叉车主体这一部件还拥有两个辅助搬运设备，这样的设计在提高工作效率的同时也起到节约成本的目的。叉车驾驶员可以通过自动设定来控制子部件的工作模式，在子部件工作时，周围会产生安全距离的指示灯，不工作时可以将子部件放入叉车后端底部，不占用其他空间。同时为叉车增添了辅助轮，提高其工作高度且不容易侧翻，通过将其原本的柴油动力模式改变为油电混合的动力模式，符合当下绿色发展的设计理念。

### Design notes

Forklifts are widely used in the industry as traditional machinery and equipment, and this model is designed to add two auxiliary handling equipment, which provides work efficiency and cost savings at the same time. The forklift driver can control the working mode of the sub-parts by automatically setting, and the safety distance indicator will also be generated around the working of the sub-components, and the sub-parts can be placed at the bottom of the rear end of the forklift when idle, without taking up other space. At the same time, the auxiliary wheel is added to the forklift to improve its working height and is not easy to roll over. By changing its original diesel power mode to a hybrid power mode of oil and electricity, this design is in line with the current design concept of green development.

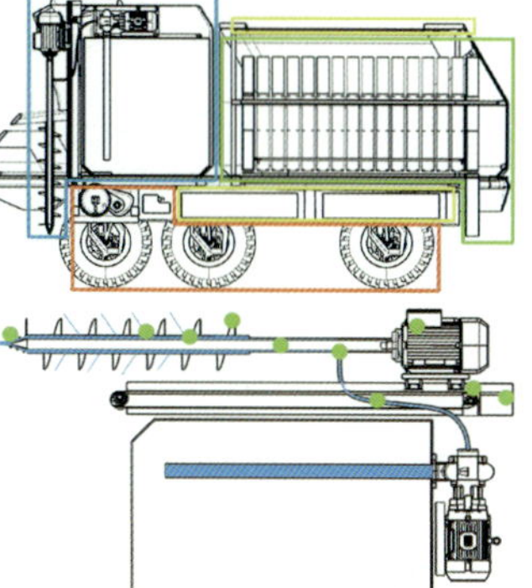

# 梭梭树自动栽植机

## Mr. Slash Automatic Planter

作　　者：张嘉颖　胡鑫鹏
指导老师：张　磊
所在院校：南通大学

### 设计说明

本产品旨在加快荒漠化治理进程并提高植树造林的机械化水平。它基于梭梭树管件防护体系，构建了一个智能监控种植后台，结合自动控制和智能化技术，设计了一套自动种植梭梭树的流程，以实现梭梭树的一体化自动栽植。该产品突破了以往传统的打孔、浇水、放苗等分段式种植工序，在打洞的同时，水流会通过浇灌主孔和侧孔喷出，利用高压水流迅速冲散沙土，减少掘进阻力，形成湿土层，这为梭梭树的生长创造了更有利的条件。储苗箱位于机器上层，可以整体抽出，以便放置防护管件和幼苗。放置完毕后，苗箱整体通过落苗管移至下层。随后，下层的幼苗和管件通过传送机构输送，依次到达下苗口，并通过下苗管完成种植过程。

### Design notes

This product is designed to accelerate the desertification control process and increase the mechanization level of afforestation. Based on the protection system of haloxylon pipe fittings, it constructs an intelligent monitoring and planting background, combines automatic control and intelligent technology, designs a set of automatic planting process of haloxylon, so as to realize the integrated automatic planting of haloxylon. This product breaks through the traditional planting process of drilling, watering, seedlings and other segments, and at the same time, water will be poured through the main hole and side hole, using high pressure water to quickly disperse sand, reduce the digging resistance, and form a wet soil layer, which creates more favorable conditions for the growth of haloxoul tree. The seedling storage box is located on the upper level of the machine and can be drawn out in one piece to accommodate the protective fittings and seedlings. After placement, the box as a whole is moved to the lower layer through the drop tube. Subsequently, the seedlings and pipe fittings of the lower layer are transported through the conveying mechanism, successively reaching the lower seedling entrance, and the planting process is completed through the lower seedling tube.

金
奖

银
奖

铜
奖

优秀奖

106

# 智能黄豆采摘运输设备创新设计

## Innovative Design of Intelligent Soybean Picking and Transportation Equipment

作　　者：张　飞

指导老师：朱云峰

所在院校：南京工业职业技术大学


### 设计说明

基于中国当下对大豆采摘智能设备的市场需求，在整合智能探测装置、无线通信和智能控制技术基础上，设计了这款集无人采摘与短途运输于一体的智能黄豆采摘车，由采摘台、运输设备、清理设备、贮运箱等部分构成，希望能够给传统的黄豆采摘模式提供一个智能化转型思路，助力我国智能农机发展。

### Design notes

Based on China's current market demand for soybean picking intelligent equipment, based on the integration of intelligent detection device, wireless communication and intelligent control technology, We design this unmanned picking and short transportation of intelligent soybean picking car, by the table, transportation equipment, cleaning equipment, storage box, hope to be able to give traditional soybean picking mode an intelligent transformation, power intelligent agricultural machinery development in our country.

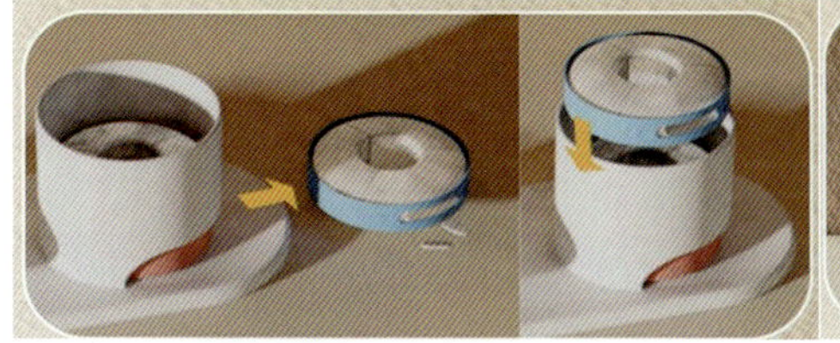

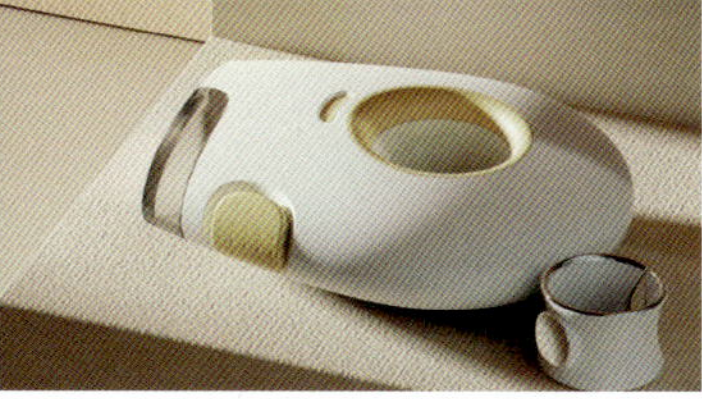



# 适老化智能药盒设计

## Design of an Anti-Aging Intelligent Medicine Box

作　　者：黄　辰
指导老师：陈　默
所在院校：南京工业大学

### 设计说明

“愈环”是一款为老龄群体设计的便携式智能药盒，旨在为用户提供更加灵活、柔和的服药体验，兼顾便利性、人性化、多功能，改善用户用药体验，引导用户形成长期稳定用药习惯，从而改善用药不依从现象，提升用户基本健康水平。其基本功能包括：子母化设计，将药盒分为主体与指环两部分，可方便老人携带与出行，震动提醒方式更加人性化；自然化设计，药盒包装直接印刷使用步骤，只需旋转即可查看操作方式，关注初次用户体验；智能化设计，通过早、中、晚分隔的方式方便用户直接取用，减少情境性障碍可能性。

### Design notes

The "illess ring" is a portable smart pill box designed for the elderly. It provides users with a flexible and gentle medication experience that is convenient, user-friendly and multifunctional. The "illess ring" aims to help users form long-term and stable medication habits, improve their medication adherence and enhance their basic health level. Its functions include: The child-mother structure, the pill box consists of two parts-the main body and the ring. The ring can be detached and carried around by the elderly for travel or daily use. The vibration reminder mode is more humane and discreet ; Natural design, the pill box uses printing technology to display the instructions on the surface of the box. Users only need to rotate the box to view the operation mode, which improves the initial user experience ; Intelligent design, the pill box has separate compartments for morning, afternoon and evening doses. This makes it easier for users to access their pills according to their schedule and reduces the possibility of situational obstacles.

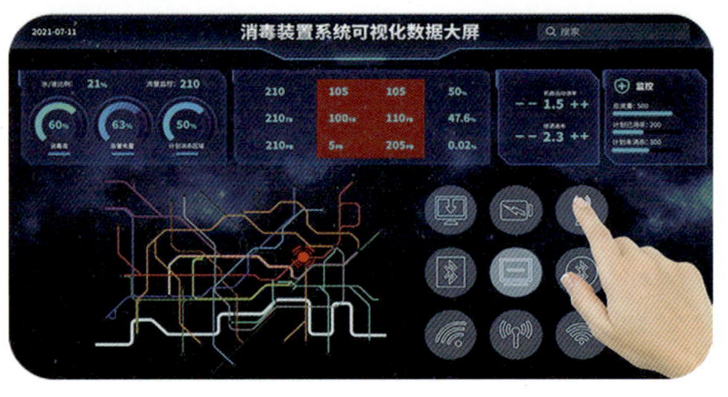

# DEFENDER——地铁消毒机器人

## DEFENDER——Subway Disinfection Robot

作　　者：瞿　敏　杨生智
指导老师：严　波
所在院校：河海大学

### 设计说明

目前国内大多数地铁消毒采用人工喷洒消毒剂的方式，这种方法对人体可能产生一定的伤害，并且存在喷洒不均匀和区域喷洒不到位的问题。为了改善这一状况，设计了该款能够自动运行的地铁消毒机器人。该机器人能够确保消毒工作的精确性和高效性，提高乘客乘坐地铁时的安全。

### Design notes

At present, most of the subways in China are disinfected by manual spraying of disinfectants, which may cause certain harm to the human body, and have problems of uneven spraying and inadequate regional spraying. In order to improve this situation, this subway disinfection robot that can operate automatically was designed. The robot can ensure the accuracy and efficiency of the disinfection work and improve the safety of passengers when taking the subway.

# 用于杂交水稻授粉制种的无人机设计

## Design of UAV for Hybrid Rice Pollination and Seed Production

作　　者：刘国烨　权金拓　吴姝欣
甘宇航　刘晓璟
指导老师：丁　一
所在院校：南京理工大学

### 设计说明

授粉制种是杂交水稻生产流程中的一个关键环节。在生物学领域，杂交水稻面临着花期短、花粉寿命短、柱头露出率低、花粉贮存技术难实现等困境。这些困难导致了对机械辅助授粉技术的迫切需求，即要求生产者在花粉活性的极短时间内快速完成花粉的传播工作。本产品利用无人机产生的风场，通过调整旋翼与水平面的倾角，改变旋翼与空气的相互作用，从而控制风向和力量的方向，以扩大无人机产生的水平风场范围。在这样的高扰动环境下，母本植株的柱头露出率得以提升，同时父本的花粉在受到扰动后脱落，并在风场中分布，最终到达母本并完成授粉过程。该设计为杂交水稻的制种提供了一种更为专业且针对性强的策略，它不仅减少了人工投入，还有效提高了制种的效率。

### Design notes

Pollination seed production is a key link in the production process of hybrid rice. In the biological field, hybrid rice is faced with difficulties such as short flowering period, short pollen life, low stigma exposure and difficult pollen storage technology. These difficulties have led to an urgent need for mechanically-assisted pollination techniques, which require producers to quickly disperse pollen within a very short period of time when pollen is active. The production makes use of the wind field generated by the UAV, changes the interaction between the rotor and the air by adjusting the inclination angle between the rotor and the horizontal plane, and then controls the direction of the wind and force to expand the range of the horizontal wind field generated by the UAV. In such a highly disturbed environment, the stigma exposure rate of the mother plant was improved, and the pollen of the father fell off after the disturbance and was distributed in the wind field, finally reaching the mother and finishing the pollination process. This design provides a more professional and targeted strategy for hybrid rice seed production, which not only reduces the manual input, but also effectively improves the efficiency of seed production.

金奖
银奖
铜奖
优秀奖

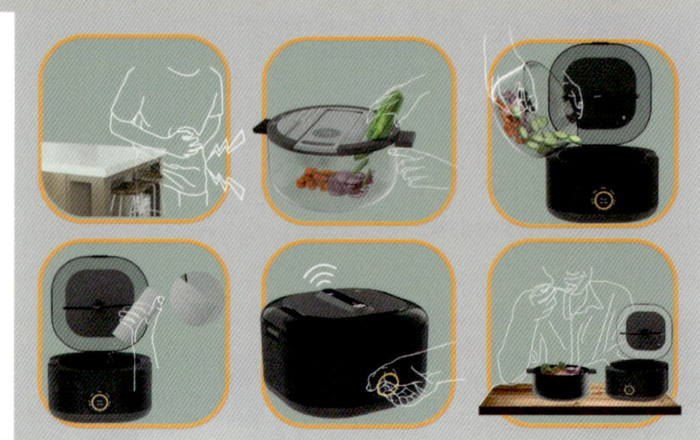

# BLINDCOOKER——面向盲人群体的厨房用具设计

## BLINDCOOKER——Kitchen Appliance Design for Blind People

作　　者：濮依婕　朱芙靖　余诗柯
指导老师：戚玥尔　徐　乐　张梦盈
所在院校：浙江工业大学之江学院

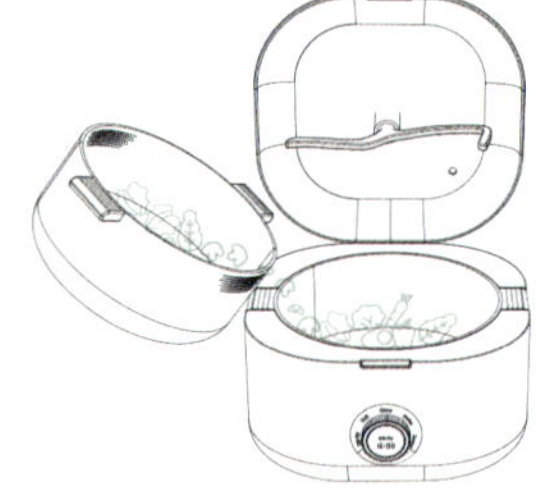

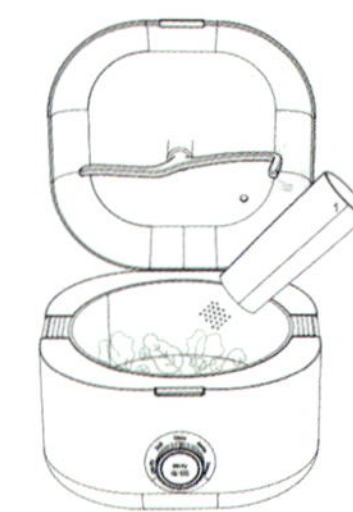

### 设计说明

该系列产品分为三个部分，切菜、烹饪和定量倒出调料，其功能包含了盲文指引、语音提示以及触觉提醒，能有效地帮助盲人在做饭过程中解决切菜、炒菜、出锅时遇到的大部分问题，避免出现割伤、烫伤等情况，从而提高盲人的生活质量。

### Design notes

The series is divided into three parts, chopping, cooking and dosing and pouring out spices. The functions include Braille guidance, voice prompts and tactile reminders, which can effectively help blind people to solve most of the problems they encounter when chopping, frying and pouring out vegetables during the cooking process, allowing them to avoid cuts and burns, thus improving the quality of life of blind people.

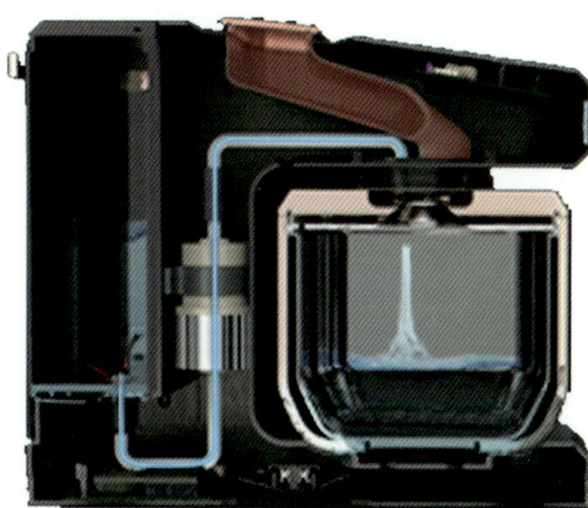





# CI 家用智控煎药机

## CI Home Intelligent Control Decoction Machine

作　　者：徐可欣　丁荥佳
指导老师：朱昱宁　唐智川
所在院校：浙江工业大学

### 设计说明

CI 智控家用煎药机，以“治愈与陪伴”为理念，是一款智能自动化程度较高，能够结合手机端 APP 设定煎煮程序的家用中药煎煮机。机器本身能够实现预约、先煎后下、多煎、自动定时加水加药等功能，达到用户启动煎煮后零操作，轻松享受每日新鲜药剂，保证最佳药效的目标。用户可在 APP 端上传个人药方，得到专业药师的指导方案及相关养生推荐，让煎药机成为你的每日健康养护伴侣。

### Design notes

CI intelligence control of household tisanes machine, in order to "cure and company," as a high degree intelligent automation, can decoct traditional chinese medicine at home combined with mobile terminal APP. The machine itself can achieve timing, appointment, fried, timing, add water and medicine, and other functions. Its goal is to realize user zero-operation after decoction starting, enjoying your daily dose of fresh medicine to ensure optimum efficacy. Users can upload personal prescriptions on the APP end, and get the professional pharmacist's guidance plan and relevant regimen recommendation, so that the decocting machine can become your daily health maintenance partner.

绿植盲盒

扫描种子上的二维码
查看种子种类

根据种植指南种植绿植

用户社区分享种植经历

消费者输入绿植兑换码

贩卖机装配绿植盲盒

制备咖啡渣肥料：
咖啡馆店员投入
每日咖啡渣

# 基于新零售的循环咖啡渣绿植贩卖系统设计

# Design of A Circular Coffee Grounds and Green Plant Sales System Based on New Retail

作　　者：王可欣　黄燕玲　朱惠婷
指导老师：李文杰　董烨楠　张乐凯
所在院校：浙江工业大学

## 设计说明

在新零售新消费模式的语境下，本设计充分利用咖啡所承载的社交属性，将咖啡渣堆肥与绿植贩卖相结合，推出含咖啡渣肥料的绿植盲盒这一衍生产品，并将这一服务融入咖啡馆的会员服务体系。该设计不仅能帮助咖啡馆就地实现咖啡渣循环处理，还能与品牌建设、用户互动结合，促进咖啡馆自身品牌营销与线上线下导通，形成用户—产品—体验的新生态，是一个将可持续与新零售品牌建设结合的全新尝试。

该系统通过软硬件结合提供服务，硬件贩卖机放置在咖啡馆内，可自动对咖啡渣进行堆肥以及绿植盲盒装配，每个绿植盲盒包含含咖啡渣肥料的土壤和一张随机的种子纸，配合软件中的种子图鉴、分享社区等功能界面，用户可以更好体验种植的乐趣。

## Design notes

Green grinder is a system designed to address current coffee grounds recycling issues. Placed in cafes, it enables a daily composting of the coffee grounds on-site. The fertilizer produced will be made into home potting kits together with soil and seed paper and vended on the machine. With this design, coffee grounds could be recycled while consumers could enjoying growing plants.

Each grow kit comes with a cup of potting soil , fertilizer and a piece of mystery seed paper, so no more additional materials for planting are required. The lid of the cup also works as a water collector. The type of seeds on each seed paper is random. Moreover, on the app ,there are functions to make the planting process more interesting.

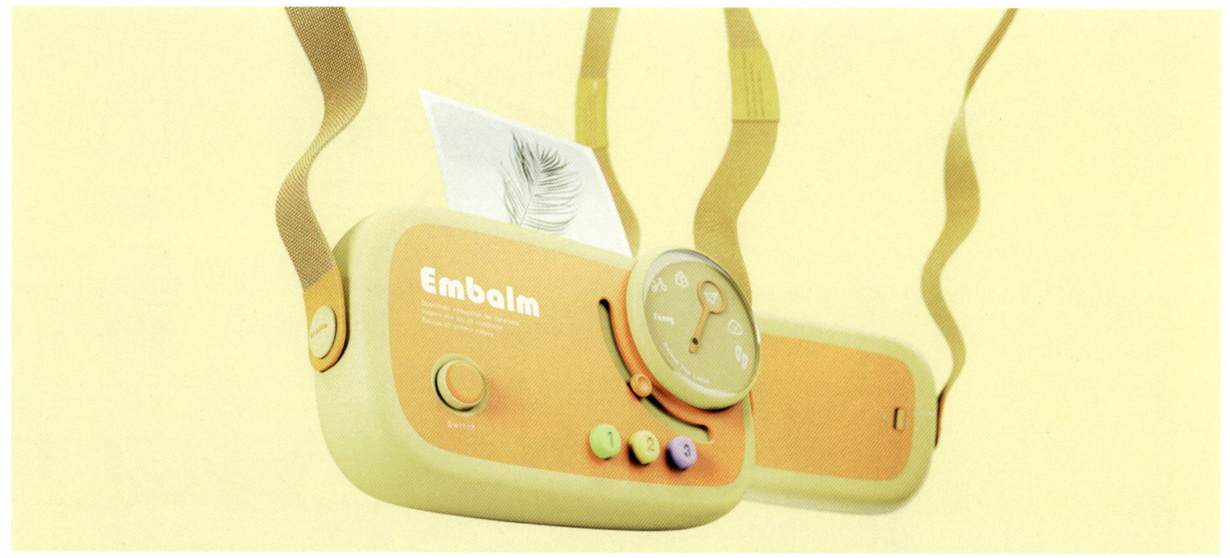

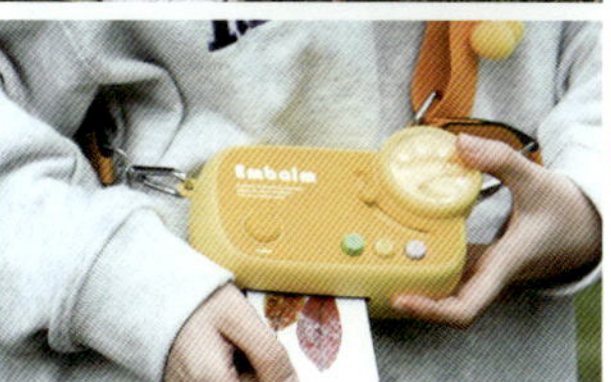

# 植物标本 DIY 塑封机

## DIY Sealing Machine for Plant Specimen

作　　者：刘　韩　孔　媛　刘华瑛
指导老师：冯　迪
所在院校：浙江工业大学

### 设计说明

6~12岁这个年龄段，孩子们对世界有了基本的认识，并且正处于积极探索世界的时期，他们对周围的环境充满了好奇心。本设计的初衷是鼓励孩子们亲手制作植物标本，记录下植物的原始形态，并将之绘制成生动有趣的画作，以此定格时间，留住四季的记忆，使植物的美丽得以长久保存。与传统的植物标本制作相比，后者需要投入更多的时间和精力，而植物标本塑封机提供了一种更加快速和便捷的解决方案，孩子们可在旅途中轻松制作植物标本，同时保持植物的自然形态不变。

### Design notes

Between the ages of 6 ~ 12, children have a basic understanding of the world and are in a period of active exploration of the world and nature, and they are full of curiosity about their surroundings. The original intention of this design is to encourage children to make plant specimens by hand, record the original form of plants, and draw them into vivid and interesting paintings, so as to freeze time, retain the memory of the four seasons, so that the beauty of plants can be preserved for a long time without fading. Compared to traditional plant preparation, which requires more time and effort, the embalm plant preparation machine provides a faster and more convenient solution, enabling children to easily prepare plant specimens on the go, while keeping the natural form of the plant unchanged.

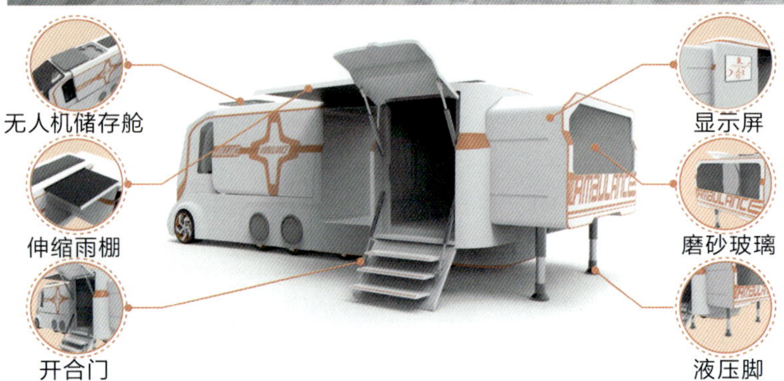


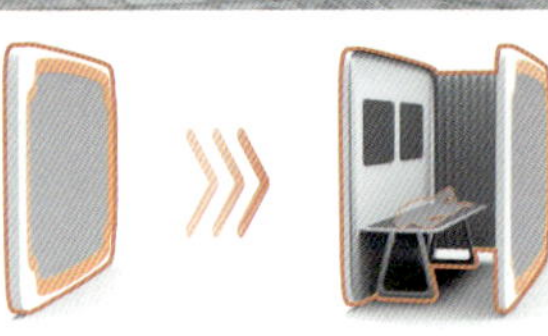

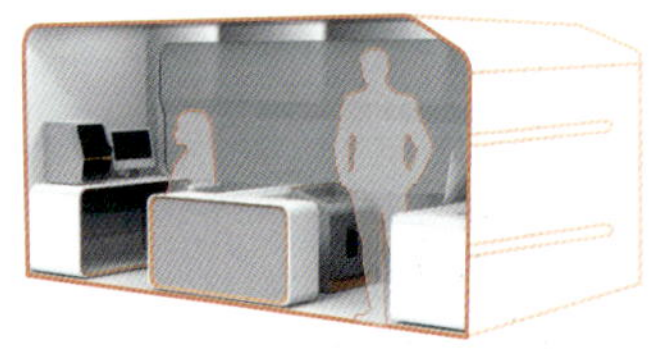



# 急救移动医疗体检车

## First Aid Mobile Medical Examination Vehicle

体检舱　　无人机　　救援舱

作　　者：方俊杰　陈美燕　帅昊宇
指导老师：俞　凯
所在院校：浙江财经大学东方学院

## 设计说明

流动医疗体检车融合了模块化整合和智能化设计理念，将紧急救援、体检服务以及远程医疗服务有效结合。它基于智能化与模块化的设计原则进行了创新设计，目的是解决现有偏远乡村地区医疗资源短缺和设备落后的问题，进而，提升这些地区居民的健康意识和整体健康水平。

## Design notes

The mobile medical examination vehicle integrates the modular integration and intelligent design concept, and effectively combines emergency rescue, medical examination services and telemedicine services. It is an innovative design based on intelligent and modular design principles, aiming to solve the problem of the lack of medical resources and badward equipment in remote rural areas, furthemore, to improve the health awareness and overall health level of residents in these areas.

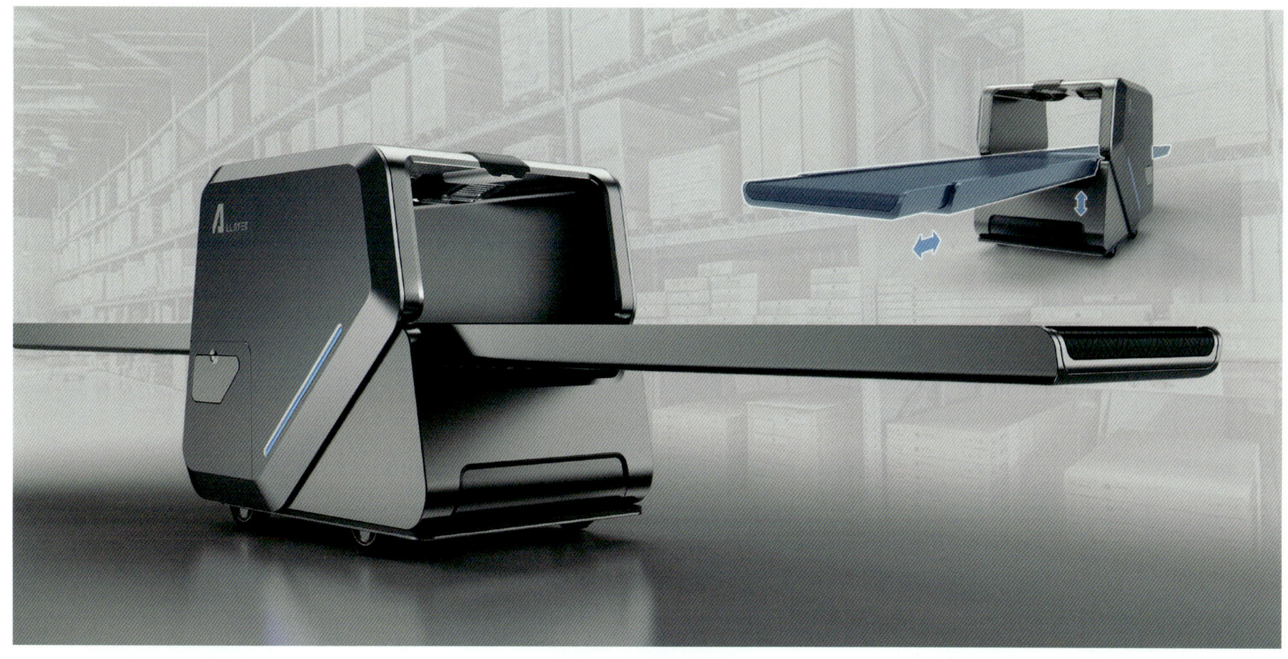

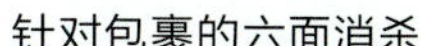

针对包裹的六面消杀

物品智能感应

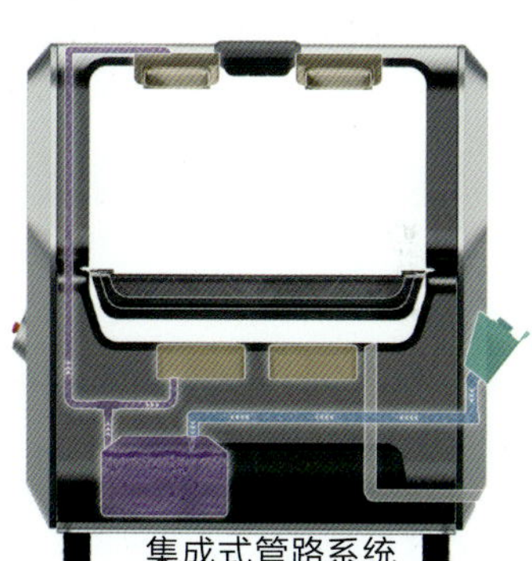

集成式管路系统

# ALLAYER——针对物流卸货口的快递消杀设备

## ALLAYER——Disinfection Equipment for Delivery

作　　者：蔡于婷
指导老师：黄　薇
所在院校：浙江工业大学

### 设计说明

ALLAYER 是一套专为大型物流卸货口设计的消杀设施，它配备了集成式专业化消杀系统，能够为快递包裹提供高强度的全面消毒，同时减少人员操作和降低试剂污染。此外，其可升降和可伸缩的传送带结构可以适应不同的卸货口环境。通过物流扫码模块系统能实现物品、站点、操作人员信息的整合，构建完整的物流溯源链条，协助疾控部门进行精准防控和快速响应。

### Design notes

ALLAYER is a sterilizationl facility designed for large logistics drop-offs, equipped with an integrated, specialized sterilization system that provides high-intensity, comprehensive disinfection of express packages while reducing personnel operations and reagent contamination. In addition, its lifting and retractable conveyor structure can be adapted to different discharge port environments. Through the logistics scanning module, the system can integrate the information of items, sites and operators, build a complete logistics traceability chain, and assist the disease control department to carry out accurate prevention and control and rapid response.

# TORRENT——新型手推式草饼打包割草机

## TORRENT —— Hand-Pushed Straw Cake Packing Mower

作　　者：马文静　罗丹妮　杨朝松
李旺林
指导老师：张　萍　祝　莹
所在院校：合肥工业大学

### 设计说明

这是一款新型的手推式割草机，它通过内置的齿轮转动机构将收集到的杂草压缩成规则的草饼，并通过位于草箱底部的排草结构进行排出，极大地简化了割草过程中的杂草处理。其推手与机身的一体化设计以及机身的透明视窗，为用户提供了舒适且有趣的使用体验。

### Design notes

This is a new type of hand mower, which compresses the collected weeds into regular grass cakes through the built-in gear transmission mechanism, and exhausts them through the grass discharge structure located at the bottom of the box, greatly simplifying the weed treatment during the mowing process. The integrated design of the push-hand and the body, as well as the transparent windows of the body, provide users with a comfortable and fun experience.

森林对抗赛

公路竞速跑道

极限挑战跑道

越野极限跑道



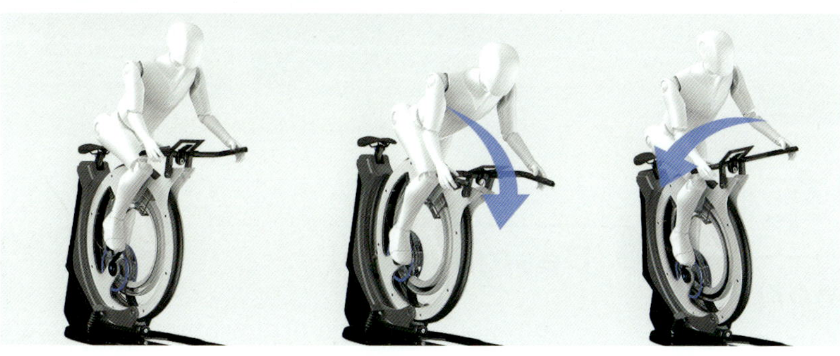

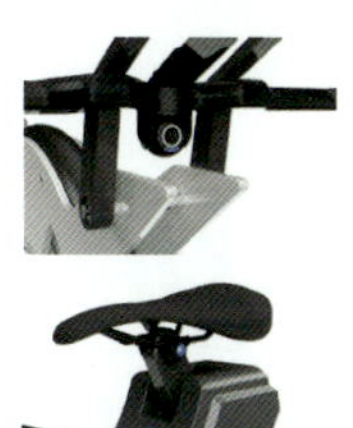

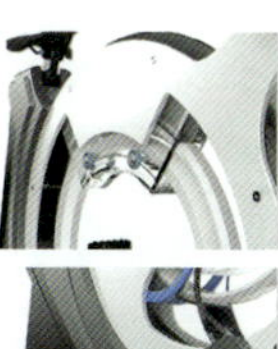

# CLRCLE——健身车

## CLRCLE—— Fitness Bike

作　　者：李炜彬　高梦琦　徐子庚
　　　　　江龙武　夏　天
指导老师：林　伟
所在院校：福州大学

### 设计说明

跑步机和健身车是比较常见的健身器材，由于市面上的健身器材样式单调，普通人难以坚持，最终放弃。因此，设计一款集创新、趣味性和安全性于一体的娱乐型健身车，以“有趣”为核心概念，提高人们的锻炼积极性和持续性。

### Design notes

People are paying more and more attention to health, especially under the influence of the epidemic, treadmills and exercise bikes have become extremely popular. However, due to the general boring tradition of fitness equipment on the market, ordinary people are difficult to adhere to and eventually give up. Therefore, we design this entertainment bike characterized by a collection of innovation, fun and safety, with "fun" as the core concept, to improve people's exercise enthusiasm and sustainability.

# 两栖浒苔清理车

## Amphibious Enteromorpha Cleaning Vehicle



作　　者：王佳豪
指导老师：汪少烽
所在院校：福州大学

### 设计说明

两栖浒苔清理车将两栖载具与浒苔清理设备进行创新结合，解决了海陆不同场景下浒苔难以收集处理的问题。

将海陆两栖载具与浒苔清理设备进行创新结合，拓展工作范围，优化浒苔收集方式，增加浒苔多样处理功能，提高清理的效率，让机械化工作变得更加全面。海面浒苔清理工作与陆地的清理工作可以连贯起来，加强工作线的完整度，又不影响处理浒苔的其他方式，将前期的步骤结合成为收集处理的集合体，也为进行两栖工作的载具提供了更多的可能性和创新性的拓展。

### Design notes

Amphibious Enteromorpha cleaning vehicle innovatively combines amphibious vehicles with Enteromorpha cleaning equipment, which solves the problem that Enteromorpha is difficult to collect and deal with in different sea and land scenes.

Innovatively combine amphibious vehicles and Enteromorpha prolifera cleaning equipment, expand the scope of work, optimize the collection mode of Enteromorpha prolifera, increase the diversified processing functions of Enteromorpha prolifera, improve the cleaning efficiency, and make the mechanized work more comprehensive. The sea surface Enteromorpha cleaning work and land cleaning work can be connected to strengthen the integrative of the work line without affecting other ways of dealing with Enteromorpha prolifera. The early steps are combined into an aggregate of collection and treatment, which also provides more possibilities and innovative expansion for amphibious vehicles.

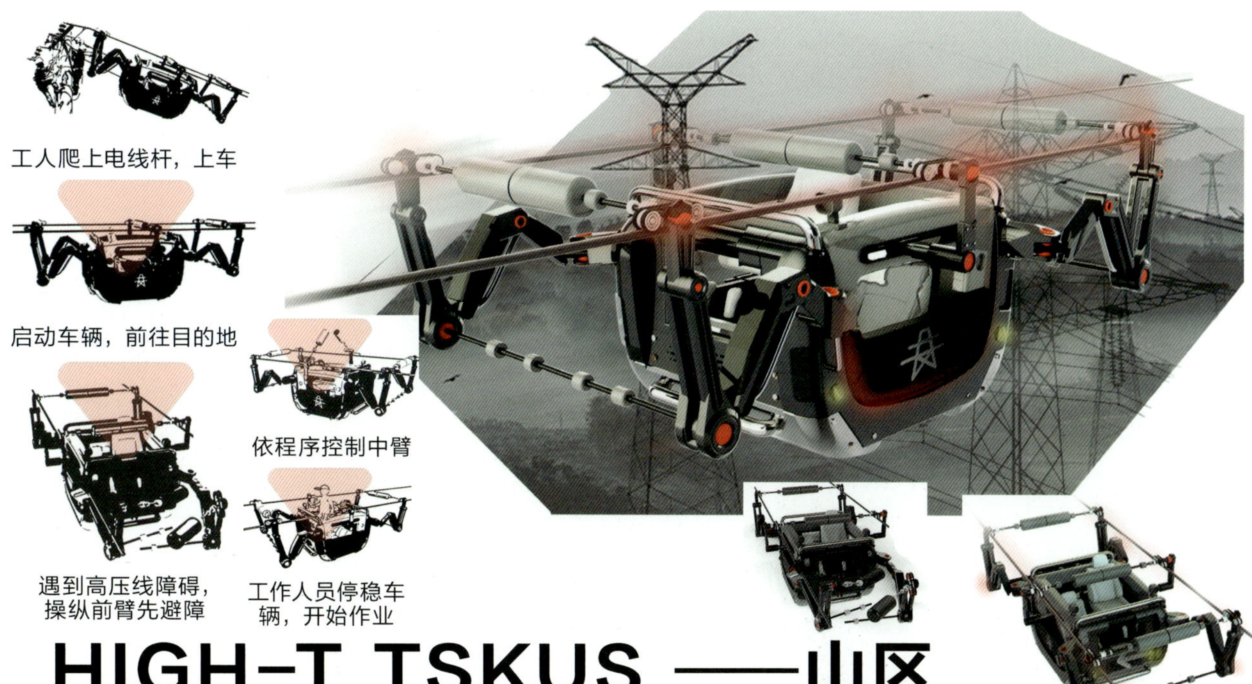


# HIGH-T TSKUS ——山区高压线作业人员专用缆车设计

## HIGH-T TSKUS——Design of A Dedicated Cable Car for High-Voltage Line Operators In Mountainous Areas

作　　者：庄　彬
指导老师：唐　刚
所在院校：厦门大学嘉庚学院

### 设计说明

TSKUS 是一款专为山区高压线缆作业人员设计的载具。考虑到高空作业人员的安全性和工作效率，我们设计了这款不仅便于操作而且辅助作业的设备，以帮助高空作业人员更加安心和高效地完成工作。它有三大优势，其一，它能够有效解决高空作业人员在维修工作中频繁移动的困难；其二，它确保作业人员在高空中安全工作，无需担忧诸如高处坠落、触电等各类职业风险；其三，它减少了作业人员反复上下高压线塔和长途跋涉的麻烦。该载具配备了三对机械臂，用于越过各种线缆障碍物，同时车身内部设有两个可升降座椅，并且在车身后部配有两架梯子，便于工作人员到达更高位置进行线路的维护和保养。

### Design notes

TSKUS is a vehicle designed for high-voltage cable operators in mountainous areas. With safety and efficiency in mind, we designe this device that is not only easy to operate, but also assists the work of the aerial workers to help them complete their work with greater peace of mind and efficiency. There are three major advantages, First, it can effectively solve the difficulties of frequent movement of aerial workers in maintenance work; Second, it ensures operators work safely in the high air without worrying about occupational risks such as falling from heights and electric shock; Third, it reduces the trouble of operators repeatedly going up and down the high-voltage tower and traveling long distances. The vehicle is equipped with three pairs of robotic arms for crossing various cable obstacles, two adjustable seats inside the body, and two ladders at the rear of the body to allow crews to reach higher positions for line maintenance.

工具区形态

维修工具箱

取水端形态

电动气泵

NFC 识别

工具箱

出水口

水位线

# 城市呼吸健康骑行共享智能服务触点——骑行浮标产品设计

## URBAN RESPIRATORY HEALTH CYCLING SHARED INTELLIGENT SERVICE TOUCHPOINT —— Cycling Buoy Product Design

作　　者：苏晋龙
指导老师：欧阳芬芳
所在院校：华侨大学

### 设计说明

用户在骑行过程中往往会遇到骑行车状态不佳以及缺水问题，为了解决用户在骑行过程中途维护和生命补给问题，以树洞为产品造型，设计了骑行浮标，内含工具箱存放维修设备，以及取水装置。在造型上，树洞本身有空间的意义，并且是绿色环保的，结合品牌理念，使用科技蓝涂装，以达到科技服务自然的产品效果。

### Design notes

Users often encounter problems such as poor vehicle condition and water shortage during cycling. To solve the problem of life supply and mid-way maintenance for users during cycling, a cycling buoy is designed with tree holes as the product intention. In terms of shape, the tree hole itself has the meaning of space, and is green and environmentally friendly. Combined with the brand concept, it is painted with technological blue to achieve the effect of technology serving nature.

# 下肢残障人群模块化可穿戴设计

## Modular Wearable Design for People With Lower Limb Disabilities

作　　者：陈安冉
指导老师：屈新波
所在院校：郑州轻工业大学

## 设计说明

这款产品打破传统假肢的概念，主要针对下肢截肢且保存关节的患者，减少一定的产品承重性，对关节起到了保护作用。打破假肢需要定制的概念，可自行进行调试，对于胖瘦高矮的人群都适合，并具有内部平衡感应器，膝盖配置可进行拆卸调节，加入按摩板块，方便出行，且具备一套属于自己的系统化设计，配有不同的配件，以供在不同的场合进行选择。

每个人的身体都是有差别的，对于假肢这样的产品，一般都需要进行定制，但是定制的假肢又会对关节产生一定的挤压，对健全的部分肢体造成二次伤害。本产品打破定制的传统，改为可调节的，可以调整宽度，也可以调整高度，针对重量也进行了特别的设计，使用模块化设计，通过芯片感应来进行运行，同时加入 App，让产品成为了一个系统。

## Design notes

This product breaks the traditional concept of artificial limbs, mainly for patients with lower limb amputation and joint preservation, reduces the bearing capacity of the product, and plays a certain role in protecting joints. It breaks the concept that artificial limbs need to be customized and can be debugged by itself. It is suitable for fat, thin and tall short crowd, and has an internal balance sensor. The knee configuration can be disassembled and adjusted, and the massage plate is added to facilitate travel. It also has its own systematic design with different accessories.

Everyone's body is different. Generally, products such as artificial limbs need to be customized, but customized artificial limbs will squeeze joints to some extent, which will cause secondary damage to some healthy limbs. This product breaks the tradition of customization and is adjustable, with adjustable width and height. It has also been specially designed for weight, using modular design, running through chip induction, and adding App to make the product a system.

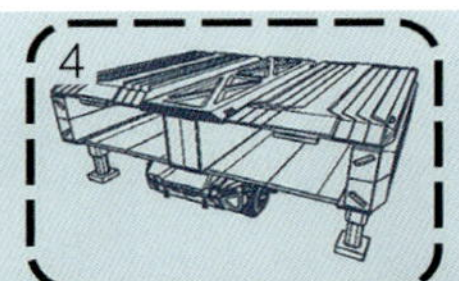

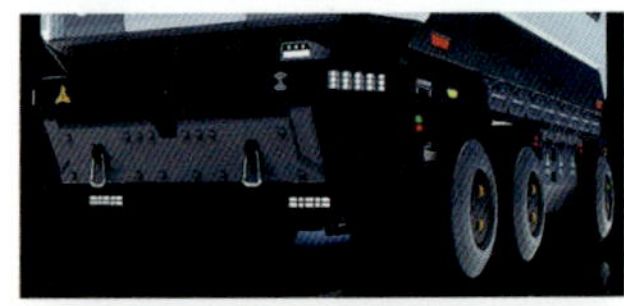

# 灾区智能移动医疗车

# Intelligent Mobile Medical Vehicles In Disaster Areas

作　　者：樊逢清　王钰齐　杨文鹏
　　　　　付连铖　李　坤
指导老师：周莉莉　薛富强
所在院校：郑州经贸学院

## 设计说明

智能移动医疗车主要用于在地震、洪涝、疫情等灾害情况下，当医疗基础设施遭受自然灾害破坏或无法满足需求时提供支援。在重大自然灾害发生之后，智能移动医疗车能够伴随医疗救援队伍前往灾区，快速建立流动医院，并执行医疗物资的运输任务。在救援队伍行进途中，智能移动医疗车可以利用雷达扫描周围环境，为救援车队提供道路安全预警信息，并且与前导车辆协同，实现精准的车队内部无人驾驶。

## Design notes

Smart mobile medical vehicles are mainly designed to provide support in the case of earthquakes, floods, epidemics and other disasters, when basic medical facilities are damaged by natural disasters or cannot meet the needs. In the aftermath of major natural disasters, intelligent mobile medical vehicles can accompany medical rescue teams to disaster areas, quickly establish mobile hospitals, and carry out the transportation of medical supplies. On the way of the rescue team, the intelligent mobile medical vehicle can use radar to scan the surrounding environment, provide road safety early warning information for the rescue team, and cooperate with the leading vehicle to achieve accurate driverless inside the team.

之前

之后

把针头一端直接扎在墙上
由防滑垫一端固定纸张等

# U 型图钉

## U-Shaped Thumbnails

作　　者：朱浩然　李盛文　白浩男　孙福安
指导老师：孙许方
所在院校：中原工学院

### 设计说明

普通图钉在从盒中抓取时，十分容易被随机的针头扎伤，这款 U 形图钉将两个图钉合在一起，相互保护了对方的针头，在抓取时就不用担心会受伤了。并且在使用 U 形图钉时还可以避免刺穿纸张，它另一端加有防滑薄垫，把针按压到墙上的压力足够固定纸张，特别是在固定相片这种特殊材料时优势更加明显！

### Design notes

We are easily pricked by random needles of ordinary pushpin when grabbing from the box, and this U-shaped pushpin combines the two pushpins together to protect each other's needles, so you don't have to worry about getting hurt when grabbing. And when using U-shaped pushpins, you can also avoid piercing the paper, because it is added with a non-slip thin pad at the other end, and the pressure of pressing the needle against the wall is enough to fix the paper. The advantage of this pushpin is even more obrious when tackling specid material like photo.

# 星球探测车

## PLANETARY EXPLORATION

作　　者：杨淑然　高雪纯　匡品品
　　　　　谢苗苗
指导老师：王　峥
所在院校：河南工业大学

### 设计说明

这款星球探测车是一款用塑料制成的手板模型玩具车，适合三岁及以上儿童使用，是孩子们日常玩耍的伴侣，在设计上采用活力橙与神秘黑相结合。这款产品在激发儿童的创造力和想象力方面具有很大的助益，属于益智型玩具。

### Design notes

The planet rover is a plastic hand-board model toy car for children aged three and up. It is not only a companion for children's daily play, but also a combination of vibrant orange and mysterious black in its design. This product is of great help in stimulating children's creativity and imagination, and belongs to the educational toy.

# 基于老旧社区生活污水管道清理设备产品设计

# Design of Cleaning Equipment for Domestic Sewage Pipelines in Old Communities

作　　者：王东海　刘　伟　侯有朋
马佳伟　陈倩丽
指导老师：王　璐　郭永欣
所在院校：河南城建学院

## 设计说明

本产品是对现有老旧社区生活污水管道进行疏浚的设备，改变外观及喷头作业方式，在产品功能上加入收纳装置，使产品更加一体化，方便人工操作。高压水射喷头加入轴承装置，由直射水流式改良为螺旋式喷水，从而达到更大面积的清洗管道内壁，并避免高压直射水流对老旧管道的损耗。

## Design notes

By improving the appearance and nozzle operation mode of the existing products, this design betters the existing pipeline dredging equipment in old communities of the city. The storage device is added to the product function to make it more integrated and convenient for manual operation. The high-pressure water jet nozzle is improved from direct water flow type to spiral water jet by adding bearing device. Thus, the inner wall of the pipeline can be cleaned in a larger area, and the loss of the old pipeline caused by the high-pressure direct water flow can be avoided. At the same time, the design and operation are more in line with ergonomics and operation logic.

金
奖

银
奖

铜
奖

优秀奖



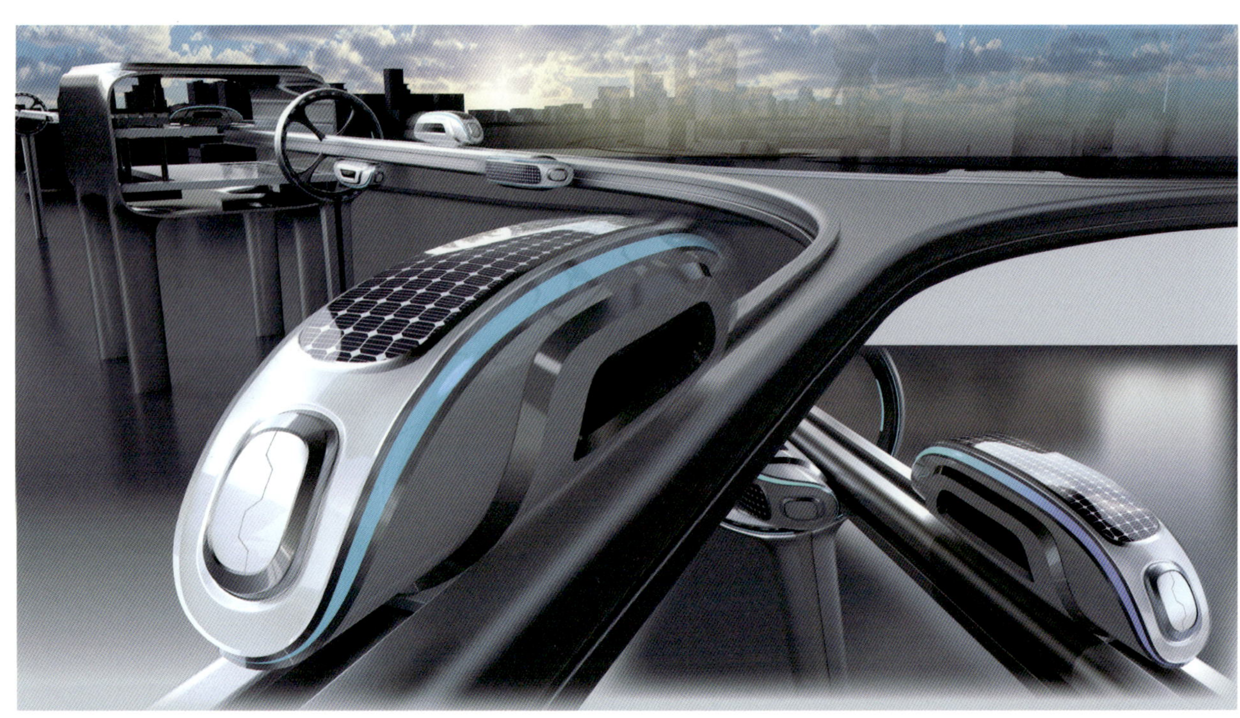

# 基于光伏能源的低空快递运输系统

## Design of Low altitude Express Transportation System Based on Photovoltaic Energy

作　　者：段宇浩
指导老师：杜妍洁
所在院校：湖北美术学院

### 设计说明

该设计方案通过对光伏能源、物流行业以及国内外相关案例进行研究，总结并分析其优劣之处，再对未来可能的运输方式进行创新设计。从绿色可持续的角度出发，分析交通运输的本质需求，通过对我国智慧交通现状进行剖析，提出基于光伏能源的低空快递运输系统新路径，探索低空领域的交通运输发展，为未来的交通运输发展提供新的思路。

### Design notes

This topic first summarizes and analyzes the advantages and disadvantages of photovoltaic energy, express transportation industry, and related cases at home and abroad, and then carries out innovative design for possible transportation modes in the future. Starting from the characteristics of the era of green and sustainable development, this paper analyzes the essential needs of transportation, analyzes the current situation of smart transportation in China, proposes a new path for low altitude express transportation systems based on photovoltaic energy, explores the development of transportation in the low altitude field, and provides new ideas for the future development of transportation.

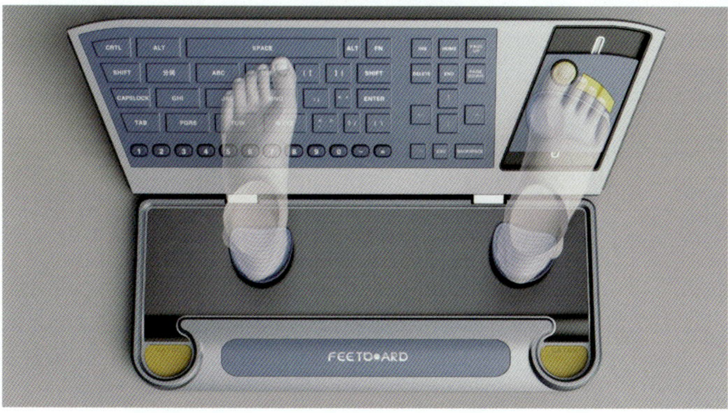

# FEETBOARD
# ——脚用键鼠

## FEETBOARD——Keyboard and Mouse

作　　者：詹浩然　钟美惠　刘凌风
指导老师：汤　军
所在院校：武汉理工大学

足跟垫的上部分
可以在一定角度内旋转

### 设计说明

这是一款专为上肢残疾者而设计的脚用键鼠，他们在使用传统键盘鼠标时都会遇到很大的困难，这款脚用键鼠根据足部特点进行设计，改变了键位布局与键位数量、以及鼠标的操作方式等，使上肢残疾者可以用双脚轻松地进行操作。同时考虑到长时间使用的情况，增加了供足跟搁置的版块，让使用者久用而不易劳累。通过这款键鼠，上肢残疾者仍能用双脚操作电脑。在这个数字经济日益发展的时代，这类人群能够借此拓宽收入来源，改善生活质量。

### Design notes

This is a foot use keyboard and mouse specially designed for people with upper limb disabilities. It is designed according to the characteristics of the foot, which changes the layout and number of keys, as well as the operation mode of the mouse and many other aspects to fit people with uper limb disabilities easily. It adds a plate for heel placement, so that users can use it for a long time without being tired. With this keyboard and mouse, people with upper limb disabilities can still operate computers with their feet. In this era of growing digital economy, this group of people can take this opportunity to broaden their sources of income and improve their quality of life.

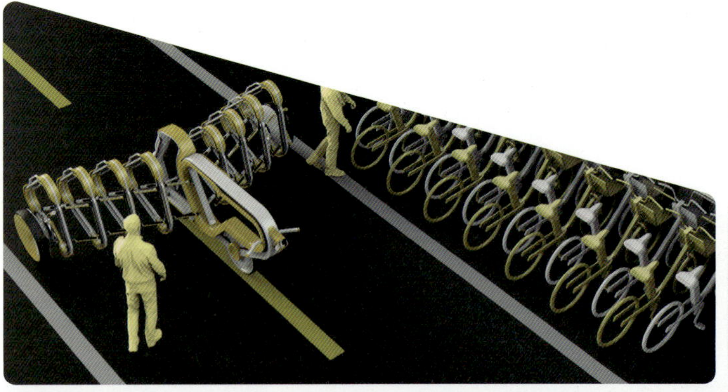

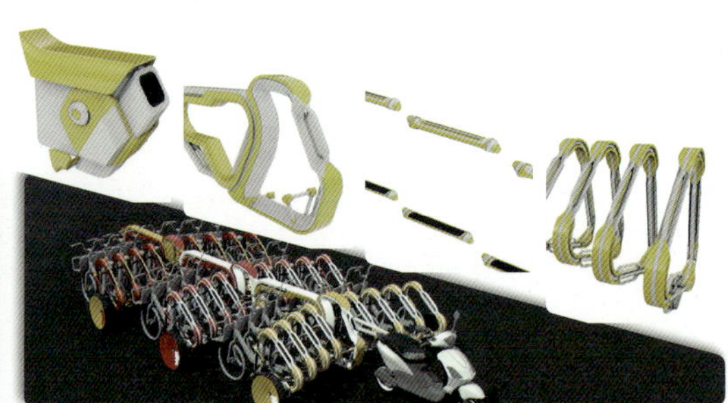

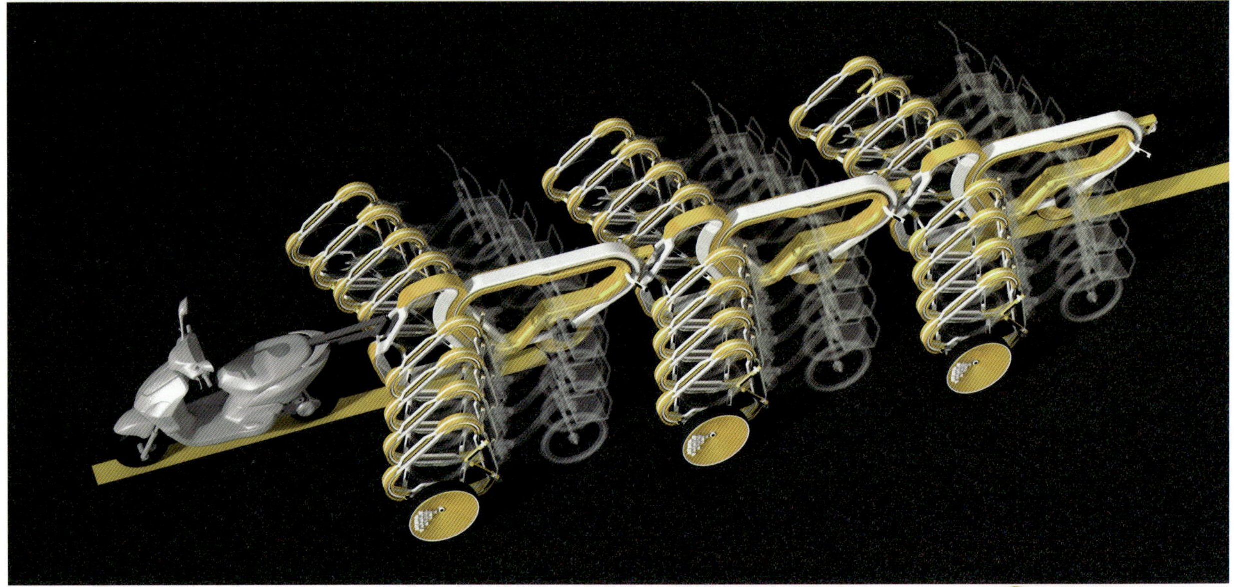

# 共享单车运输车

## Transportation of Shared Bicycle

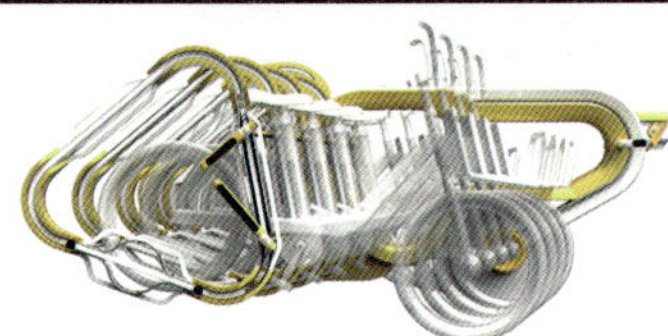

将共享单车后轮固定，前轮着地

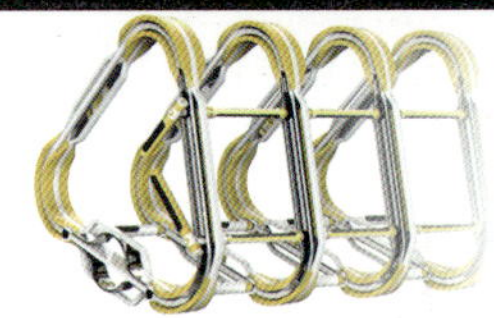

自行车后轮用四杆固定，分别分布在三角形车架的两边，底下的两杆固定

作　　者：李福轩
指导老师：吕杰锋
所在院校：武汉理工大学

## 设计说明

本设计是一款方便快捷解决城市共享单车运输的无动力共享单车运输车。在牵引车的牵引下，采用后轮固定，前轮着地的模式运输共享单车，装载过程和卸载过程操作简单，节约时间，节省人力物力；一辆共享单车运输车一次可运输八辆共享单车，并且采用模块化设计，可多辆运输车同时并存，提高运输效率。

## Design notes

A unpowered shared bicycle transport vehicle that can easily and quickly solve the problem of urban shared bicycle transportation is designed. Under the traction of the tractor, the shared bicycle is transported in the mode of fixed rear wheels and front wheels free on the ground. The loading process and unloading process are simple, saving time, manpower and material resources; One transporter can transport eight shared single vehicles at a time, and the modular design can allow multiple transporters to coexist at the same time, improving transportation efficiency.

1. 向上滑动观察视窗

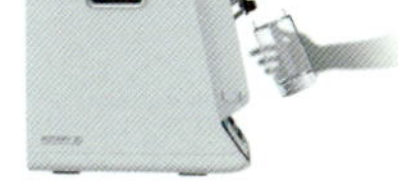
2. 将培养瓶放入设备中

3. 通过显示屏对细胞培养瓶的转速进行设置

4. 通过显示屏对设备的湿度、二氧化碳浓度、湿度、pH 值进行设置

5. 通过显示屏进行显微成像，观察细胞的实时生长状况

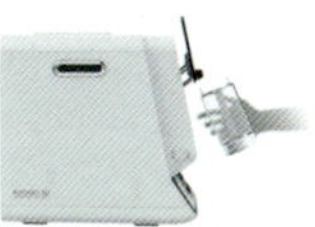
6. 取出细胞培养瓶，对细胞培养瓶中的细胞进行研究

# 旋转式三维细胞培养系统

## Rotary Three-Dimensional Cell Culture System

作　　者：卢鸣豪
指导老师：康红娜
所在院校：南华大学

### 设计说明

本产品通过旋转式细胞培养瓶能模拟细胞的天然微环境（3D 环境）。与细胞在平坦的培养皿或培养瓶上以单层生长的2D 环境相比，细胞的天然微环境（3D 环境）能够提供自然的生长动力学和细胞附着。该系统具有二氧化碳、温度、湿度、pH 值的监测和控制、细胞培养瓶旋转转速控制，细胞显微成像、灭菌等功能，是一个具有多功能，简化细胞培养操作步骤，提高三维细胞培养的效率，实用性、便捷性的产品。

### Design notes

This product can simulate the natural microenvironment of cells (3D environment) through a rotating cell culture bottle. Compared to the 2D environment where cells grow in a monolayer on a flat culture dish or bottle, the natural microenvironment of cells (3D environment) can provide natural growth kinetics and cell adhesion. This system has functions such as monitoring and controlling carbon dioxide, temperature, humidity, and pH, controlling the rotation speed of cell culture bottles, cell microscopy imaging, sterilization, etc. It is a multifunctional product that simplifies the operation steps of cell culture, improves the efficiency, practicality, and convenience of three-dimensional cell culture.

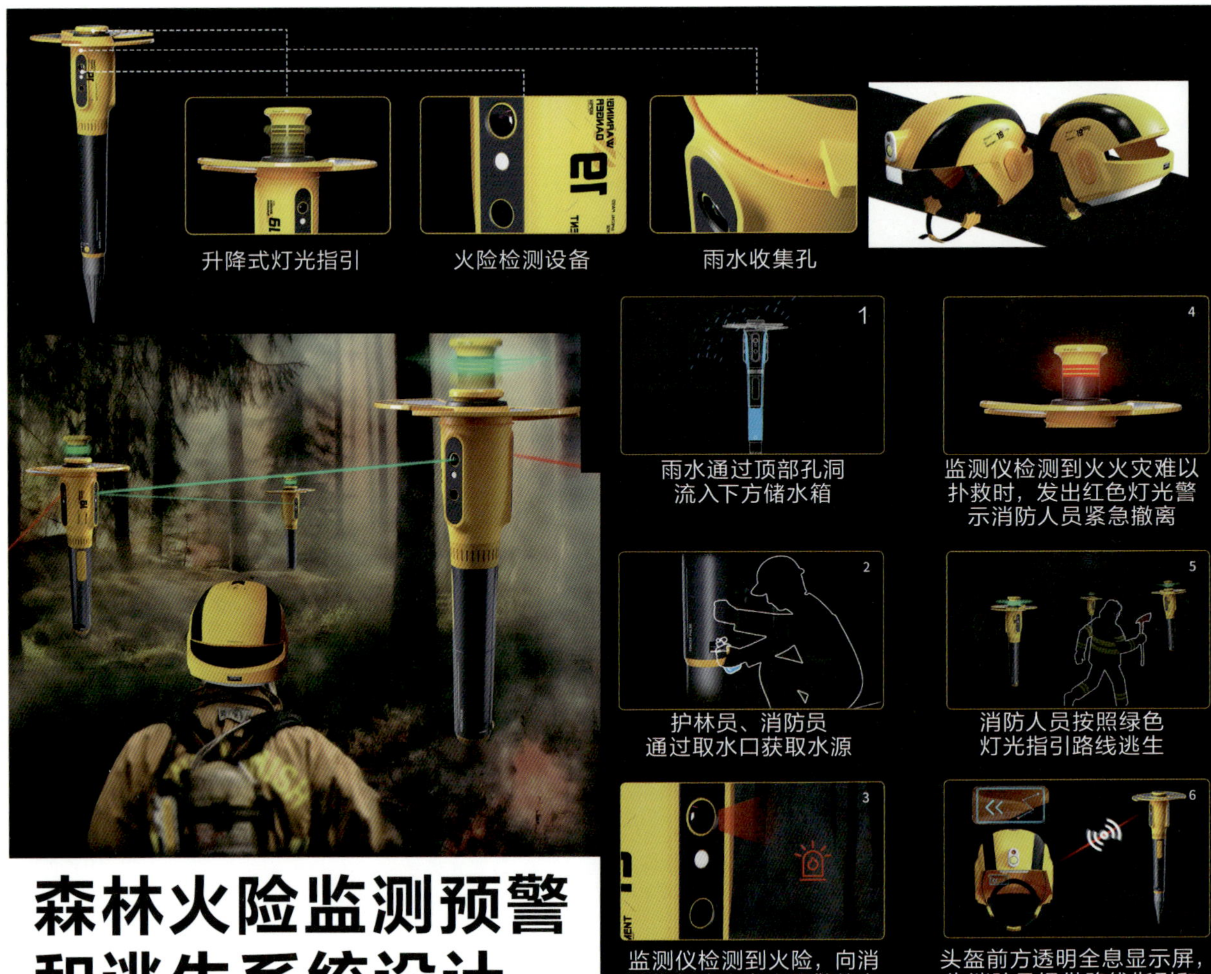


# 森林火险监测预警和逃生系统设计

## Forest Fire Monitoring, Early Warning, and Escape System Design

作　　者：田皓瑞
指导老师：张　黎
所在院校：广东工业大学

### 设计说明

森林火险监测预警和逃生系统是由火险监测仪和智能防护头盔组成。火险监测仪通过烟感检测器、GIS 地理信息系统等技术，全天候对森林进行火险监测，能及时发现森林火灾，快速向消防指挥中心发送预警信息。火险监测仪发出指示语音和闪烁灯光，为消防人员提供安全逃生路线指引。火险监测仪通过雨水收集装置，为护林和消防人员提供一定的水源补给。智能防护头盔前方所带透明全息显示屏，为消防员提供火灾信息显示和逃生路线指引等功能。

### Design notes

The forest fire risk monitoring, early warning and escape system is composed of a fire risk monitor and an intelligent protective helmet. The fire risk monitor monitors forest fires around the clock through technologies such as smoke detectors and GIS geographic information systems. Detect forest fires in time and quickly send early warning information to the fire command center. The monitor emits an instructional voice and flashes lights to provide firefighters with safe escape route guidance. The transparent holographic display screen on the front of the smart helmet provides firefighters with functions such as escape route guidance.

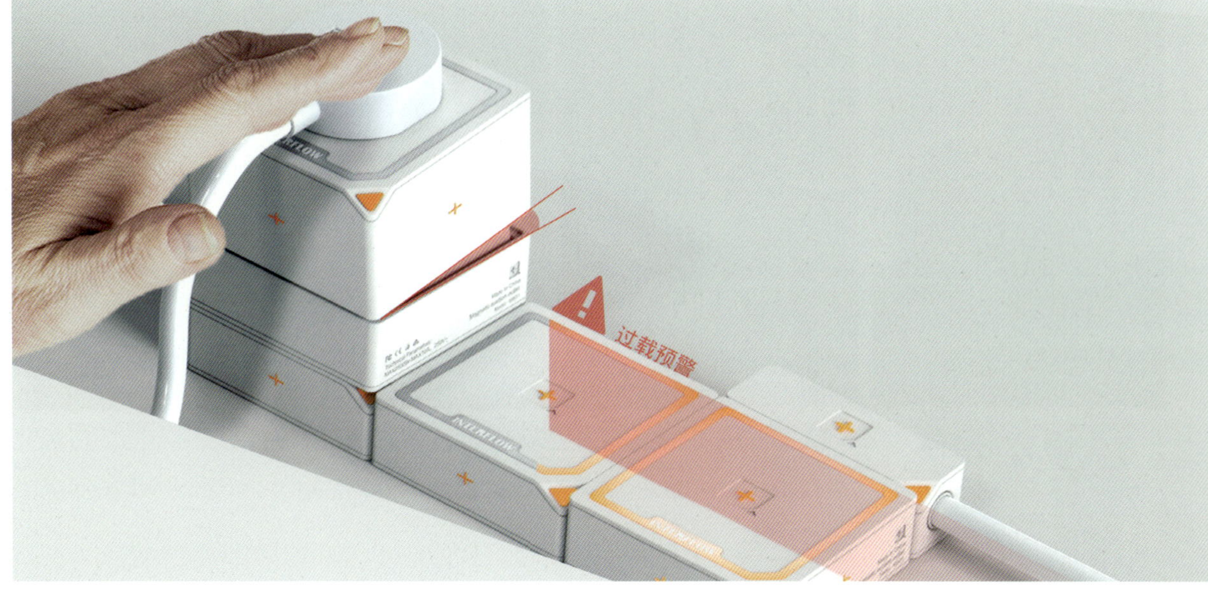


# INTERFLOW 交流电——适老型家用磁组插座设计

拔取断电

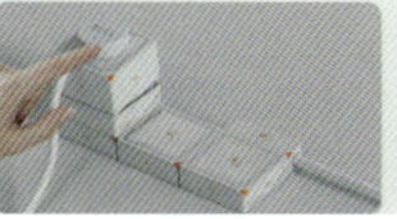
单手按压断电

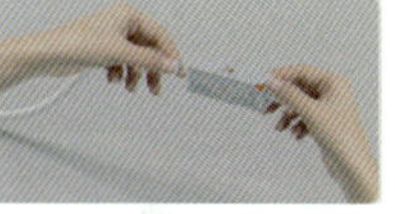
可延长电源线



## INTERFLOW——Modular Socket Design with Magnetic Suction for The Elderly

作　　者：徐艺玮
指导老师：刘　蔚
所在院校：东莞城市学院

### 设计说明

本款磁吸插座产品将无线接电技术与传统插座相结合，意在搭建传统产品与新兴技术之间沟通的桥梁，减少传统电器因技术不匹配所造成的被迫淘汰。是基于适老化理念在安全性、易用性、易学性上进行的创新设计。通过对目标用户行为方式、生理及心理方面的调研分析，将设计点聚焦在提升容错率、安全警示及用电引导上，并采用老年群体最熟悉的接断电方式，在最小改变现有条件、降低学习成本的基础上满足目标用户更多需求。

### Design notes

This magnetic socket product combines wireless electricity connection technology with traditional sockets, aiming to build a bridge of communication between traditional products and emerging technologies, and reduce the forced elimination of traditional electrical appliances caused by technology mismatch.

Based on the concept of proper aging, innovative design is carried out in terms of safety, ease of use and learnability. Through the studies of the target user behaviors, physiological and psychological aspects, the design focuses on increasing the rate of fault tolerance, safety warning and guidance, and adopts the most familiar modes of electrictity conection and disconneotion to the elderly group,has the smallest change on the basis of the existing condition, reduces the learning cost of target users and to meet more demand of them.

# 重塑 共生

## Reshaping Symbiosis



作　　者：范怡丽
指导老师：梁　敏　张兆梅
所在院校：广州美术学院

### 设计说明

现代社会在经济高速发展的同时也出现了一系列的环境问题，本系列设计聚焦塑料废弃物对海洋污染的问题，综合运用切割、塑形、染色等工艺手法对废弃塑料瓶和废旧 PVC 材料进行重组再造，赋予废弃塑料新的形态与生命力。探索废弃塑料在服装中循环利用的更多可能性。首先，服装在色彩上提取海洋深浅不一的蓝色，面料上以欧根纱为主面料，把欧根纱切割成起伏变化、大小不一的荷叶边，营造海浪翻涌的效果。其次，把废旧 PVC 切割为同样的形状进行多次扎染，使其呈现蓝紫渐变的效果与主题色彩相呼应。最后，把废弃塑料瓶制作成晶莹剔透的手工花进行装饰。废弃塑料与代表海洋形态的服装浑然一体，实现重塑与共生，服装整体也呈现出了大海般微波粼粼的效果。

### Design notes

With the rapid economic development of modern society, a series of environmental problems have also emerged. This series of design focuses on the pollution of plastic waste to the ocean. The waste plastic bottles and waste PVC materials are reconstituted and reconstituted by comprehensive use of cutting, shaping, dyeing and other technological techniques, so as to give new forms and vitality to waste plastics and explore more possibilities of recycling waste plastics in clothing. Different shades of ocean blue are extracted from the ocean, and organza is used as the main fabric. The organza is cut into flounces of varying sizes and undulation to create the effect of waves surging. Secondly, the waste PVC is cut into the same shape and tie-dyed many times, so that the effect of blue and purple gradient echoes the theme color. Finally, the waste plastic bottles are made into crystal clear handmade flowers for decoration. The waste plastic is integrated with the clothing that represents the shape of the ocean, realizing remodeling and symbiosis. The whole clothing also presents the effect of sparkling wave in the sea.

# 一士·趣味化灯具设计

## YISHI · Interesting Lamp

作　　者：梁宗益
指导老师：刘颍希　裴悦舟
所在院校：广州美术学院

### 设计说明

“一士”灯具是趣味化照明产品设计的可行性方案，通过对照明产品的多维度研究与设计，将趣味化与仪式感融入照明产品中，满足了新生活方式下人们对照明产品的情感需求，表达了人们对生活仪式感的美好向往。

### Design notes

"YISHI" lamps are a feasible solution for the design of interesting lighting products. Through the multi-dimensional research and design of lighting products, the fun and sense of ceremony are integrated into lighting products, which meets the needs of people in the new lifestyle. The emotional needs of lighting products express people's beautiful yearning for the ritual sense of life.

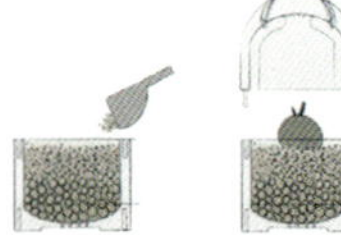

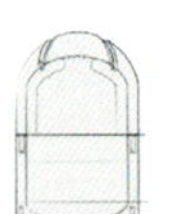
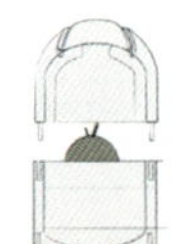

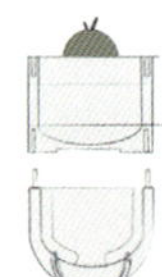
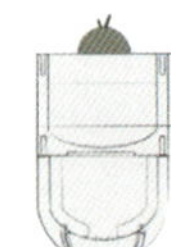

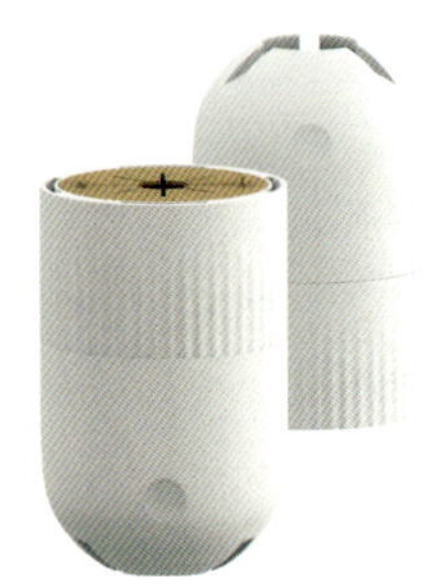

# 弹竹

## Bullet Bamboo

作　　者：吴雪凤　符思静　罗莹莹
指导老师：杨赛男
所在院校：广东省外语艺术职业学院

### 设计说明

弹竹是一款专为幼苗植物设计的便携式可降解花盆，旨在解决商家在打包植物时使用不可降解材料导致的环境污染和资源浪费问题。另外，在运输过程中，弹竹能够有效保护植物避免受到碰撞损伤。该产品采用透气且环保的竹纤维可降解材料制成，可在幼苗生长期作为培育杯使用。当植物生根后，可将花盆直接插入土中，它会在土壤中完全降解。这样不仅环保，还能极大地提升用户的便捷体验。

### Design notes

Bullet bamboo is a portable biodegradable pot designed for seedlings to solve the problem of environmental pollution and resource waste caused by the use of non-biodegradable materials when packaging plants. In addition, during transportation, bamboo can effectively protect plants from collision damage. The product is made of breathable and environmentally friendly bamboo fiber degradable material and can be used as a breeding cup during the growth period of seedlings. When the plant takes root, the pot can be inserted directly into the soil, where it will completely degrade. This is not only environmentally friendly, but also greatly improves the user's convenient experience.

# Yè 椅

## Yè Chair

作　　者：谢兆文　张　竟　薛雨尔
指导老师：陈振益
所在院校：五邑大学

### 设计说明

这款椅子的设计灵感源自树木及其枝叶的自然形态。其骨架模仿了树枝的连接方式，而靠背则呈现出树叶般的曲线弧度，整体造型流畅且细节丰富，赋予了产品一种雕塑般的美感，为室内空间增添了现代感的美学氛围。在功能性方面，椅子的靠背设计扩展了前后翻折的弧度，增强了对使用者的包裹感，同时提供了水平向上的承托空间，兼具了椅子扶手的功能。这种可翻转的靠背与稳固的结构设计，使得人们可以以多种姿势使用椅子，包括正坐、侧坐或反坐，自由伸展和放松，确保了舒适的坐感，并适用于不同的用户和场景。材质上，椅子采用天然木材搭配皮革，提供多种颜色和材质选择。这是一款既注重外观设计又兼顾实用功能的椅子，带给用户独特的使用体验。

### Design notes

The design of this chair is inspired by the natural form of trees and their branches and leaves. Its skeleton mimics the connection of branches, while the backrest presents a leaf-like curve, the overall shape is smooth and rich in details, giving the product a sculptural beauty, adding a modern aesthetic atmosphere to the interior space. In terms of functionality, the backrest design of the chair expands the curvature of the front and back folding, enhances the sense of wrapping to the user, and provides a horizontal upward supporting space, with the function of the chair armrest. This reversible backrest and solid structural design allow people to use the chair in a variety of positions, including sitting, side or reverse, free to stretch and relax, ensuring a comfortable sitting feeling, and suitable for different users and scenarios. In terms of material, the chair is made of natural wood and leather, and is available in a variety of colors and materials. This is a chair that pays attention to both appearance design and practical functions, bringing users a unique experience.

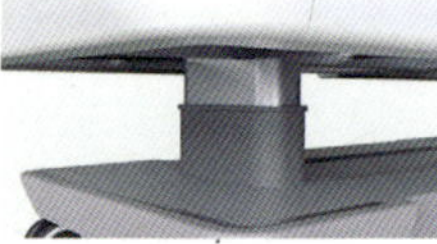
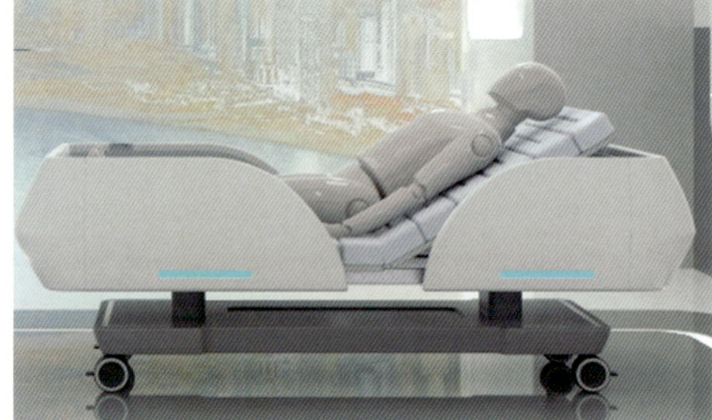
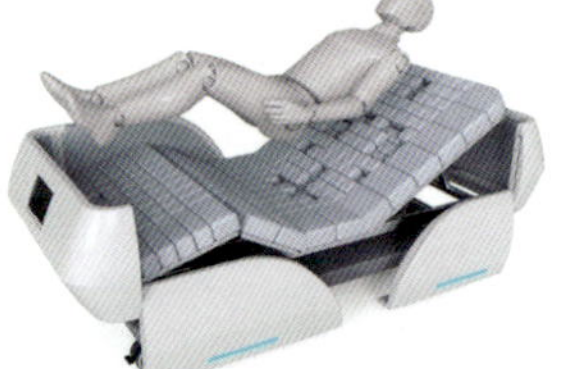
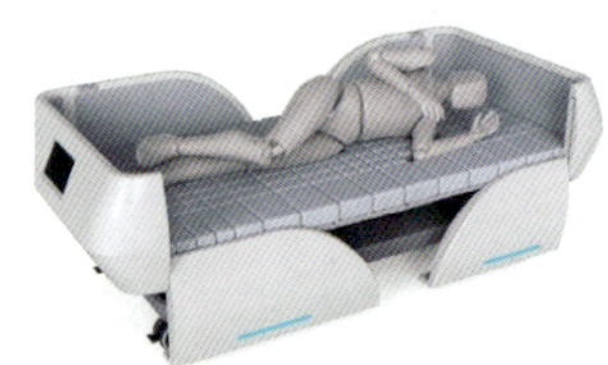

# ICU 防压疮病床设计

## Design of Anti-Pressure Sore Bed in Icu

作　　者：巫晓庆
指导老师：罗向兼
所在院校：广东技术师范大学

### 设计说明

ICU（重症加强护理病房）患者是压疮发生的高危人群，这是因为久卧导致患者同一部位长时间受压。而压疮会增加患者的痛苦，死亡率也会增加。因此压疮已经变成一个全球性的健康问题。

因此，本设计通过两个方面实现预防压疮的发生。第一方面，大动作：多角度翻身。病床床板的翻折实现病人多个体位的改变，避免同一部位长时间受压，达到全身减压。第二方面，小动作：精细化床垫。气垫床的单元分割，使得气垫能精细化放气，易受压部位也精准减压，从而更加有效地预防压疮。另外，病床具备 AI 人工智能系统，自动检测患者情况，方便医生的观察和治疗。

病床的自动翻身功能实现全身减压，并配合气垫床的精准放气实现易受压部位的精准减压，双重预防 ICU 里的患者压疮的发生。

### Design notes

ICU patients are at high risk of developing pressure ulcers because prolonged lying down causes pressure on the same part of the patient body for prolonged periods of time. Pressure ulcers increase patient suffering and increase mortality. As a result, pressure ulcers have become a global health problem.

Therefore, this design achieves the prevention of pressure sores through two aspects. In the first aspect, the big move: turning over from multiple angles. The folding of the bed board realizes the change of multiple body positions of the patient, avoids long-term pressure on the same part, and achieves decompression of the whole body. In the second aspect, small actions: refine the mattress. The unit division of the air mattress enables the air mattress to deflate finely, and the parts that are prone to pressure are also precisely decompressed, so as to prevent pressure sores more effectively. In addition, the hospital bed is equipped with AI system, which can automatically detect the patient's condition and facilitate the doctor's observation and treatment. The automatic turning function of the hospital bed realizes the decompression of the whole body, and cooperates with the precise deflation of the air bed to realize the precise decompression of the vulnerable parts, and double prevents the occurrence of pressure sores in ICU patients.

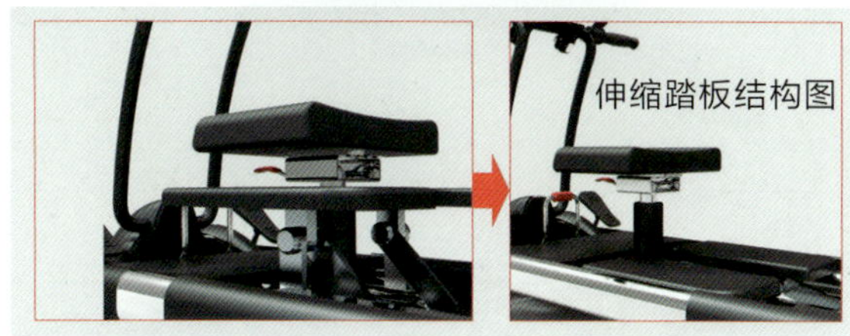


# 陪伴——可伸缩老人代步轮椅车设计

## COMPANION — Retractable Elderly Leisure Transport

作　　者：陈依漫　陈桂钦　伍颖欣　魏晓涵
指导老师：郭　武　邝　芸　何心怡　吴志聪
所在院校：广东岭南职业技术学院

### 设计说明

这是一款针对腿脚不便的老人设计的生活轮椅车，它兼具轮椅和代步车的功能。当老人一个人外出时，可以当代步车使用，也可以两个老人一起结伴出行，此时只需要拉动红色手柄，启动液压杆，即从座位下伸出一个小座位和相应脚踏板。可以在老人单独出行遇到紧急情况时，方便旁人代驾送老人就医，以节省时间，赢得抢救机会。

本设计结构紧凑而巧妙，在进行座位切换时，轮椅车能自动适应座位变化带来的踏板和空间变化。当车缩短成单人驾驶模式时，踏板和小座位自动缩进车体内，当车伸长成俩人模式时，踏板和小座位随动伸出，小座位高度可调。不需要额外的操作，只需要启动液压杆顶出即可，方便高效，人性化十足。

### Design notes

This is a life wheelchair designed for elderly people with limited mobility, which combines the functions of a wheelchair and a commuting vehicle. It can be used as a walking bike by one elder person or shared by too with simply pulling the red handle, activating the hydraulic rod, and extending a small seat and corresponding foot pedal from under the seat. When the elderly travel alone and encounter emergency situations, it is convenient for others to drive them to seek medical treatment, saving time and winning the opportunity for rescue. This design has a compact and clever structure, and the wheelchair can automatically adapt to the pedal and space changes caused by seat changes during seat switching. When the car is shortened to single person driving mode, the pedals and small seats automatically retract into the body. When the car is extended to two person mode, the pedals and small seats extend along with the movement, and the height of the small seats is adjustable. No additional action is required, just activate the hydraulic rod to push it out, which is convenient, efficient, and user-friendly.

# 背负式多功能消防破拆工具设计

## Design of a Backpack Type Multifunctional Firefighting Dismantling Tool



作　　者：朱鸣飞　韩舒阳　陈羽森　杨泽远
指导老师：李纳墨
所在院校：桂林电子科技大学

### 设计说明

这是一款针对消防人员救援设计的背负式多功能消防破拆工具，造型简约，颜色采用消防破拆器材通用的红黑色，对比强烈，强烈的视觉冲击力防止在混乱的场景中丢失器材。部件主要有可更换破拆钳以及电钻，把手设计最突出的特点是快速切换和稳定性，三角形结构防止把手晃动，合并后的垂直握把契合电钻把持方式，后方的旋转把手在上方为破拆钳的手持动作，切换到下方为电钻的手持方式，在摇杆处增加电路转换方式，避免两种驱动开关混乱导致工具无法正常使用。散热口开阔能够更好进行散热，防止内部零件高温损坏延长寿命。

### Design notes

This is a piggyback multi-functional fire demolition tool designed for firefighter rescue, with a simple and stable square design. The color adopts the red and black which is common to fire demolition equipment, the color contrast is strong, and the visual impact prevents the loss of equipment in the chaotic scene. The main functions are replaceable demolition pliers and electric drills. The most prominent feature of the handle design is fast switching and stability, the triangular structure prevents the handle from shaking, the combined vertical grip fits the electric drill holding mode, and increase the circuit conversion mode at the rocker to avoid the confusion of the two drive switches and cause the tool to not be used normally. The open heat dissipation port can better dissipate heat, prevents high temperature of internal parts damage its life span.

# UFO 童年趣味驱蚊夜灯

## UFO Childhood Fun Mosquito Repellent Night Light

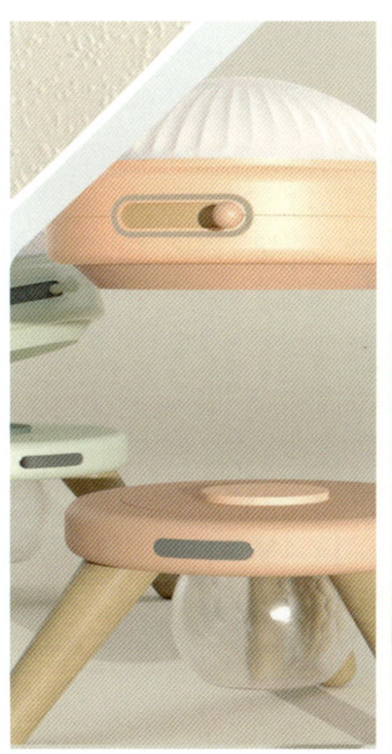

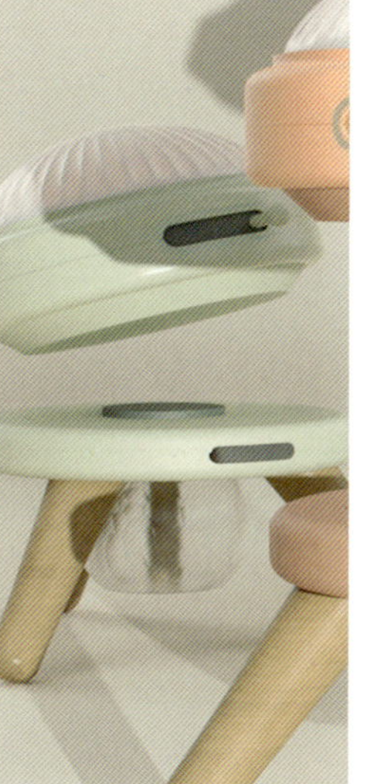

作　　者：尚　磊　李文凯　杜佳凝
杨雅琳　张　帅
指导老师：都　江　刘　聪
所在院校：昆明理工大学

### 设计说明

这是一款为年轻人群设计的驱蚊小夜灯。在使用时它会发出微光，独特的 UFO 造型设计带给用户趣味性和压力的舒缓，为入眠困难的人群提供一个温馨自然的环境，增加体验感和舒适感。

### Design notes

This is a mosquito repellent night light designed for young people. When in use, it emits a faint light, and the unique UFO design provides users with fun and stress relief, creating a warm and natural environment for those who have difficulty to fall asleep, increasing the experience and comfort.

# 基于马家窑彩陶文化的火锅餐具设计

## Design of Hot Pot Tableware Based on Majiayao Colored Pottery Culture

作　　者：张梦莹　闫星云　邓晶晶　冯新雅
指导老师：李　丽
所在院校：兰州理工大学

### 设计说明

此款火锅餐具设计从甘肃马家窑彩陶文化中汲取元素，设计上保留了传统器物的基本属性，贴合当下国际工业制品简约的审美趋向，同时符合工业化生产方式，以食器设计为载体，以传播甘肃传统文化为目标，选取马家窑文化中的彩陶文化进行创新设计实践。

### Design notes

Drawing design elements from Gansu Majiayao painted pottery culture, retaining the basic attributes of traditional utensils in the design, it is designed from the meaning of modern aesthetics, in line with the simple aesthetic trend of current international industrial products, and at the same time in line with industrialized production methods, allowing existence of the past Majiayao painted pottery not only for awakening the new life today, but also for facing the future.

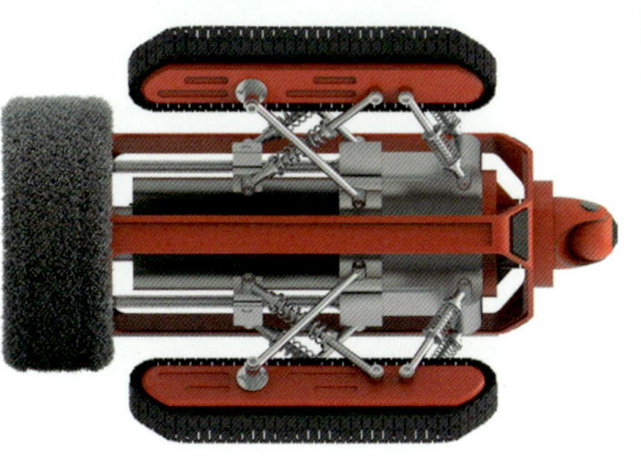

# 模块化多维态势适应型智能窄管清洁机器人

## Modular Multi-Dimensional Adaptive Intelligent Narrow Tube Cleaning Robot

作　　者：杨梦石　胡若晨
指导老师：徐　悬　冯　明
所在院校：北京理工大学

### 设计说明

大型建筑物通风管道、室内多用途狭管的清洁与消毒问题愈发被人们所重视。产品以一种管道检测机器人为技术原型进行设计开发，尾部端口兼容多型清洁与探查接口，为通风管道多型狭管清洁与消毒提供了更有力的保障。同时，产品形态依据应用空间可自动调节，同时对同类型竞品造型呆板笨重、功能单一、效果粗糙等问题进行了优化。该产品造型简洁大方，符合现代工业设计的审美习惯。

### Design notes

In the post-epidemic era, the problem of cleaning and disinfection in ventilation ducts of large buildings and indoor multi-purpose narrow pipes has been paid more and more attention. The product is designed and developed based on a patented pipeline inspection robot for the technical prototype. The tail port is compatible with multiple cleaning and exploration interfaces, which provides a more powerful guarantee for the convenience of cleaning and disinfection of multiple narrow pipes of ventilation pipes. At the same time, the shape of the product according to the application space can be self-adaptive, self-adjustment, meanwhile drawing from the same type of competitive products stiff and heavy, single function, rough effect and other problems were optimized. The product modeling is simple and generous, hard and bright, in line with the semantics of modern industrial design and aesthetic habits.

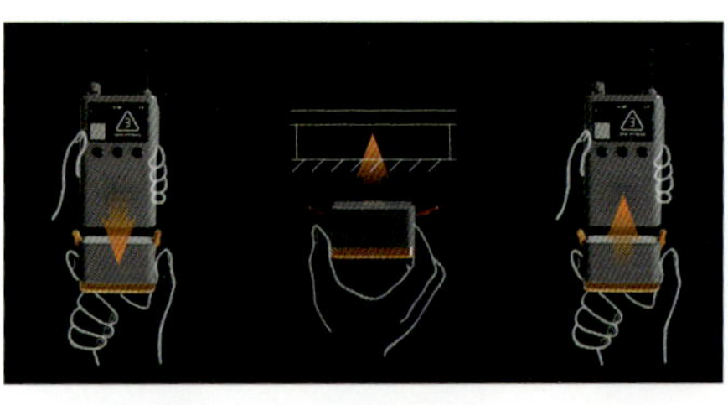

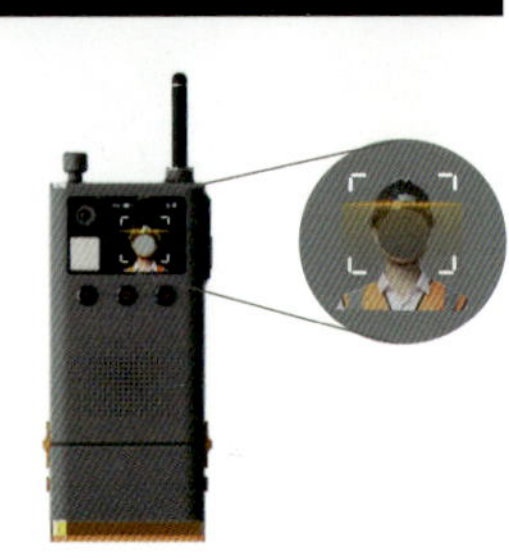

# SOS——高铁上道作业人员安全警示系统

## SOS —— Safety Warning System for Workers Working on High Speed Rail

作　　者：滕雪苗　俞迪恺
指导老师：吴雪松　李子龙
所在院校：湖南大学

### 设计说明

SOS 是一个铁路工人安全保护系统，它结合了工具带、列车接近传感器和多功能终端，以保障铁路工人的安全、舒适和工作效率。

它由三部分构成：一个紧凑的工具带，一个火车接近传感器和一个多功能终端。在开始工作之前，对讲机下面的磁吸式传感器可以被移除并放置在铁路轨道上。当高速列车接近时，传感器将发出警报，让工人离开轨道。所有的工具，包括对讲机，都可以放在单腿工具带上，它还可以作为护膝，保护那些必须长时间单膝工作的工人。45度角的工具通道使工人更容易获得工具。

### Design notes

SOS is a safety protection system that combines a tool belt, a train approaching sensor and a multifunctional terminal to improve the safety, comfort and efficiency of railway workers.

It combines three equipment requirements: a compact tool belt, a train approaching sensor and a multifunctional terminal. Before beginning of work, a magnetic suction sensor under the intercom can be removed and placed on the railway track. When a high-speed train is approaching, the sensor will give out a warning for workers to leave the tracks. All tools, including the walkie-talkie, can be carried on the one-legged tool belt, which also serves as a knee guard to protect workers who must work on one knee for extended periods of time. The 45-degree angled tool access makes it easier for workers to access tools.

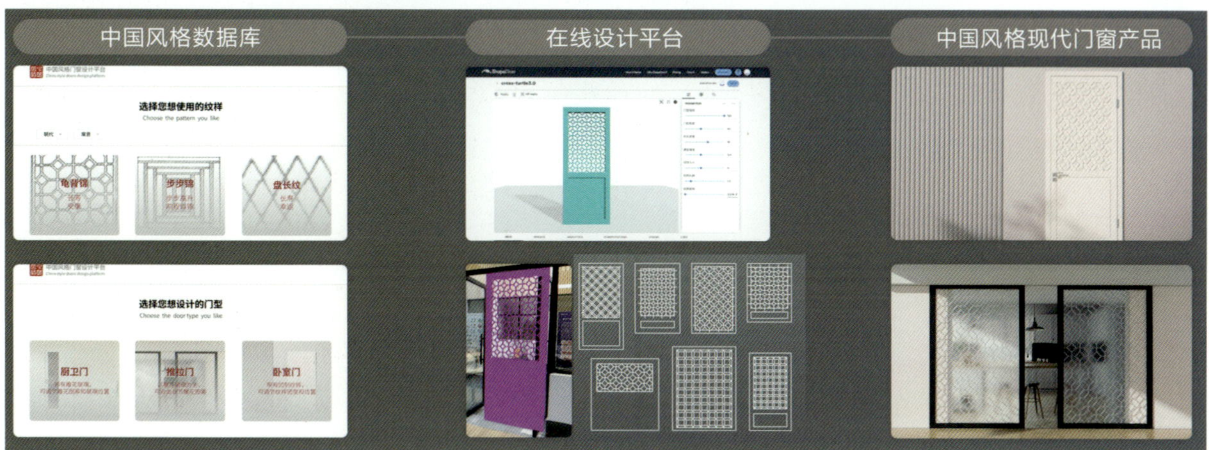


# 中国风格门窗的生成式设计

## Generative Design of Chinese Style Doors and Windows

作　　者：李　昊　张力扬
指导老师：季　铁　郭寅曼
所在院校：湖南大学

### 设计说明

以中国传统格扇门窗的几何格心纹样为研究对象，深入研究其演化逻辑、形制构成、使用场景和文化寓意，结合现代形式工艺进行家居门窗产品的参数化定义与智能生成，最终实现在线设计，为门窗设计师提供了既有文化依据又快捷高效的设计方式。通过这个设计，门窗设计师可以轻松地在网页上调节参数（如长、宽、纹栏大小、纹样密度等）生成中国风格门板，并利用 AR 技术实时查看生成效果，体验新的门窗定制流程。

### Design notes

We take the geometric lattice pattern of traditional Chinese doors and windows as the research object, deeply study its evolution logic, form composition, use scenario and cultural symbolism, combine with modern form process to define and intelligently generate the parameters of home door products. Finally, it is deployed to the web platform to realize online design, providing door and window designers with a cultural basis and a fast and efficient design method. Through this design, door and window designers can easily adjust the parameters (such as length, width, pattern bar size, pattern density, etc.) on the webpage to generate Chinese style door panels, and use AR technology to view the generated results in real time to experience the new door and window customization process.

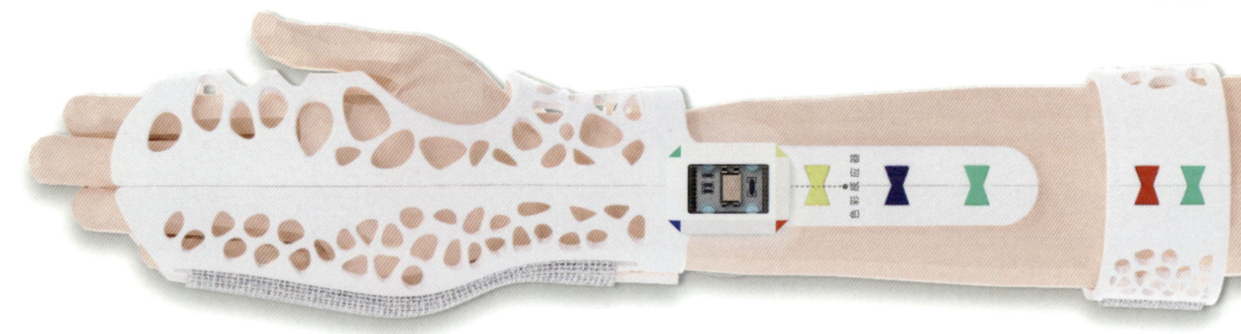

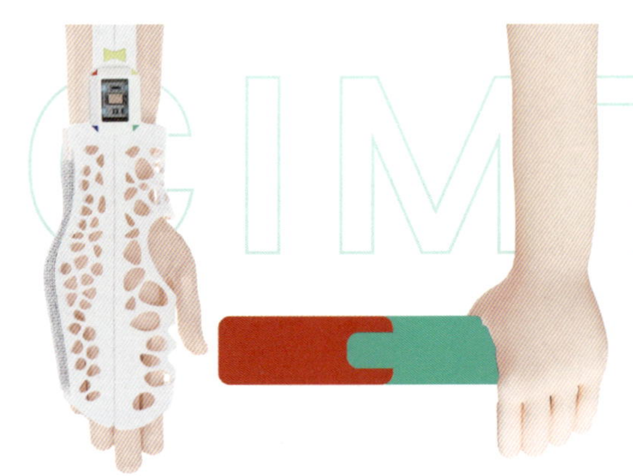

# 基于 CIMT 与 HABIT 的偏瘫儿童辅助臂设计

## Design of Assistive Arms for Children with Hemiplegia Based on CIMT and HABIT



作　　者：韦佳伶
指导老师：欧　静
所在院校：湖南大学

### 设计说明

本设计课题面向因上肢健全但单侧上肢功能受损而出现习得性废用现象的偏瘫儿童。结合优势互补的两个临床作业疗法：CIMT（constrained induced movement therapy-限制健侧诱导患侧）与HABIT（hand-arm bimanual intensive training-手－臂双侧徒手强化训练），以音乐游戏和榫卯拼接玩具为载体的辅具设计，更好地实现两个疗法的同时为二者的结合建立桥梁。让7-11岁的偏瘫儿童通过拼接榫卯结构的乐器，以游戏的方式单手进行简化版弦类、鼓类以及指挥等活动进行手指、手腕、手肘、手臂、肩膀的综合训练。

### Design notes

This design subject is aimed at children with hemiplegia who have learned disuse due to a healthy upper limb but impaired unilateral upper limb function. Combined with two complementary clinical occupational therapies: CIMT (constrained induced movement therapy- limiting the healthy side to induce the affected side) and HABIT (hand-arm bimanual intensive training- hand-arm intensive training), the auxiliary equipment design based on music games and mortise and tenon splicing toys can better realize the two methods at the same time to build a bridge for their combination. Let 7~11 years old hemiplegic children through the tenon and tenon structure of musical instruments, play one-hand simplified version of string, drum, and conduct activities for fingers, wrist, elbow, arm, shoulder comprehensive training.

外部滚轮方便移动 两种运动需求轻松切换 居家环境融合

与桌面类家居结合使用

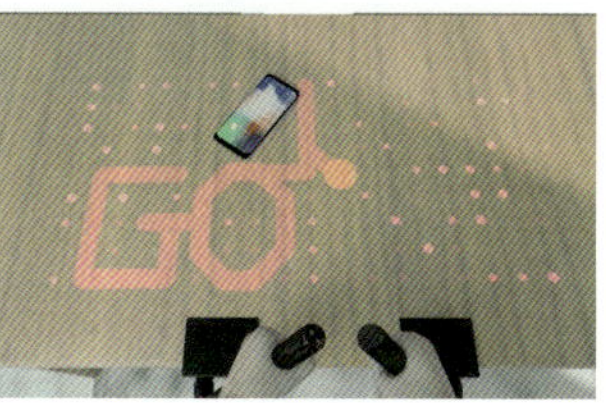


实时训练交互

# 健间单车

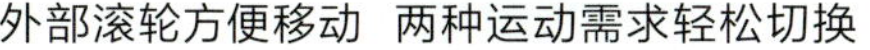

## BITWEEN

作　　者：陈罗千绘　甘陈宇
指导老师：刘　胧　樊　中
所在院校：同济大学

### 设计说明

健间单车，旨在为生活和工作繁忙中的年轻初级健身爱好者提供一种新型的健身工具。它允许用户在日常间隙时间探索健身的可能性。BITWEEN 具备“日常坐姿健步模式”和“健身站姿登山模式”，既能减轻长时间久坐和不良坐姿带来的负面影响，又可通过高效锻炼有针对性地增进身体健康。其紧凑设计和外围滚轮使得模式切换、携带和收纳都更加便捷。此外，通过智能 app 监测用户的久坐时间，并提供定时提醒；利用即时反馈和智能模拟来预期成果，增强积极的健身动力；通过设定日常活动化的健身小任务和周期性总结，使健身成为生活的一个不可或缺的部分，从而激发成长型思维，并帮助用户摒弃“缺乏意志力”的借口。与 BITWEEN 携手，我们可以更好地利用时间，丰富生活，不断成为更好的自己。

### Design notes

BITWEEN (bike-between) is designed to provide a new way of fitness for young primary fitness enthusiasts in busy life and work. It allows users to explore fitness possibilities in daily intervals, such as seat time. BITWEEN has a "daily sitting walking mode" and a "fitness standing climbing mode" that can both reduce the negative effects of prolonged sitting and poor posture, and improve health through targeted exercise. Its compact design and external roller make mode switching, carrying and storage more convenient. In addition, smart apps monitor users' sedentary time and provide regular reminders; use instant feedback and AI simulations to anticipate outcomes and enhance positive fitness motivation; by setting daily activity-based fitness tasks and periodic summaries, make fitness become an integral part of life, thereby stimulating growth thinking and helping users discard the excuse of "lack of willpower". Together with BITWEEN, we can make better use of our time, enrich our lives, and constantly become better ourselves.

# 全国大学生工业设计大赛
# 优秀作品集 · 2022

Collection of Award-winning Works of
China Universities Industrial Design Competition
2022

# 优秀奖

# Excellent Award

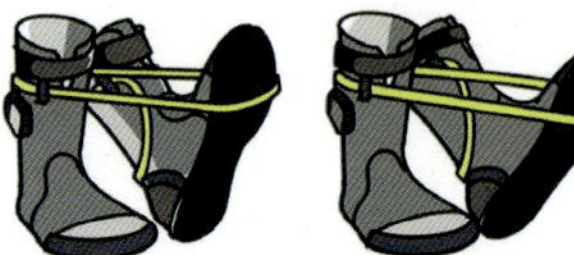

1. 弹力带侧脚横摆

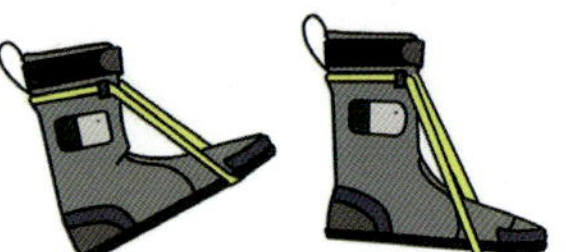

2. 弹力带前脚掌下压

3. 运动过程监测

# 老年人居家健身辅助产品与训练服务

## Home Fitness Assistance Products and Training Services for the Elderly

作　　者：许娜琪
指导老师：崔艺铭
所在院校：北京服装学院

## 设计说明

本设计是一款专为老年人设计的下肢肌肉训练靴，该训练靴特别考虑了下肢弹性带的方向性阻力，针对性地强化小腿内外侧及大腿肌群的训练，旨在增强老年人的行动能力。此外，该设计与视频训练内容的结合，促进老年用户与同龄人建立健身社交网络，进而通过完成训练任务获得奖章、排名提升及实物奖励，增强他们的社会参与感和自我价值感。

## Design notes

This design proposes a lower limb muscle training boot specifically designed for the elderly, which takes into account the directional resistance of the lower limb elastic band and specifically strengthens the training of the inner and outer calf and thigh muscle groups, aiming to enhance the mobility of the elderly. In addition, the design combines with video training content to promote the establishment of fitness social networks between elderly users and their peers. By completing training tasks, they can receive medals, rankings, and physical rewards, enhancing their sense of social participation and self-worth.

# 梭——个人通勤出行载具设计

## SHUTTLE——Design of personal commuting vehicles

作　　者：尹文昊
指导老师：李　盈　张　弛
所在院校：北京服装学院

### 设计说明

随着城市化进程的加速，人们面临的通勤挑战也日益增加，包括通勤距离的延长、交通拥堵和通勤疲劳，这些因素共同降低了日常通勤的效率。针对这一问题，人们迫切需要一个既独立又私密，同时兼具舒适、安全与高效性的通勤空间。本设计通过优化交通通勤体验，特别针对长距离城市通勤者设计了一种个人化交通载具，并深入研究了载具空间与用户之间的互动关系，提高用户在车内的休息、娱乐及办公效率，进而增强城市空间的人文关怀并提升社会和经济效益。

### Design notes

With the acceleration of urbanization, people are facing increasing commuting challenges, including extended commuting distances, traffic congestion, and commuting fatigue, all of which collectively reduce the efficiency of daily commuting. In response to this issue, people urgently need a commuting space that is both independent and private, while also providing comfort, safety, and efficiency. This design proposes an innovative travel solution specifically for long-distance urban commuters by optimizing transportation commuting efficiency. Based on the investigation of the actual average passenger capacity during the morning and evening rush hours of urban transportation (only 1.25 people), a personalized transportation vehicle was designed, and the interaction between the vehicle space and users was deeply studied to improve the efficiency of users' rest, entertainment, and office work in the vehicle, enhance the humanistic care of urban space, and improve  the social and economic benefits.

金奖

银奖

铜奖

优秀奖

# 紧急救援餐车设计

## Design of Rescue Rolling Kitchen

作　　者：付子哲
指导老师：李光亮
所在院校：北京理工大学

### 设计说明

本设计系统地研究了自然灾害救援车辆的理论基础、人体工程学及产品创新相关理论，通过对救援餐车当前状况及未来趋势的调查研究，深入分析了用户需求及相关的人体工学尺寸，能为突发自然灾害的地区提供一种灵活性较高的便捷餐车。该产品通过优化餐品制备流程，为受灾人群提供一个更安全、健康的就餐环境。

### Design notes

This study systematically explores the theoretical basis, human factors engineering, and product innovation theory of natural disaster rescue vehicles. Through investigation and research on the current situation and future trends of rescue food trucks, it deeply analyzes user needs and related ergonomic dimensions. This study designs an internal layout plan for a dining car, providing a flexible and convenient dining car for areas affected by sudden natural disasters. This design optimizes the food preparation process to provide a safer and healthier dining environment for the affected population.

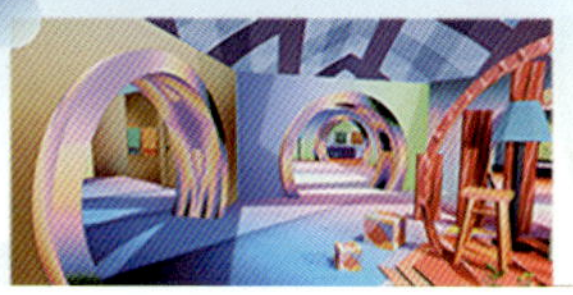



# MEGO 元宇宙校园个人展馆

## MEGO Metaverse Campus Personal Pavilion

作　　者：史昊冉
指导老师：安　丛
所在院校：北京信息科技大学

### 设计说明

MEGO 元宇宙校园个人展馆是为大学生提供的一个可互动交流、展示的主题平台，激发学生的强互动性与自主学习性。平台内部设计了五个特色主题展馆，分别是“活·悦”生活馆、“巢·舍”建造馆、“技·异”科技馆、“缘·结”社交馆、“忆·愿”毕业季馆，每个展馆都配备了具有独特性格特点的虚拟角色，供学生选择并通过这些角色探索各个展馆。该平台允许学生在元宇宙环境中展示个人日常生活，提供校园生活咨询，构建理想宿舍场景，探索从工业到信息技术的发展脉络，与不同领域的人建立联系，回顾校园生活记忆，并撰写或查看留言祝福等，从而丰富学生的线上校园体验，加深他们对校园文化的归属感。

### Design notes

The MEGO Metaverse Campus Personal Exhibition Hall is a themed platform for students to interact, communicate, and showcase, stimulating their strong interactivity and self-directed learning. There are five themed pavilions, namely <Living - Joy> Living Pavilion, <Nesting - House> Building Pavilion, <Technology - Difference> Technology Pavilion, <Karma - Ties> Social Pavilion and <Remembering - Wishes> Graduation Season Pavilion, each exhibition hall is equipped with unique virtual characters for students to choose from and explore various and other exhibition halls through these characters. This platform allows students to showcase their daily lives in the metaverse environment, provide campus life consultation, construct ideal dormitory scenarios, explore the development trajectory from industry to information technology, establish connections with people from different fields, review campus life memories, and write or view messages of blessings, thereby enriching students' online campus experience and deepening their sense of belonging to campus culture.

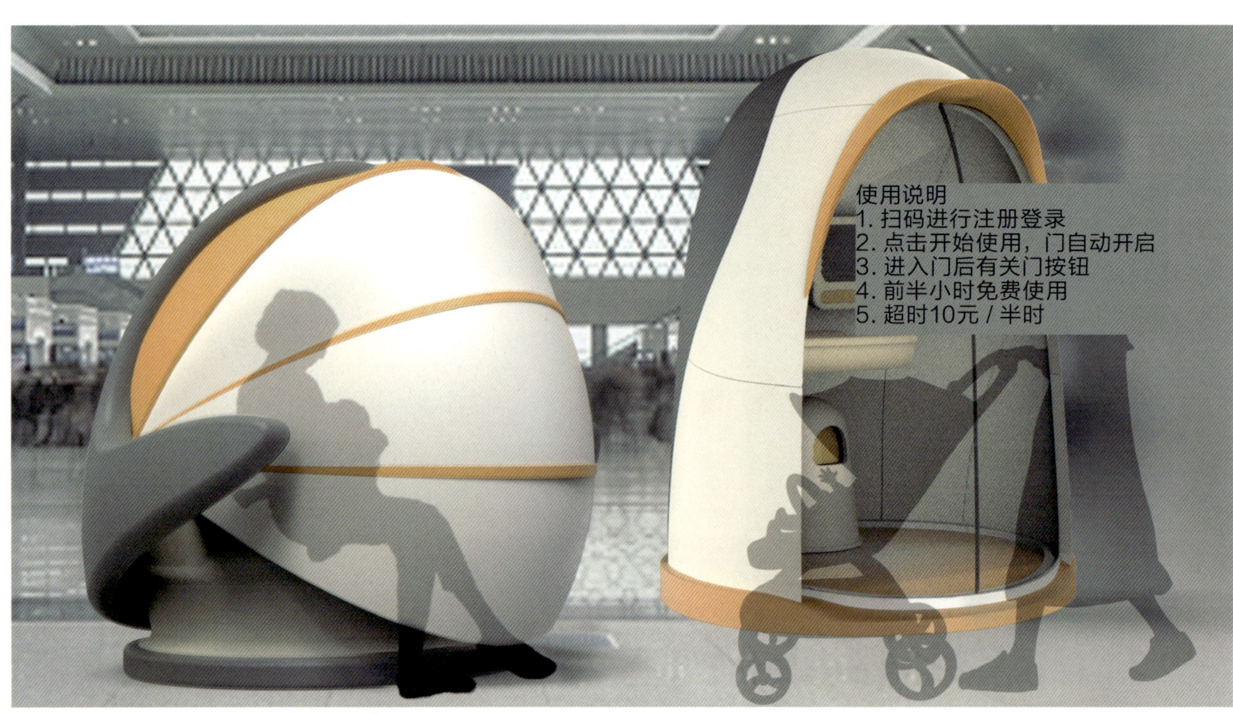




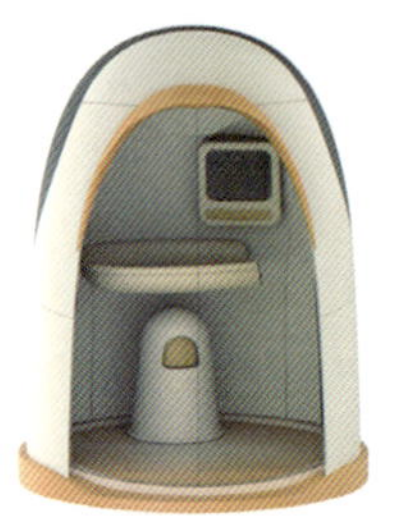
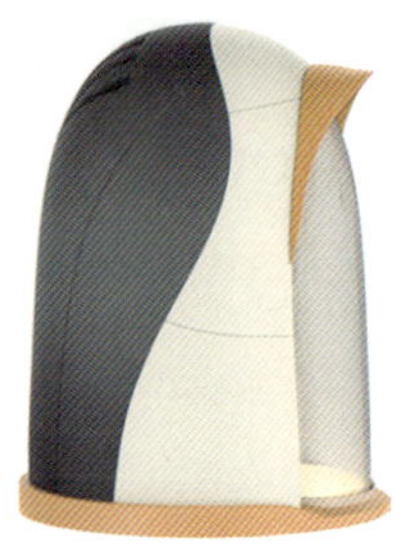

# 温暖的它

## WARMIT

作　　者：王耘琪　常佳楠
指导老师：苏　艺
所在院校：北京服装学院

### 设计说明

本产品为一款集婴儿照料、智能定位导航与城市文化展示于一体的智能移动哺乳椅，采用企鹅仿生设计理念，为哺乳期女性在公共场所提供一个温馨、私密且便捷的喂养和婴儿护理环境。该空间的清洁室使用电致变色玻璃，便于清洁并在使用时变成磨砂状态以提供隐私保护。室内配备了尿不湿、棉柔巾、湿巾等母婴应急物资及婴儿安抚设备。

### Design notes

This product is designed as an intelligent mobile space that integrates baby care, intelligent positioning navigation, and urban culture display. It adopts the penguin biomimetic design concept and provides a warm, private, and convenient feeding and baby care environment for lactating women in public places with a rounded shape. The cleaning room of this space uses electrochromic glass, which is easy to clean and turns into a frosted state during use to provide privacy protection. Indoor facilities are equipped with emergency supplies such as diapers, cotton wipes, and wet wipes for mother and baby, as well as baby comfort equipment.

# 循途——老年及残障人士过街助行设备

# TRACING THE PATH——Elderly and Disabled Pedestrian Crossing Equipment

作　　者：张　龙
指导老师：陈净莲
所在院校：北京林业大学

## 设计说明

由于老年人和残障人士受到个人身体条件的限制，难以提高过街速度，因此对辅助设施的需求变得迫切。为此，设计了一款专为老年和残障人士过街助行设备，使该群体能够安全快速地抵达马路对岸。

## Design notes

Due to personal limitations on their abilities, elderly and disabled individuals find it difficult to improve their crossing speed, making the need for auxiliary facilities urgent. The current flat crossing facilities have not fully met the needs of accessible groups, and the design of pedestrian overpasses and underground passages undoubtedly increases their crossing difficulties. Therefore, this design proposes a street crossing assistance platform specifically designed for elderly and disabled individuals to focus on community roads, enabling them to safely and quickly reach the opposite side of the road.

# RIDGE-X 串列式双座电动越野概念车

## RIDGE-X Tandem Concept Electric Off-road

作　　者：慕雨森
指导老师：张　雷　刘志国
所在院校：清华大学美术学院

### 设计说明

RIDGE-X 概念车针对未来越野运动普及及人们对极限挑战和刺激体验的需求而设计。该车采用串列式车身布局，实现了驾驶和领航两种独立而独特的乘坐体验于一体的创新设计。

### Design notes

The RIDGE-X concept car is designed for the future popularization of off-road sports and people's demand for extreme challenges and thrilling experiences. The car adopts a serial body layout, achieving an innovative design that combines driving and navigation as two independent and unique riding experiences.

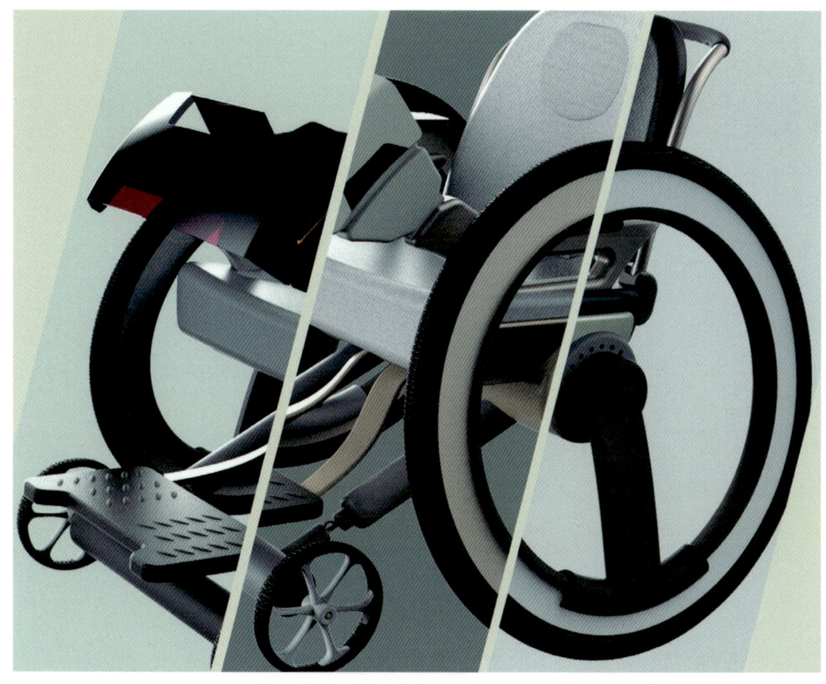

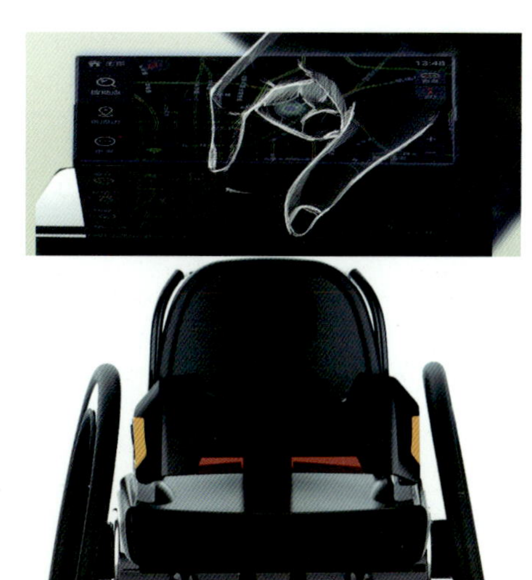



# 针对高位截瘫患者心理健康设计的轮椅

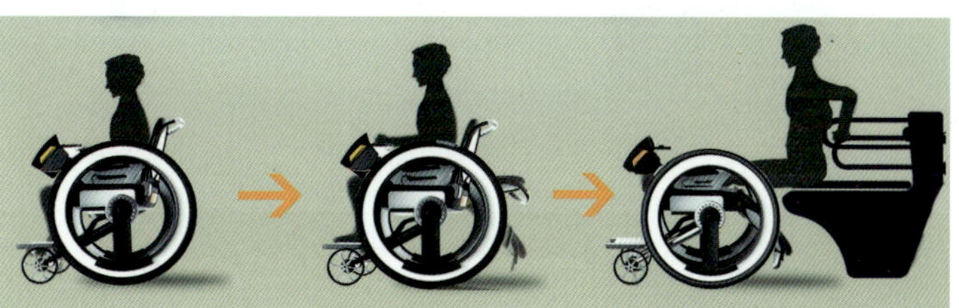



## A Wheelchair Designed for the Mental Health of Patients with High Paraplegia

作　　者：周颀媛
指导老师：刘永翔　边　鹏
所在院校：北方工业大学

### 设计说明

高位截瘫患者群体心理状况极为脆弱，针对这一情况，我们设计的轮椅采用了智能交互系统，一定程度上缓解患者的心理和精神状况。该轮椅具备半自动功能，轮椅靠背根据人体脊柱的自然弧度设计，能够更合理地贴合患者背部，使患者更加舒适。根据不同人的身高、体格将轮椅的座椅设计为可调节装置。交互台面使用全息投影系统，使用更智能方便，同时关注患者心理健康。

### Design notes

The psychological status of patients with high-level paraplegia is extremely fragile. In response to this situation, the wheelchair we designed adopts an intelligent interaction system, which to some extent alleviates the psychological and mental conditions of patients. This wheelchair has semi-automatic function, and the backrest of the wheelchair is designed according to the natural curvature of the human spine, which can fit the patient's back more reasonably, making the patient more comfortable. Design wheelchair seats as adjustable devices based on the height and physique of different individuals. The interactive tabletop uses a holographic projection system, which is more intelligent and convenient to use, while also paying attention to the patient's mental health.

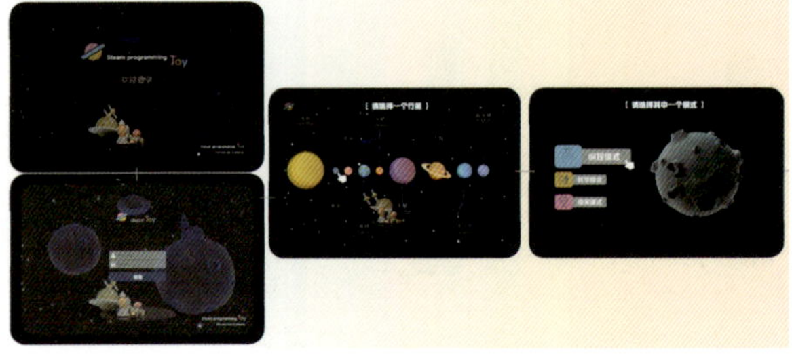

# 儿童 steam 编程玩具设计说明

## Design Instructions for Children's Steam Programming Toys

作　　者：徐嘉琦
指导老师：杨　新　孙博文
所在院校：北京理工大学

### 设计说明

本产品是一款以宇宙为主题专为学龄前期儿童设计的 steam 编程玩具。通过互动式学习，该玩具不仅能增进儿童对宇宙科学的认识，还能在游戏的过程中引入编程概念，激发对编程代码的兴趣和理解。该 steam 玩具围绕科学、数学、工程、艺术、技术五大方面锻炼儿童的认知和创造能力。

### Design notes

This product is a steam programming toy designed specifically for pre-school children, with the theme of the universe. Through interactive learning, this toy not only enhances children's understanding of cosmic science, but also introduces programming concepts during the game process, stimulating interest and understanding of programming code. This steam toy focuses on five aspects of science, mathematics, engineering, art, and technology to exercise children's cognitive and creative abilities.

# 灾后应急折叠屋设计

## Emergency Folding House Design

作　　者：孙东东
指导老师：魏泽崧　易　晓　石　彭
所在院校：北京交通大学

### 设计说明

本设计，为地震灾害后的人们提供既保障私密性又不失公共性的暂时避难场所。该避难所由四面防水帆布构成，可堆叠放置，便于运输和存放。避难所装备有三个太阳能模块，可提供250千瓦时的电能，满足基本能源需求。内部设有排风扇和6盏LED照明灯以保证内部空气流通与光照，另有2盏LED灯专为室外照明设计。为了增加使用者的便利性，内部还配置有可折叠椅，适用于室外休息。底部为存储紧急物资的方形空间，物资足以支撑一名成年人在无外援情况下生存10天以上。

### Design notes

This plan designs a temporary shelter to provide people after earthquake disasters with a temporary shelter that ensures both privacy and commonality. The shelter is composed of four waterproof canvas sheets that can be stacked for easy transportation and storage. The shelter is equipped with three solar modules that can provide 250 kwh of electricity to meet basic energy needs. Internally it is equipped with exhaust fans and 6 LED lighting fixtures to ensure internal air circulation and lighting, and 2 LED lights specifically designed for outdoor lighting. In order to increase user convenience, there are also foldable chairs inside, suitable for outdoor rest. The bottom is a square space for storing emergency supplies, which are sufficient to support an adult to survive for more than 10 days without external assistance.

金奖 银奖 铜奖 **优秀奖**

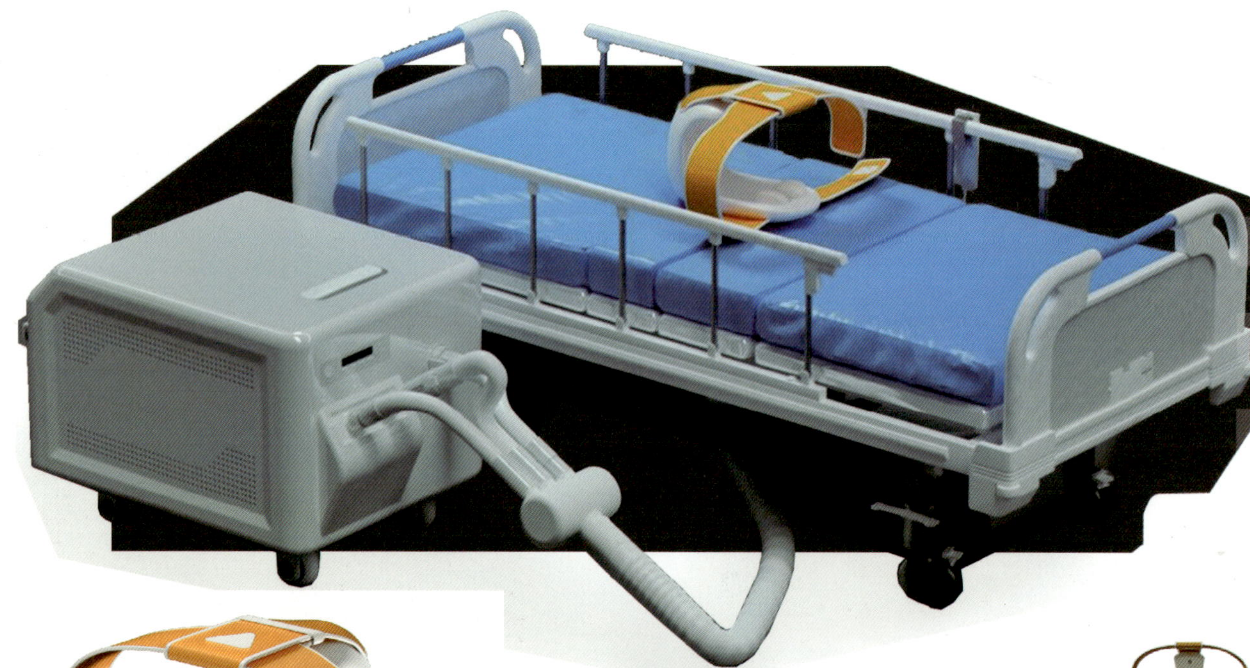



# 床上如厕智能护理产品系统

## Bed Toilet Intelligent Nursing Product System

作　　者：翟　静
指导老师：胡　鸿
所在院校：北京工业大学

### 设计说明

这是一款针对中重度失能老人的如厕清理产品，该产品集成了自动感应技术、智能冲洗系统、智能充气垫和热风烘干器，与可穿戴设备配合使用，安置在带有便孔位的家用护理床上，组成床上如厕智能护理产品系统。为中重度失能老年人提供一个方便、卫生的床上如厕解决方案，从而提升了他们的生活质量，也减轻了陪护人员的负担。

### Design notes

This project proposes a toilet cleaning product for elderly people with moderate to severe disabilities. This product integrates automatic sensing technology, intelligent flushing system, intelligent inflatable cushion, and hot air dryer, and is used in conjunction with wearable devices. It is placed on a home care bed with a toilet hole position and combined with box equipment to form a bed toilet intelligent care product system. Providing a convenient and hygienic bed toilet solution for elderly people with moderate to severe disabilities has improved their quality of life and reduced the burden on caregivers.

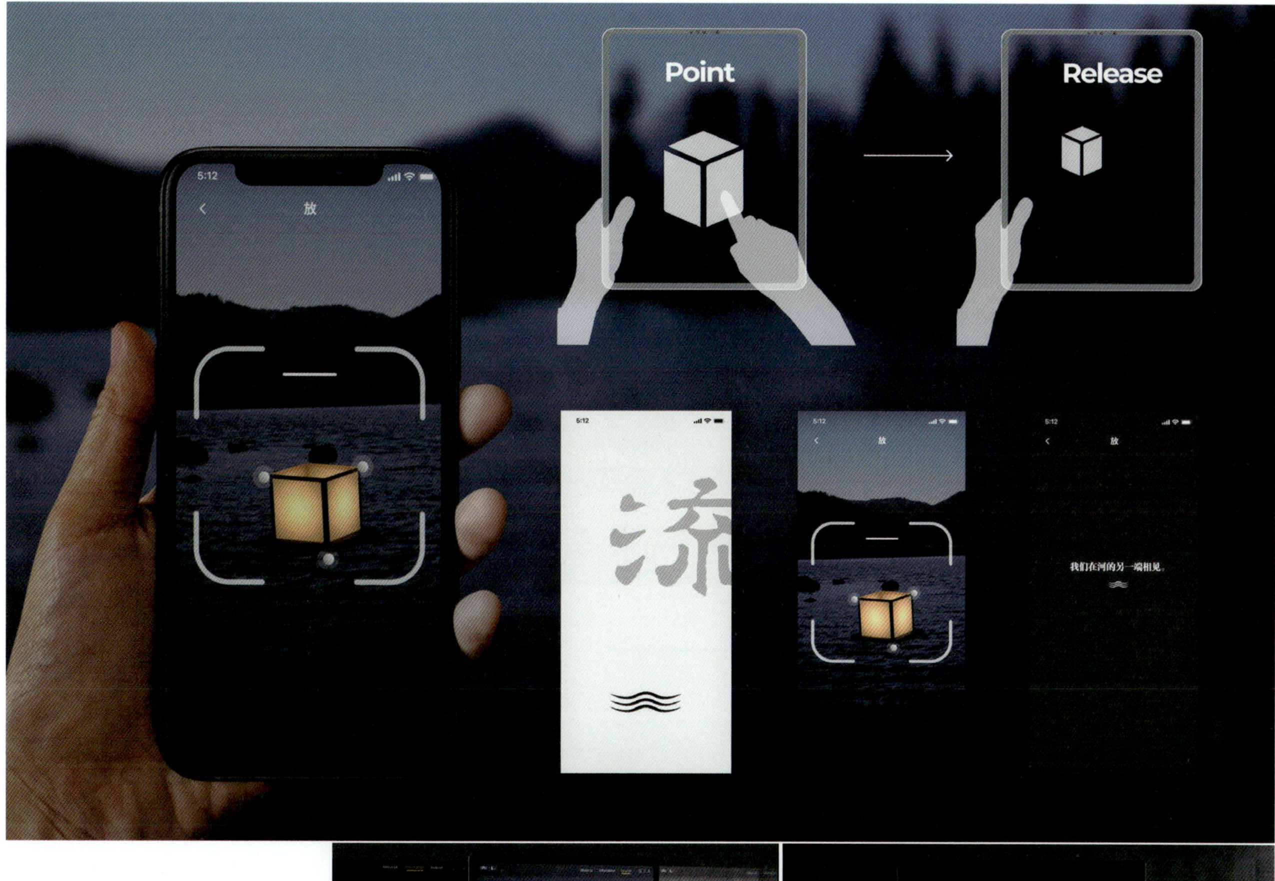




# 流
## Liu

作　　者：石玙睿　张子鹏　伊俊翼
指导老师：陈　路　尹　虎
所在院校：北京航空航天大学

## 设计说明

“流”是一个以河灯为主题，串联了纪念馆、手机程序和网站的线上祭祀平台。为了让放河灯的传统祭祀行为与环境保护寻得平衡，“流”以全新的交互方式使得绿色祭祀成为可能。用户可在纪念馆的交互式屏幕或通过手机应用放置一盏电子河灯以纪念所思之人，使点灯思人不再拘泥于地点。

## Design notes

"Liu" is an online sacrificial platform with river lanterns as its theme, connecting memorial halls, mobile programs, and websites. In order to find a balance between the traditional ritual of releasing river lanterns and environmental protection, Liu has made green worship possible in a new interactive way. Users can place an electronic river lamp on the interactive screen of the memorial hall or through a mobile application to commemorate the person they are thinking of, or browse the sacrificial messages left by users from all over the world on the website. In this way, lighting a lamp to think of people is no longer limited to the location, and any place can become a temple for ancestor worship.

# LIONDANCE——醒狮系列潮玩

## LIONDANCE ——Series Trendy Play



作　　者：李双言　刘文瀚　邹仁耀
　　　　　张　鑫
指导老师：申华平
所在院校：北京理工大学

### 设计说明

醒狮又称瑞狮，意为吉祥如意，是驱邪避害的吉祥瑞物。其来源于中国传统舞狮文化中的南狮，广泛流行于岭南地区，属广府文化。该设计旨在提取醒狮文化中的造型元素，设计一系列 IP 形象，并衍生相关的文化创意产品。以刘备狮、关羽狮、张飞狮、马超狮、黄忠狮作为主要造型进行设计，提取红、黑、绿等特征色彩元素，保留狮缨、球缨、额镜以及狮角等特征部位，使得形象更具有辨识度。同时，在狮耳处附以如意纹、虎斑纹等纹理，狮嘴处融入玩偶形象，使得该 IP 在保留醒狮原有神态的基础上增加俏皮感。该设计将舞狮这一传统文化与现代潮流文化相融合，进而发扬其以狮作舞、提振精神的文化内涵。

### Design notes

The lion, also known as the auspicious lion, means good luck. It is a auspicious symbol for warding off evil and avoiding harm. It originates from the southern lion in traditional Chinese lion dance culture and is widely popular in the Lingnan region, belonging to the Guangfu culture. This design aims to extract styling elements from lion culture, design a series of IP images, and derive related cultural and creative products. Design with Liu Beishi, Guan Yushi, Zhang Feishi, Ma Chaoshi, and Huang Zhongshi as the main shapes, extracting characteristic color elements such as red, black, and green, while retaining characteristic parts such as lion tassels, ball tassels, forehead mirrors, and lion horns, making the image more recognizable. At the same time, textures such as Ruyi patterns and tiger stripes are attached to the lion's ears, and a doll image is incorporated into the lion's mouth, adding a playful touch to the IP while retaining the lion's original demeanor. This design integrates the traditional culture of lion dance with modern trend culture, and further promotes its cultural connotation of lion dance and awakening spirit.

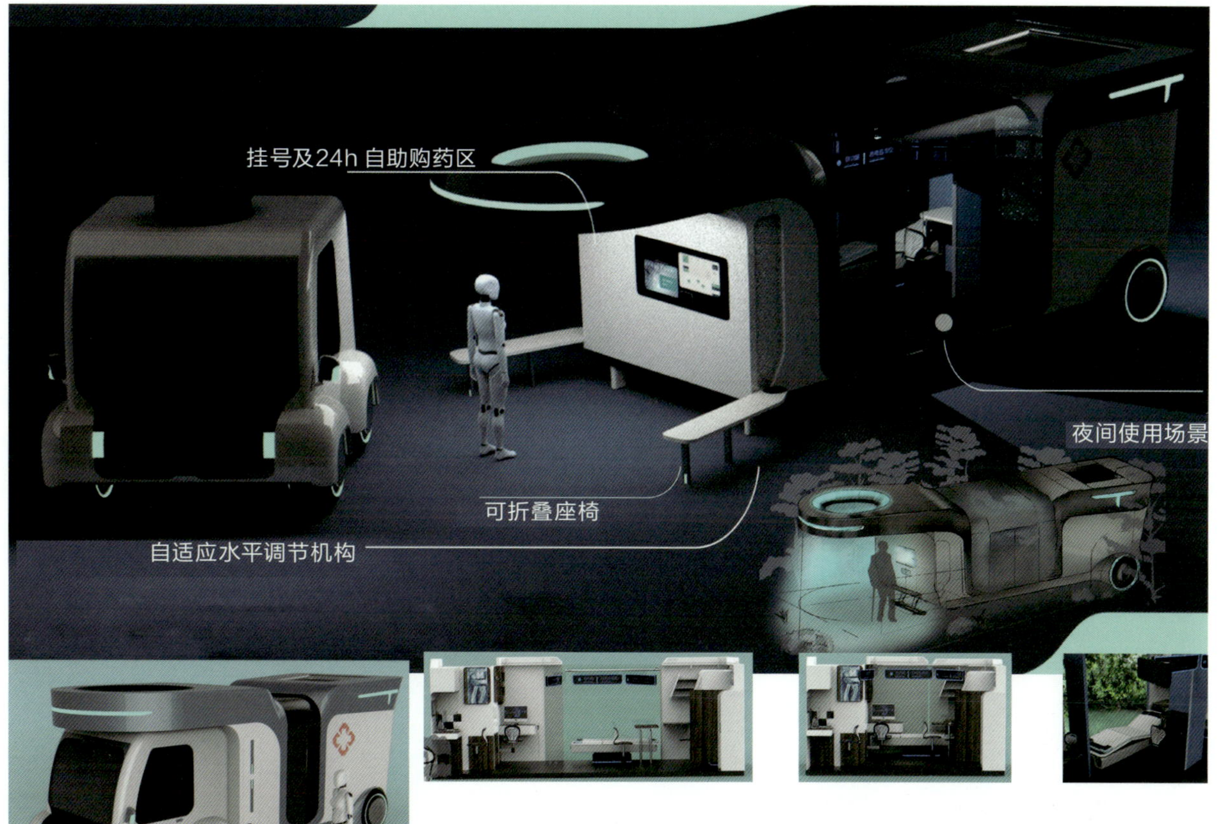






# 乡村流动医疗车设计

## Design of a Rural Mobile Medical Vehicle

作　　者：武晓宇
指导老师：柳　沙
所在院校：中国农业大学

## 设计说明

农村空巢老人普遍面临经济来源有限、医疗保障不足及精神关怀缺失等问题。针对这一现状，设计了一款流动医疗服务车，为农村老年人提供定期的健康检查和医疗巡诊服务，车体可延伸扩展更多的就诊空间，车辆前端配备自助购药机和候诊休息区，提供更加便捷和舒适的医疗服务体验，从而丰富乡村老年人的健康服务选择。

## Design notes

As of 2020, the number of empty nest elderly people in rural areas of China has reached 50 million, accounting for nearly half of the rural elderly population. This group generally faces problems such as limited economic sources, insufficient medical security, and lack of spiritual care. In response to this situation, this design proposes a mobile medical service vehicle that provides regular health checks and medical patrols for rural elderly people, with special attention to common and multiple diseases among the elderly, to meet their needs for health examinations, disease prevention, and healthcare. The design features include a vehicle extension mechanism to expand more medical space during parking, as well as a self-service medication machine and waiting area equipped at the front of the vehicle, aiming to provide a more convenient and comfortable medical service experience, thereby enriching the health service choices of elderly people in rural areas.

# 模块化音乐编曲器

## Modular Music Arranger

作　　者：林珈羽　黄惠翔
指导老师：张　扬
所在院校：北京化工大学

### 设计说明

本设计提出的模块化音乐编曲器旨在服务于青少年和成年音乐初学者。该编曲器通过模拟积木拼搭的方式，引导用户探索音乐构成原理，用户可以选择喜爱的曲目，并根据个人对音乐的理解与感知，创作出属于自己的旋律。产品采用简化的模块化编程方法，通过不同模块的组合排列，鼓励创作者通过自主实验去探索音乐的规律。

### Design notes

The modular music arranger proposed in this design is aimed at serving teenagers and adult music beginners. This arranger guides users to explore the principles of music composition by simulating building blocks. They can choose their favorite tracks and create their own melodies based on their personal understanding and perception of music. The product adopts a simplified modular programming method, and through the combination and arrangement of different modules, encourages creators to explore the laws of music through independent experiments.

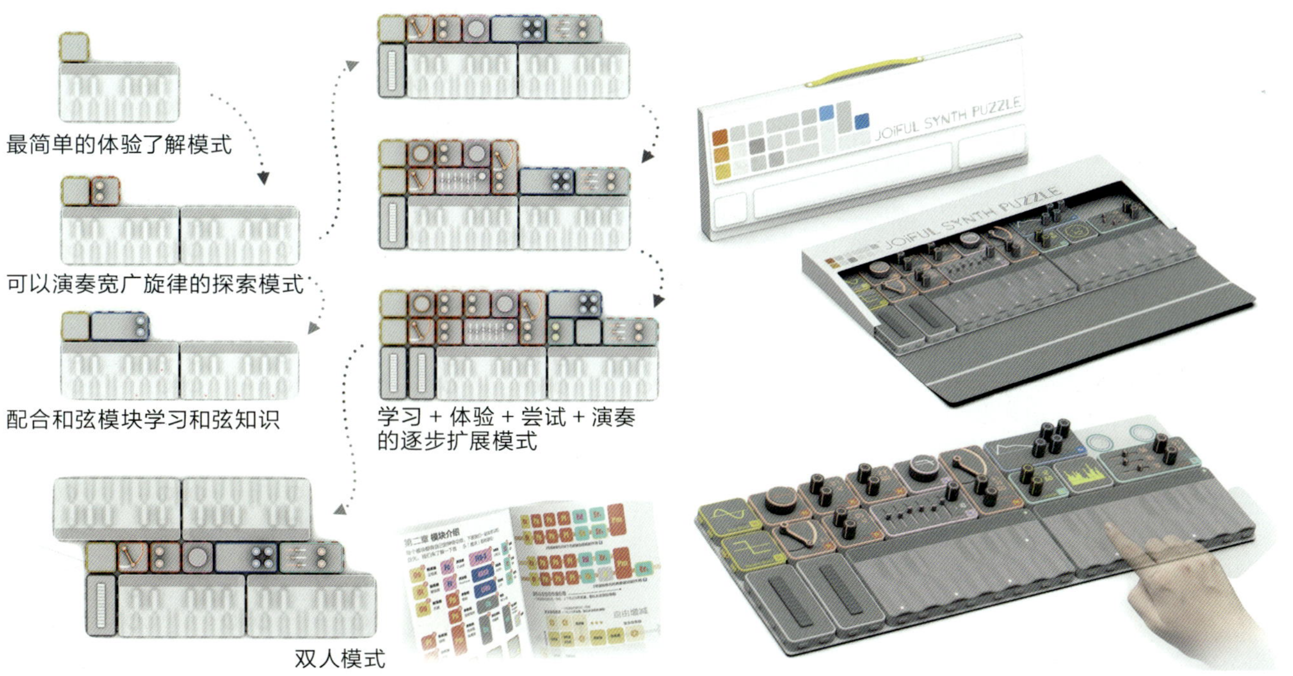




# 音乐魔块

## JOiFUL Synth Puzzle

作　　者：黄昊成
指导老师：杨　新
所在院校：北京理工大学

### 设计说明

现有合成器多为成人和专业用户设计，对青少年而言，其复杂性构成了学习的障碍，不利于他们音乐创造力和逻辑思维的培养。音乐魔块通过将合成器的基本原理与电路设计相结合，引入了一种创新的学习方法，使学生能够简单、灵活地进行拼装和组合，适应各个学习阶段的需要。借助可视化的参数显示和色彩提示，即便是零基础也能进行创作。音乐魔块借助于电路中串联与并联的无限可能性，突破了传统合成器设计的局限，促进青少年科学思维的发展，激发学生探索和创造独特音乐的兴趣。

### Design notes

Existing synthesizers are mostly designed for adults and professional users, and for teenagers, their complexity poses a learning barrier, which is not conducive to the cultivation of their musical creativity and logical thinking. The music magic block introduces an innovative educational method by combining the basic principles of synthesizers with circuit design. Providing a series of basic components, through the logic of deconstruction and reconstruction, enables students to easily and flexibly assemble and combine to meet the needs of various learning stages. With the help of visual parameter display and color prompts, even basic zero can be used for creation. By leveraging the infinite possibilities of series and parallel connections in circuits, music magic blocks break through the limitations of traditional synthesizer design, promote the development of STEAM thinking in young people, and stimulate their interest in exploring and creating unique music.

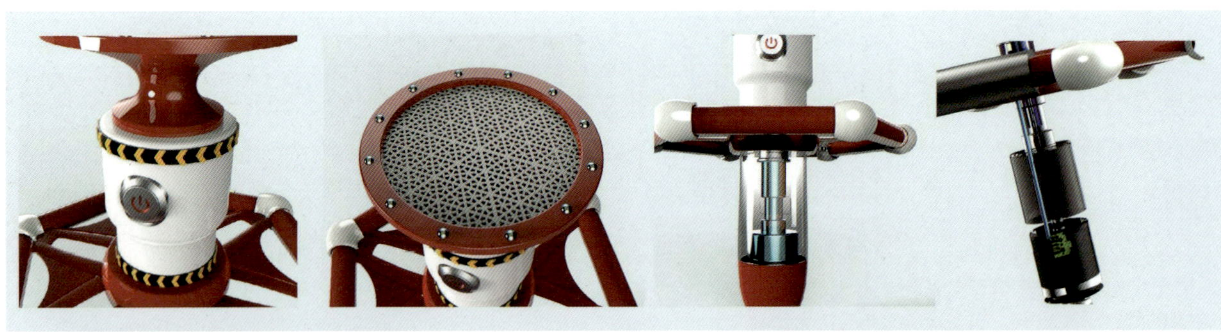

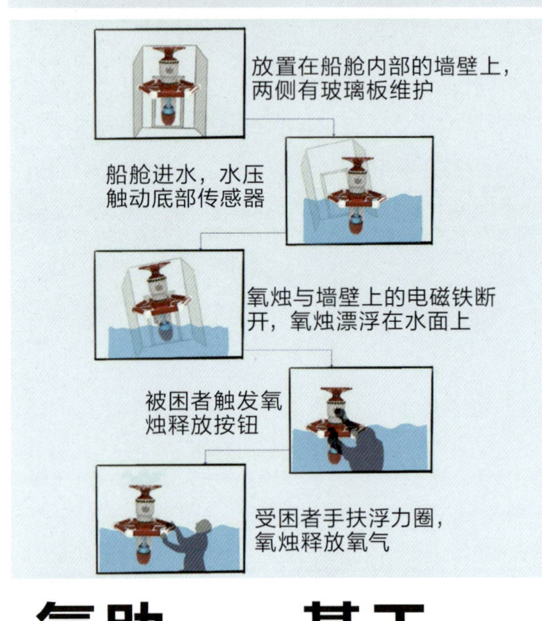

# 氧助——基于氧烛原理的海难应急求生装置

# OXYGEN CANDLE—Emergency Survival Device for Maritime Accidents Based on Oxygen Candle Principle

作　　者：崔宇航　崔益玮　陈颖颖
黄沐晗
指导老师：朱天阳
所在院校：北京化工大学

## 设计说明

在应对突发沉船紧急情况时，本设计通过运用氧烛制氧技术为受困者提供氧气支持，创造较大的生存空间。氧烛技术的关键特性之一是其放热反应，该反应不仅促成了氧气的生成，同时也为受困人员提供了重要的热源，有助于保持体温，可以延长受困人员在船舱中的存活时间，从而提高救援的存活人数。

## Design notes

In response to the emergency situation of a sunken ship, this design uses oxygen candle technology to provide oxygen support for the trapped and create a larger living space. One of the key characteristics of oxygen candle technology is its exothermic reaction, which not only promotes the generation of oxygen, but also provides an important heat source for trapped personnel, helping to maintain body temperature and prolonging the survival time of trapped personnel in the cabin, thereby increasing the number of survivors rescued.

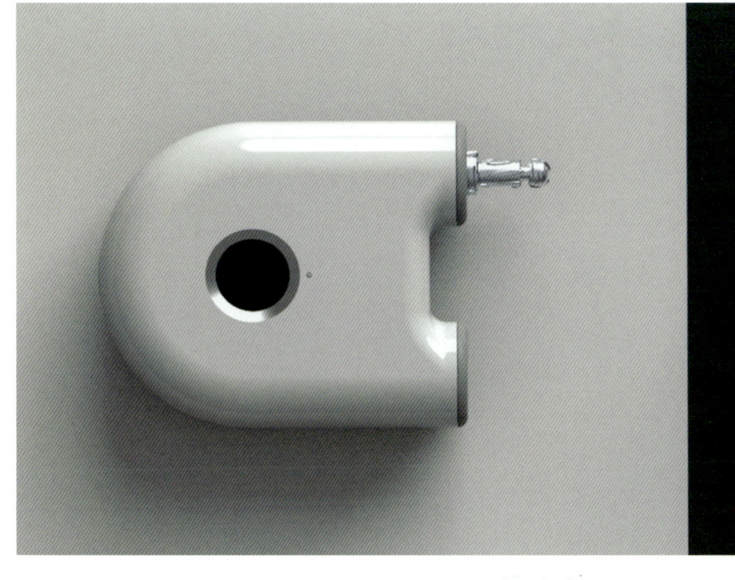

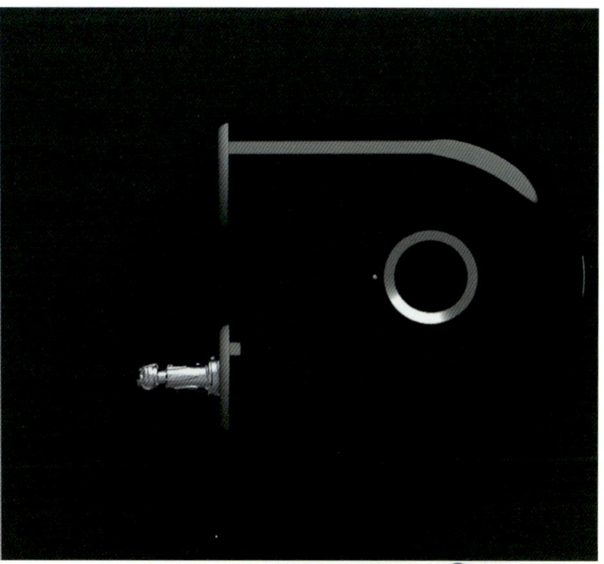

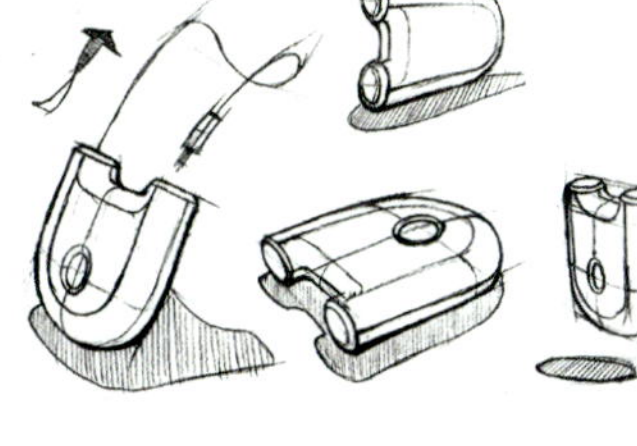

金奖
银奖
铜奖
优秀奖



# 嗒咔——智能指纹挂锁

# TAKA —— Intelligent Fingerprint Hanging Lock

作　　者：田蕴轩　郭雨阳　王林萍
段京超　仁青多吉
指导老师：郗小超　马　洋
所在院校：北京邮电大学

## 设计说明

当用户携带大件行李独自外出，特别是在高铁站、汽车站等人流量大的公共场所时存在许多不便。因此，我们设计了一款智能抽绳锁，该产品通过与手机端互联，能够将行李灵活锁定在座椅扶手等固定物体上，可实时查看行李位置及状态。抽拉锁绳能调节长度、指纹识别可快速开锁、手机端实时显示位置与状态。同时，本产品小巧轻便、易于携带，给用户提供安心便捷的出行体验。

## Design notes

When users carry large luggage and go out alone, especially in public places with high traffic such as high-speed railway stations and bus stations, it can easily cause many inconveniences. Therefore, we have designed an intelligent drawstring lock that can flexibly lock luggage onto fixed objects such as seat armrests through interconnection with the mobile phone, and view the position and status of luggage in real-time. The specific implementation methods include adjusting the length of the drawstring, quickly unlocking with fingerprint recognition, and displaying the position and status in real-time on the mobile phone. Meanwhile, this product is compact, lightweight, and easy to carry, providing users with a safe and convenient travel experience.

# 便捷式破拆钳——城市应急避险工具设计

## CONVENIENT DEMOLITION TONGS —— Design of Urban Emergency Shelter Tools

硝化纤维火药解析

爆速：6300m/s(含氮13%)
爆轰气体体积：841L/kg(含氮13.3%时)
自燃点：170℃
原理：$2KNO_3 + S + 3C ==== K2S + N_2\uparrow + 3CO_2\uparrow$
优点：能量密度高
缺点：易燃易爆，应避光贮存

作　　者：武鑫宇
指导老师：刘永翔
所在院校：北方工业大学

### 设计说明

本设计以开发一种既便携又高度可靠的破拆工具为目标，我们从内部结构到外观设计全面进行了产品开发。产品采用硝化纤维火药驱动技术，以铝合金作为动力传导结构，将火药的爆炸冲击力安全、无损地传至钳口，实现剪切。设计成功达到了最终预期，实现了一款兼具便携性、长续航力、高可靠度的应急破拆钳。

### Design notes

The original intention of this project is to develop a portable and highly reliable dismantling tool. Therefore, we have comprehensively developed the product from internal mechanisms to external design. The product adopts nitrocellulose powder driving technology and uses aluminum alloy as the power transmission structure to safely and non-destructively transmit the explosive impact force of the powder to the jaws, achieving shear. The design has successfully met the final expectations, achieving an emergency breaking pliers that combines portability, long endurance, high reliability.

# LIME LIGHT & TIMER——灯具设计

## LIME LIGHT & TIMER —— Lamp Design

作　　者：葛诗清　张子甜　于欣莹
指导老师：孙博文　李光亮
所在院校：北京理工大学

### 设计说明

本产品将计时器与照明工具结合，旨在帮助用户睡前合理安排时间。与传统以数字显示时间不同，我们设计采用用光线的渐暗，表达时间的流逝，让时间不再是冰冷的数字，而是人们可以确切感知的变化。因此，本产品不仅帮助人们珍惜宝贵的睡前时刻，而且促进了更充分的休息，增强了人们对光线和时间价值的认识。同时，LIME 通过使用低色温灯创造了一个更自然的发光环境，让人们获得更好的睡眠体验。

### Design notes

This product combines a timer with lighting tools, aiming to help users arrange their time properly before bedtime. Unlike traditional digital display of time, our design uses the gradual dimming of light to express the passage of time, so that time is no longer a cold number, but a change that people can accurately perceive. Therefore, this product not only helps people cherish precious bedtime, but also promotes more adequate rest and enhances people's understanding of the value of light and time. Meanwhile, LIME creates a more natural luminous environment by using low color temperature lamps, allowing people to have a better sleep experience.

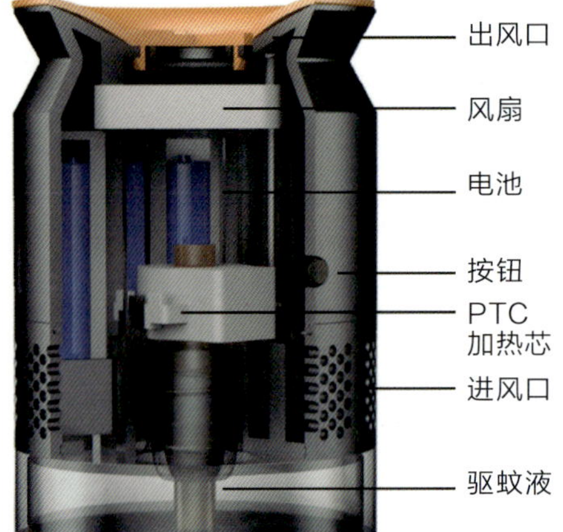




# 关注你的学习——智能驱蚊产品

## CARE FOR YOUR STUDY —— Intelligent Mosquito Repellent

作　　者：陈楚茹　张海文　田淳西
　　　　　侯家祺　肖润泽
指导老师：郗小超
所在院校：北京邮电大学

### 设计说明

这是一款专为学生设计的智能驱蚊器，通过定时功能避免传统驱蚊器造成的资源浪费和安全隐患。其简洁小巧的外观与智能静音的运行模式结合，并在驱蚊技术上做出了创新，改善高温挥发技术的不足之处，引入气流扩散技术，配合精密的能耗计算功能，提供一种对环境友好的使用体验。

### Design notes

This design introduces an intelligent mosquito repellent designed specifically for students, aiming to avoid resource waste and safety risks caused by traditional mosquito repellents through advanced and proactive timing functions. Its simple and compact appearance combined with intelligent silent operation mode provides students with a safe and comfortable learning environment. Unlike traditional mosquito repellents, this product has made innovations in mosquito repellent technology, improving the shortcomings of high-temperature volatilization technology. It introduces airflow diffusion technology, combined with precise energy consumption calculation, to provide an environmentally friendly user experience.

# SNOW BALL——室内虚拟场景滑雪体验机

## SNOW BALL —— Indoor Virtual Scene Skiing Experience Machine

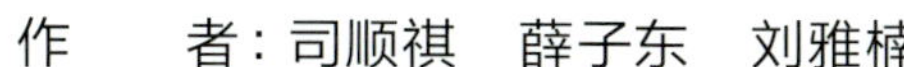

作　　者：司顺祺　薛子东　刘雅楠
张诗雨　王一冰
指导老师：张　帆
所在院校：天津工业大学

### 设计说明

本设计是一款室内虚拟场景滑雪体验机，解决人们因各种原因无法切身体会到滑雪乐趣的问题。该设备灵感来源于宇航员使用的三维滚环训练方法，并结合虚拟现实技术，尽可能地模拟真实的滑雪体验。使用者可以在滑雪机上自由活动，并可以完成一系列滑雪动作。

### Design notes

We have designed an indoor virtual skiing experience machine that addresses the issue of people not being able to experience the joy of skiing firsthand. The inspiration for this device comes from the three-dimensional rolling ring training method used by astronauts, combined with virtual reality technology, to simulate the real skiing experience as much as possible. Users can freely move on the ski machine and complete a series of skiing movements.

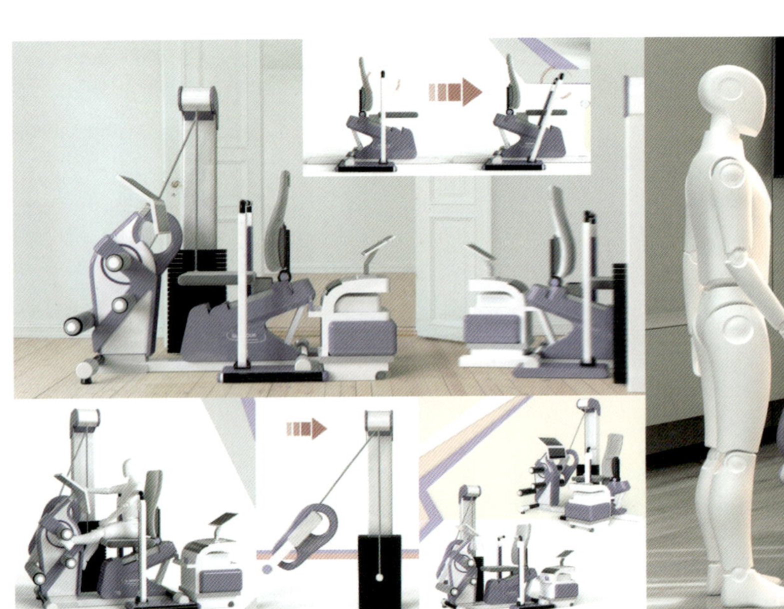

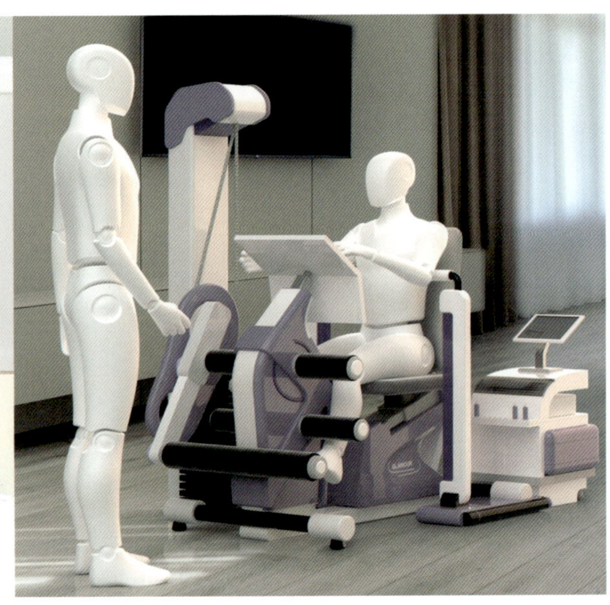

# GLAMOUR——产妇产后恢复医疗健身器械

## GLAMOUR — Medical Fitness Equipment for Postpartum Recovery

作　　者：张　娜　赵海燕
指导老师：杨冬梅　张新新
所在院校：河北工业大学

### 设计说明

本产品是一款帮助产妇进行产后恢复的医疗健身设备，它集成了盆底肌检测、盆底肌生物与物理修复、腹直肌修复、胸部康复、上下肢舒缓减脂等功能。该设备能够根据用户的生理、心理或环境需求，进行整体的大小和位置调整；产品构成了一个综合的智能交互系统，检测与运行记录可以形成个人数据库，用户每次使用均可与云端医生进行交流，并采取医生建议，获得专属定制个人修复方案。产品使用过程中搭配音乐、游戏等方式，使用过程不再枯燥。该设计致力于为产妇提供一个最舒适的产后恢复环境和方法，帮助恢复健康，达到身心的最佳状态，更显女性魅力。

### Design notes

This product is a medical and fitness equipment that helps postpartum women recover. It integrates functions such as pelvic floor muscle detection, pelvic floor muscle biological and physical repair, rectus abdominis repair, chest rehabilitation, upper and lower limb relaxation and fat reduction. The device can adjust its overall size and position according to the user's physiological, psychological, or environmental needs. At the same time, according to the changes in required functions, various details can be converted and adjusted. The product forms a comprehensive intelligent interaction system, and the detection and operation records can form a personal database. Users can communicate with cloud doctors every time they use it, and take doctor's advice to obtain exclusive customized personal repair plans. During the use of the product, music, games, and other methods are used to make the process no longer boring. This design is committed to providing postpartum women with the most comfortable recovery environment and methods, helping them recover their physical health, achieve a perfect state of mind and body, and thus more feminine charm.

五段电极片，
进行精准检测

微间隙重启
按钮设计

亲肤布料，
舒适透气

APP 界面设计





# TELENUDGE——居家远程办公头环

## TELENUDGE—— Home Telecommuting Head Ring

作　　者：王柏捷　顾佳奇　唐雨琛
指导老师：叶　武
所在院校：天津大学

### 设计说明

TELENUDGE 利用头环监测脑电波来分析员工在远程工作环境下的工作状态，并将其上传至平台。员工可通过 app 查看自己及同事的工作状态，这些状态以一种柔和且灵动的方式进行可视化呈现。这一机制使员工意识到自己的工作状态被共享，并了解同事的工作情况，从而促使他们调整自己的居家办公节奏，这不仅有助于展示自己的工作状态，同时也能获得相对自由灵活的时间安排。通过选择佩戴或不佩戴头环来明确界定工作与休闲的状态。

### Design notes

TELENUDGE uses a headband to monitor EEG waves and analyze the work status of employees in remote work environments, and uploads them to the platform. Employees can view themselves and their colleagues' work status through the app, which is visualized in a soft and dynamic way. This mechanism makes employees aware that their work status is shared and understand the work situation of colleagues, thereby prompts them to adjust their home Telecommuting rhythm. This not only helps to showcase their work status, but also allows for relatively free and flexible time arrangements. Clearly define the state of work and leisure by choosing to wear or not to wear a headband.

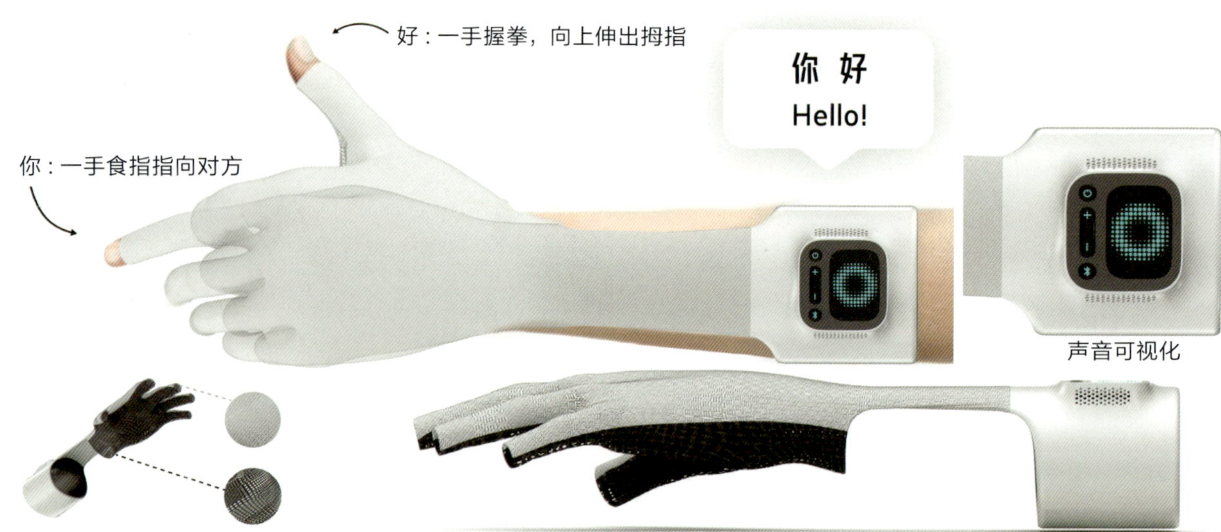


# SMART FINGER ——智能手套

## SMART FINGER ——Intelligent gloves



作　　者：廖孟哲　张淋悦
指导老师：张赫晨　邹　强
所在院校：天津大学

## 设计说明

为了改善言语障碍者与他人的日常交流，SMART FINGER 被设计为一款基于柔性传感器技术的智能可穿戴设备，基于柔性传感器技术，将用户的手语转换成语音实时播放，有效地提高言语障碍者的社交参与度。产品为蓝白色，增加视觉舒适度，手部选用柔性织物与网面的结合，对传感器等电子元件进行保护的同时，保证了使用者的舒适度。通过单元后移减轻手臂挥动的不适，设置开机、蓝牙连接和音量调节功能，以快速适应特定场景的需求；且产品支持无线充电，满足防水需求。配备手机端 app，进行多种语言、多种语音和不同语速的切换，进行语言记录、个性化手势定制及人工纠错功能配合完成深度学习。

## Design notes

In order to improve the daily communication between people with speech disorders and others, SMART FINGER is designed as an intelligent wearable device based on flexible sensor technology. Based on flexible sensor technology, it converts the user's sign language into speech for real-time playback, effectively improving the social engagement of people with speech disorders. The product adopts a blue and white tone design to increase visual comfort. The part of palm is made of a combination of flexible fabric and mesh, which protects electronic components such as sensors while ensuring user comfort. The design of intelligent hardware components takes into account comfort during use, reduces discomfort from arm swinging by moving the unit backwards, and sets up power on, Bluetooth connection, and volume adjustment functions to quickly adapt to specific scene requirements; And the product supports wireless charging, meeting waterproof requirements. Equipped with a mobile app for switching between multiple languages, voices, and different speech speeds, language recording, personalized gesture customization, and manual error correction functions to complete deep learning.

# 基于机械臂工程的流程一体化西瓜收获机

# Process Integration Watermelon Harvester Based on Robotic Arm Engineering

作　　者：黄　涵　李晓晟　降泓宇
指导老师：杨冬梅
所在院校：河北工业大学

## 设计说明

本项目创新性地开发了一套一体化作业装备，包含了西瓜的切割、采摘、运输和收集全过程，简化西瓜作业流程并显著提升了作业效率。通过整合市场上的人工采摘技术、西瓜收割机、收获机和收集机等不同设备功能，同时利用机械臂技术替代人工，实现了全程的智能化机械服务。通过重新配置的模块分区，极大地优化了功能结构。

## Design notes

This project innovatively develops an integrated operation equipment that includes the entire process of watermelon cutting, picking, transportation, and collection, simplifying the watermelon operation process and significantly improving operation efficiency. By integrating different equipment functions such as manual harvesting technology, watermelon reapers, harvesters, and collectors in the market, and using robotic arm technology to replace manual labor, intelligent mechanical services throughout the entire process have been achieved. Through reconfigured module partitioning, the functional structure has been greatly optimized.

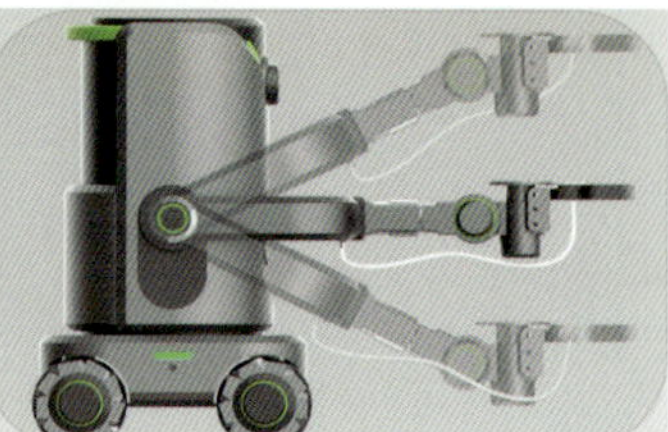



# TBRO——智能行道树喷涂设备

## TBRO —— Smart Equipment of Street Tree Spraying

作　　者：朱家谊　管孝天　吾丽江·吉恩斯
指导老师：李　鹏
所在院校：天津工业大学

### 设计说明

本设计聚焦于城市树木的生态保护与修复，提供一种高效且高质量的树木白漆喷涂解决方案，解决人工喷涂的不均匀性、低效率以及受天气变化限制无法进行作业而造成的对树木保护不充分的问题，能有效降低人为作业产生的安全隐患，提高喷漆效率。运用机械手臂自主运转、导航定位、智能测距和远程操控等技术，实现精准定位、精准喷涂，促进树木生态修复、保护，推动绿色环境发展。

### Design notes

This design focuses on the ecological protection and restoration of urban trees, providing an efficient and high-quality solution for white paint spraying on trees. By solving the problems of uneven and inefficient manual spraying, as well as the inability to carry out operations due to weather changes, insufficient protection of trees can be achieved, effectively reducing safety hazards caused by manual spraying and improving painting efficiency. This device has the characteristics of humanization and precision. By utilizing technologies such as autonomous operation of robotic arms, navigation positioning, intelligent ranging, and remote control, precise positioning and spraying can be achieved, promoting ecological restoration and protection of trees, and promoting the development of a green environment.

app 界面





# 冰上巡逻者

## Ice-spoerts Venue INSPECTOR

作　　者：袁思琪　于利洋　王依凡
王巧玲　郭海超
指导老师：孙　利　吴俭涛
所在院校：燕山大学

## 设计说明

冰球、冰壶、短道速滑等冰雪项目广受欢迎，但目前这些项目的场地维护仍然依赖大量专业人工。为了降低维护人员的学习成本，打造运动场馆维护工作的智能化，我们设计了一款冰面质量评估检测仪。该仪器可以在冰面平稳移动，便于检测每一块冰面的质量，收集冰面温度、平整度、滑度、室内温度等相关数据，并能快速直观地反馈给维护人员，从而协助他们更高效地维护场馆。

## Design notes

Ice and snow events such as ice hockey, curling, and short track speed skating are widely popular, but currently, the maintenance of these events still relies on a large number of professional manpower. In order to reduce the learning cost of maintenance personnel and create intelligent maintenance work for sports venues, we have designed an ice quality assessment and detection instrument. This instrument can move smoothly on the ice surface, making it easy to detect the quality of each ice surface. It collects relevant data such as ice surface temperature, flatness, smoothness, and indoor temperature, and can quickly and intuitively provide feedback to maintenance personnel, thereby assist them in maintaining the venue more efficiently.

# PIT 大黄蜂——未来城市外墙粉刷智能装备

## PIT HORNET —— Future City Exterior Wall Painting Intelligent Equipment

作　　者：吴连富　赵　赫　黄嘉宁
　　　　　武鸿跃　牛纪璇
指导老师：刘　渊
所在院校：华北电力大学（保定）



### 设计说明

该产品是针对新城建设与旧城改造中的缺点与弊端，结合未来智能都市所设计的一款新型外墙粉刷智能装备，在减少人工高空作业所带来危害的同时，又极大地提高城市建设与改造翻新的工作效率。该设备是车辆搭配无人机工作的形式，包含两种工作场景，第一种是在大面积空旷区域，通过车体压力泵作用将颜料、底胶、水等通过物料管道传输至喷刷无人机，可大面积高效施工；第二种是在电线密集区域，无人机直接从车载的小型物料桶中取材进行作业。无人机设计灵感来源于大黄蜂，赋予新型产品活力。此外，装备采用氢能源，减少环境污染，结合5G通信传输、搭载AI智能扫描、数据分析、红外测距等多种技术，确保了操作的高效性和智能化。有助于推动未来城市建设的效率、标准化和智能化发展。

### Design notes

This product is a new type of exterior wall painting intelligent equipment designed for the future intelligent urban fantasy, which focuses on the shortcomings and drawbacks of new city construction and old city renovation. It reduces the harm caused by manual high-altitude operations while greatly improving the efficiency of urban construction and renovation. This device includes two modes. In large open areas, the pigments, primer, water, etc. are transmitted through material pipelines to the spraying drone through the action of a vehicle pressure pump, which can efficiently carry out large-scale construction. In areas with dense power lines, drones directly extract materials from small material buckets in the vehicle for operation. The inspiration for drone design comes from hornet, giving new products vitality. In addition, the equipment adopts hydrogen energy to reduce environmental pollution, combined with 5G communication transmission, AI intelligent scanning, data analysis, infrared ranging and other technologies, ensures the efficiency and intelligence of operation. Helps to promote the efficiency, standardization, and intelligent development of future urban construction.

# 极地两栖运输车

## Polar Amphibious Transport Vehicle

作　　者：白沐凡
指导老师：赵芳华
所在院校：河北工业大学

## 设计说明

中国南极内陆考察对物资、设备和油料的需求通过地面运输得以满足，使得运输成为内陆考察活动的关键支撑。面对南极恶劣的自然环境和复杂的地形条件，后勤保障能力尤为关键。为了解决环境挑战对物资运输效率的影响并确保物资安全送达科考站，本设计提出了一种远程操控的运输车，该运输车能够适应水路及冰面条件，其顶部和尾部动力装置采用对环境友好的氢燃料驱动，并通过固定轴升降进行卸货。

## Design notes

With the arrival of the golden season for scientific research in Antarctica, the demand for materials, equipment, and oil for China's Antarctic inland exploration has been met through ground transportation, making transportation a key support for inland exploration activities. Facing the harsh natural environment and complex terrain conditions in Antarctica, logistics support capabilities are particularly crucial. In order to address the impact of environmental challenges on the efficiency of material transportation and ensure the safe delivery of materials to scientific exploration stations, this design proposes a remote-controlled transport vehicle that can adapt to waterway and ice conditions. Its top and tail power devices are driven by environmentally friendly hydrogen fuel and unloaded through fixed shaft lifting.

# 国际医生——一款变轨距高速检测车

## INTERNATIONAL DOCTOR— A Variable Gauge High-Speed Inspection Vehicle

作　　者：王泽远　杨　岚　王　璇
　　　　　陈玉山　张晓婷
指导老师：杨冬梅
所在院校：河北工业大学

### 设计说明

这是一款适用于国际高速铁路的可变轨距高速检测车，旨在克服传统大型检测车检测周期长的缺陷，进行无人化分区间平行检测作业。通过此技术，可及时发现并预防可能影响铁路行车安全的显性因素，做到防患未然。

### Design notes

This design proposes a variable gauge high-speed inspection vehicle suitable for international high-speed railways, aiming to overcome the shortcomings of traditional large inspection vehicles with long cycles and carry out unmanned parallel inspection operations between zones. Through this technology, explicit factors that may affect railway operation safety can be detected and prevented in a timely manner, by taking preventive measures.

# 儿童电动工程车设计

## Design of Electric Engineering Vehicle for Children

作　　者：黄　涵　范宇凝　高思媛
指导老师：杨冬梅　张新新
所在院校：河北工业大学

### 设计说明

这是一款玩具车，旨在通过可操控的车体构件与多方向转动，实现工程车工作的模拟仿真，促进儿童的感知能力和手眼协调能力的提升。在产品功能上，能够模拟真实工程车的载重与安全移动功能，支持多种工程车型如挖掘机、叉车、铲车等，以满足多样化操作需求。在设计原则上，重点考虑了儿童工程车的色彩丰富性、安全性、环保性及便携性，提供一个创新且符合儿童使用的电动工程车。

### Design notes

This design aims to simulate the operation of engineering vehicles through controllable body components and multi-directional motion coordination, while promoting the improvement of children's perception ability and hand-eye coordination ability. In terms of product functionality, it can simulate the load capacity and safe movement of real engineering vehicles, supporting various engineering vehicle models such as excavators, forklifts, etc. to meet diverse operational needs. In terms of design principles, the focus is on the color richness, speed safety, material environmental friendliness, and portability of children's engineering vehicles, providing an innovative and child friendly electric engineering vehicle design. In addition, the project conducts an in-depth analysis of the functional requirements of children's electric engineering vehicles, designs the layout and appearance of internal functional components, fully considers human-machine interaction, to ensure that the design ultimately meets the actual usage conditions of children's education and entertainment.

# 不占道路面修补自动化设备

## Automatic Equipment for Road Repairing without Affecting Traffic

作　　者：王　琪　贾玉婷　孟泽熙
　　　　　程　仪　孙佳欣
指导老师：张芳兰
所在院校：燕山大学



### 设计说明

当前公路养护主要依赖人工或人机协作方式进行，这在路面修补过程中需要占用大量交通资源。本设计提出了一种具有两种形态的工程车，以提高公路修补的效率并尽可能减少对交通的影响。远距离行驶时车前后臂平行翘起、底盘升起，修补路面时两臂贴合地面呈桥状，内部的装置进行坑洼和裂缝的自动修补，过往的车辆可以直接从工程车顶部开过去而不影响交通运行。在工程车的中上部位置设置有一个废料收集仓，两个原料存储仓，一个燃气罐。清理的公路废料通过废料收集管吸取到废料仓，工程车在修补完毕后可携带废料排放到指定位置。两个原料仓都配有加热保温系统，通过加热管进行输送，且顶部都设有原料填充口，方便快速补给，特别是在坑洼较多区域，减少了频繁补给的需求。燃气罐则为热风枪等设备提供所需能源，整体设计考虑了高效养护与最小化交通干扰的需要。

### Design notes

The current highway maintenance mainly relies on manual or human-machine cooperation, which requires a large amount of traddic resources in the process of road repair. This design proposes an engineering vehicle with two forms to improve the efficiency of highway repair and minimize the impact on traffic. When driving at a long distance, the front and rear arms of the vehicle are parallel and raised, and the chassis is raised. When repairing the road surface, the two arms are in a bridge shape, and the internal devices automatically repair potholes and cracks. Passing vehicles can drive directly from the top of the construction vehicle without affecting traffic operation. There is a waste collection bin, two raw material storage bins, and a gas tank set up in the upper middle position of the engineering vehicle. The cleaned road waste is collected through a waste collection pipe and transported to the waste bin. After the repair is completed, the construction vehicle can carry the waste and discharge it to the designated location. Both raw material silos are equipped with heating and insulation systems, which are transported through heating pipes. The top of both silos is equipped with raw material filling ports, making it convenient for quick replenishment, especially in areas with many potholes, reducing the need for frequent replenishment. The gas tank provides the necessary energy for equipment such as the hot air gun, and the overall design considers the needs of efficient maintenance and minimal traffic interference.

# 基于客流潮汐性的高铁客货共运系统设计

## The Design of High-Speed Railway Passenger-Freight Co-Transportation System

作　　者：潘文娟　李忆璐　孙　达
指导老师：白仲航
所在院校：河北工业大学

### 设计说明

近年来，随着我国高铁行业的快速发展，其运输业务已逐步扩展至货运领域。然而，部分地区的高铁客流量在淡季却不足50%，这种客流潮汐性现象导致了高铁运能浪费。本设计利用客流潮汐性设计客货空间转换方式，从而解决高铁运能浪费问题。该设计采用双层高铁结构，其中第二层为固定乘客区，当客流量较少时，可将第一层座椅椅背向前折叠至水平，倒凹型板块向下移动至遮盖座椅，板块上方空间即为货舱空间。同时，工作人员将货物放入模块化货舱内，称量货物及扫描信息。最后，工作人员通过面板确认信息后，货舱自动规划路径，从车厢货舱门进入车厢，驶向相应位置，后槽进行固定。当卸载货物时，相应目的地货舱解除固定，驶出车厢。

### Design notes

In recent years, with the rapid development of China's high-speed rail industry, its transportation business has gradually expanded to the field of freight transportation. However, in some areas, the passenger flow of high-speed rail is less than 50% during the off-season, and this tidal phenomenon of passenger flow leads to waste of high-speed rail transportation capacity. This design utilizes the tidal characteristics of passenger flow to design a space conversion method for passenger and freight, thereby solving the problem of waste of high-speed rail transportation capacity. The design adopts a double-layer high-speed rail structure, with the second layer being a fixed passenger area. When the passenger flow is low, the backrest of the first layer seat can be folded forward to the horizontal, and the concave plate can be moved downwards to cover the seat. The space above the plate is the cargo compartment space. At the same time, the staff puts the package into the modular cargo hold, lifts the board to weigh the package and scans the information. Finally, after confirming the information through the pans, the staff automatically plans the path for the cargo hold, enters the carriage through the cargo hold door, drives towards the corresponding position, and fixes the rear slot. When unloading the cargo, the corresponding destination cargo hold is released from the hold and is driven out from the carriage.

# GUICON 葡萄收获机设计

## Design of Grape Harvester

作　　者：高思媛　何志娇
指导老师：杨冬梅
所在院校：河北工业大学

## 设计说明

GUICON 葡萄收获机是一款实现鲜食葡萄和酿酒葡萄采摘，采用机械臂和振动棒技术、模块化运输以及伸缩结构设计的两用葡萄收获机。面对中国葡萄种植业的规范不一和复杂的种植布局，此收获机采取双模式操作，机械臂用于鲜食葡萄的精细采摘，而振动棒适用于酿酒葡萄的大批量收集。模块化的设计让采摘和运输功能视情况结合和分离，促进区域内庄园的协同收获，适用于个体化和分散化的种植模式，促进区域内庄园的协同作业，有效提高了工作效率。

## Design notes

GUICON grape harvester is a dual-purpose grape harvester that achieves the harvesting of fresh and wine grapes, using robotic arm and vibration rod technology, modular transportation, and telescopic structure design. Faced with the inconsistent standards and complex planting layout of China's grape cultivation industry, this harvester adopts a dual mode operation, with a robotic arm used for fine harvesting of fresh grapes, and a vibrating rod suitable for large-scale collection of wine grapes. The modular design facilitates the combination and separation of harvesting and transportation functions according to actual situations, promotes collaborative harvesting of estates in the region, adapts to individualized and decentralized planting modes, promotes collaborative operation of estates in the region, and effectively improves work efficiency.

# 高压电线巡检装备

## High Voltage Wire Inspection Equipment

作　　者：张琬悦
指导老师：姚小清
所在院校：华北电力大学（保定）

### 设计说明

这是一款高压电线巡检装备，通过技术革新，优化了传统的高压电线巡检流程。传统人工巡检不仅耗时、危险性高，而且在必要时还需断电进行检查，带来了极大的不便。相比之下，高压电线巡检机器人显著提升了巡检效率，用户只需将位置信息输入机器人的芯片，机器人便能自动定位。当达到预设高度后，机器人通过顶部的挂置装置稳固地挂接在电线上，并由后置驱动电机推动进行高压电线巡检作业。这种自动化的巡检方式不仅提高了检查效率，还大大降低了作业风险。

### Design notes

This project is a high-voltage wire inspection equipment. Through technological innovation, the traditional high-voltage wire inspection process has been optimized. Traditional manual inspection is not only time-consuming and high-risk, but also requires power outage for inspection when necessary, which brings great inconvenience. In contrast, high-voltage wire inspection robots have significantly improved inspection efficiency. Users only need to input location information into the robot's chip, and the robot can automatically locate. After reaching the preset height, the robot is securely attached to the wire through the top mounting device and is driven by a rear drive motor for high-voltage wire inspection operations. This automated inspection method not only improves inspection efficiency, but also greatly reduces operational risks.

# 芦苇画交互座椅

## Reed Painting Interactive Seats



作　　者：张轩维　叶淏天　申梦嫣　廉喆琦
指导老师：裴卉宁　黄晓光
所在院校：河北工业大学

## 设计说明

本设计是一款芦苇画交互座椅，融合了雄安地区的传统技艺——芦苇画与亚克力光画艺术工艺及压力传感器等现代技术。这款座椅设计用于公共和展览空间，通过用户“坐下休息”的动作激活内置光带，照亮亚克力板内的立体芦苇画，创造出装饰性和趣味性并存的交互方式，有效引导人们关注与欣赏芦苇画工艺。同时，也展现了传统技艺和现代科技的融合，体现了雄安传统技艺与城市未来高新技术共同发展的方向，标志着对城市文化标识构建的成功尝试和积极探索。

## Design notes

This design creates a reed painting interactive seat that integrates traditional techniques from the Xiong'an region, such as reed painting and acrylic light painting art, as well as modern technologies such as pressure sensors. This seat is designed for public and exhibition spaces, activating the built-in light strip through the user's "sit down and rest" action, illuminating the three-dimensional reed painting inside the acrylic board, creating a decorative and fun interactive way to effectively guide people to pay attention to appreciate the reed painting craft. At the same time, it also demonstrates the integration of traditional craftsmanship and modern technology, reflecting the direction of joint development of traditional craftsmanship and future high-tech in Xiong'an, marking a successful attempt and active exploration of urban cultural identity construction.

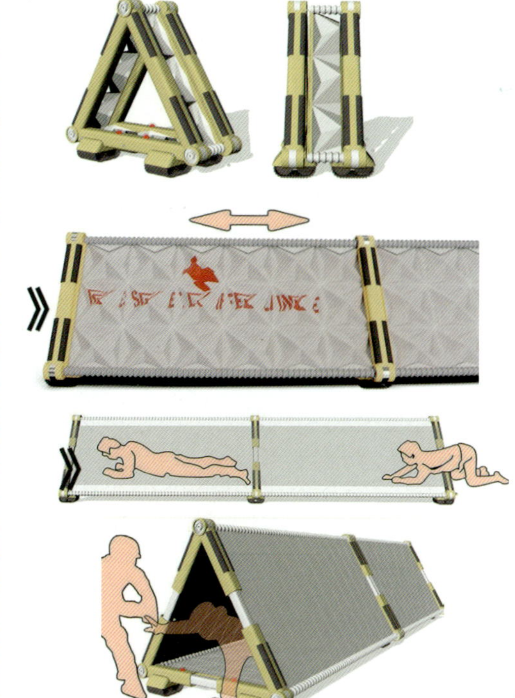

# 基于折纸结构的救援通道产品设计

## Rescue Channel Product Design Based on Origami Structure

作　　者：徐　颖　陈福源　熊小睿
路天悦
指导老师：刘月林
所在院校：燕山大学

### 设计说明

针对地震救援作业中遇到的困难和时间延长问题，本设计利用了折纸的可延展性质和应力结构原理，开发了一种适用于地震灾害废墟浅层区域的救援通道。在地震造成的不规则孔洞现场，可将通道放下，前端三角形入口横向展开，为救援人员提供进入点，采用液压技术使折纸结构纵向延展，形成一个小型的救援通道。良好的应力结构能够保证通道的坚硬度与稳定性，让营救者进入通道后将被困者救出。

### Design notes

In response to the difficulties and time extension encountered in earthquake rescue operations, this design utilizes the extensibility of origami and the principle of stress structure to develop a rescue channel suitable for shallow areas of earthquake disaster ruins. At the site of irregular holes caused by earthquakes, the channel can be lowered and the front triangular entrance can be horizontally expanded to provide an entry point for rescue personnel. Hydraulic technology is used to extend the origami structure vertically, forming a small rescue channel. A good stress structure can ensure the hardness and stability of the channel, allowing rescuers to enter and rescue trapped individuals.

# WHITEWING
# ——落叶清洁车

## WHITEWING ——
## Leaf Cleaning Vehicle

作　　者：许洪多
指导老师：徐　鸿
所在院校：河北师范大学

### 设计说明

这是一款清扫落叶专用的清洁车，旨在实现落叶的有效回收与再利用，以提升经济效益。该清洁车可通过折叠的方式对落叶进行热压缩，将松散的落叶处理为固体块状，不仅可以转化为燃料给车辆提供动力，还可以实现运输贮存和二次加工，大大地减少了资源浪费并提高了清洁车的工作效率。

### Design notes

We have designed a dedicated cleaning vehicle for the biomass waste mainly composed of fallen leaves during urban road cleaning in autumn and winter, aiming to achieve effective recovery and reuse of fallen leaves and improve economic benefits. This cleaning vehicle can compress fallen leaves by folding them into solid blocks, which not only convert them into fuel to provide power for the vehicle, but also facilitate transportation, storage, and secondary processing, greatly reducing resource waste and improving the efficiency of the cleaning vehicle.

# 高效节能蔬菜烘干机

## High Efficiency and Energy Saving Vegetable Dryer

作　　者：朱昊然　周子豪　贾　墨
严高峰　岳　洋
指导老师：姚小清　王　栋　刘明浩
所在院校：华北电力大学

### 设计说明

为满足农户对低成本烘干蔬菜的需求以及实现节能环保的目标，本设计采用了板翅式余热回收系统、太阳能空气加热器和单片机智能控温系统，显著降低了能源消耗。通过结合太阳能集热与电辅助加热，并配备余热回收系统，能够节省62.4% 的电能。

这套系统不仅满足了农户对于节能高效烘干蔬菜的需求，同时也体现了对环境友好、节能减排的设计理念。

### Design notes

To meet the needs of farmers for low-cost vegetable drying and achieve the goal of energy conservation and environmental protection, this design adopts a plate fin waste heat recovery system, solar air heater, and single-chip intelligent temperature control system, significantly reducing energy consumption. By combining solar energy collection with electric assisted heating, and equipped with a waste heat recovery system, 62.4% of electricity can be saved. This system not only meets the needs of farmers for energy-saving and efficient vegetable drying, but also embodies the design concept of environmental friendliness, energy conservation and emission reduction.

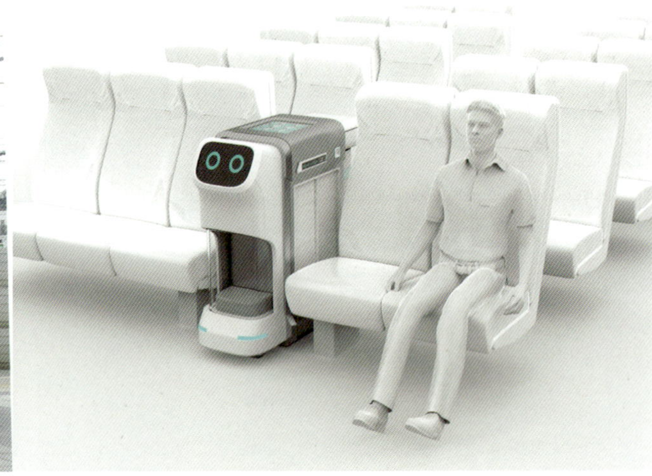

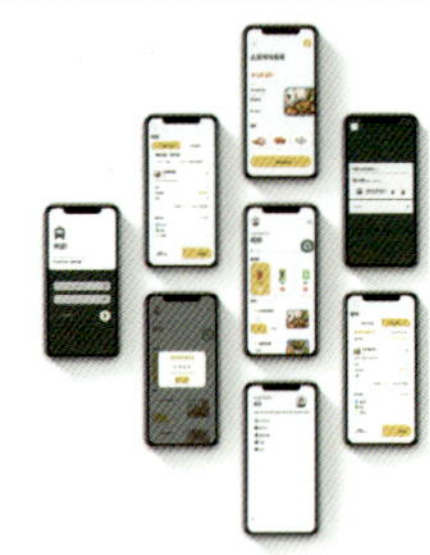

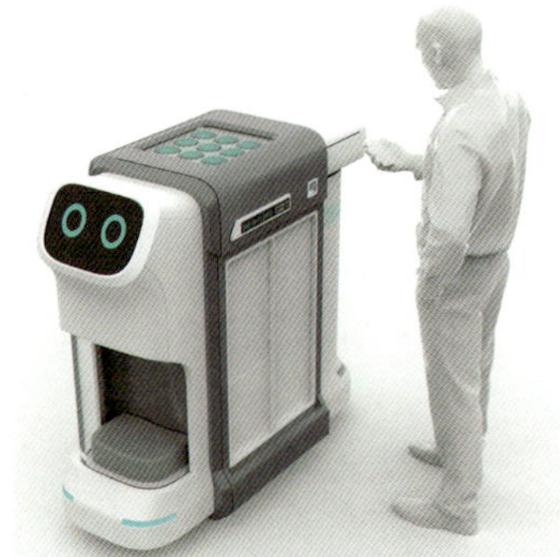



# 高铁车厢智能送餐机器人设计

## Intelligent Food Delivery Robot in High-Speed Railway Carriages

作　　者：史佳乐　刘倩倩
指导老师：赵　静
所在院校：太原理工大学

### 设计说明

为了解决高铁送餐效率低，设计了一款高铁智能送餐机器人。该机器人满足无人配送、餐盒回收等需求，可以配合自动售货机的自动补货功能、餐盒回收站和手机小程序，构建了一套完整的服务流程，提高乘客在高铁旅途中的智能化体验。该设计基于现有技术，针对高铁现存餐饮服务问题，为高铁智慧出行提供了全新的解决方案。

### Design notes

In order to solve the problems of reduced labor supply in the service industry, low efficiency of high-speed rail delivery, we designed a high-speed rail intelligent food delivery robot. This robot meets the needs of unmanned delivery and food box recycling, and can be combined with the automatic replenishment function of vending machines, food box recycling stations, and mobile mini programs to build a complete service process, enhancing the intelligent experience of passengers on high-speed rail journeys. This design is based on existing technology and provides a new solution for the existing catering service problems of high-speed rail, providing intelligent transportation for high-speed rail.

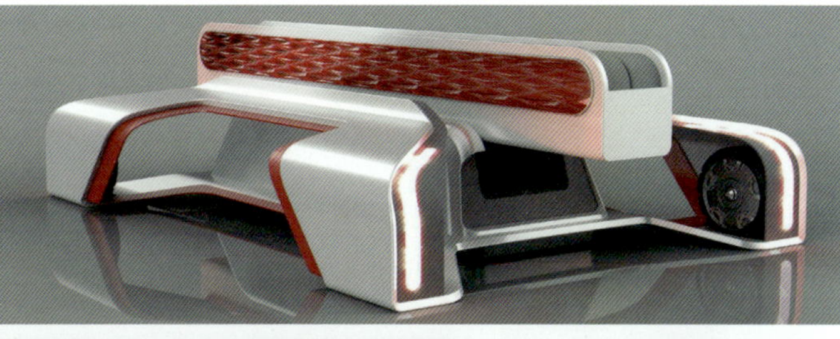

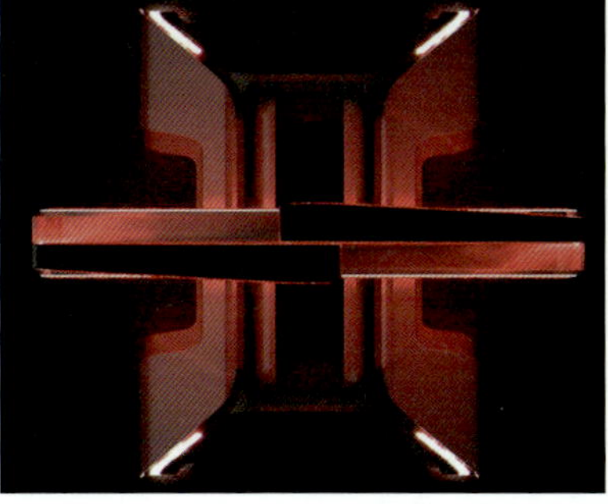

# ALERTRON——交通事故警示牌解决方案

## ALERTRON——Traffic Accident Warning Sign Solution

作　　者：韩　月　黄晓曈　王　苹
指导老师：郝建峰　刘　静　方　廷
所在院校：晋中信息学院

### 设计说明

交通中传统的三脚架式警示牌由于缺乏自动化设置功能，需要驾驶员或乘客亲自携带至适当距离放置，这显著增加了二次事故的风险。此外，事故的突发性和事后驾乘人的紧张恐惧心理都迫切需要一套解决方案。

本设计遵循以用户为中心的设计理念，通过建立典型人物个性，并深入研究用户旅程，开发了一套既适用于白天也适用于夜晚、能够迅速启动的自动警示牌设置方案。该方案使用步骤只需4步，从背部取出遥控器，设立指定距离，按压遥控器，开始自动设立，到达指定地点警示。

### Design notes

In accident scenarios where triangular warning signs need to be installed, traditional tripod warning signs lack automated setting functions and require drivers or passengers to personally carry them to an appropriate distance for placement, which significantly increases the risk of secondary accidents. In addition, the suddenness of accidents and the tension and fear of drivers and passengers afterwards urgently require a set of solutions.

This design follows the user centered design concept, establishes a typical character persona, and conducts in-depth research on user journeys to develop a set of automatic warning sign setting schemes that are suitable for both day and night, and can be quickly activated. This solution only requires four steps to use. Take out the remote control from the back, set a specified distance, press the remote control, and start automatic setting, arrive at the designated location for warning.

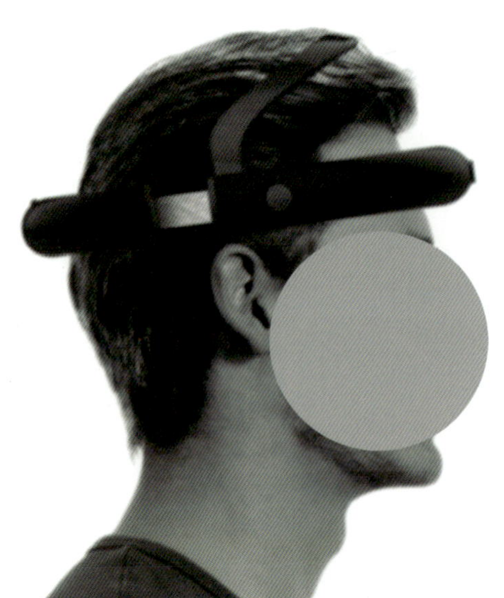

# 基于脑电信号——便携式创造力检测产品

## EEG-BASED —— Portable Creativity Testing Products

作　　者：田尚鑫　李　洋　王　杰
　　　　　焦千龙　窦世霖
指导老师：苗　秀　边　坤
所在院校：内蒙古科技大学

### 设计说明

本研究旨在基于脑电信号，探索在创造性思维活动中与创造力相关的脑电特征。我们的目标是实现对创造力的自动评估，并设计一种带有少量电极的便携式创造力脑电检测设备。通过脑电实验，我们收集了被试在进行常规想象和创造性想象时的脑电数据，并通过差异性检验来探索与人们创造力思维相关的脑电通道和频段。研究发现，在创造性思维中，位于 F3、FP1和FP2电极上的 alpha 频段的能量数值能够反映个体的创造力水平。进一步地，通过用户调研、竞品分析以及 Kano 模型，我们确定了产品的功能需求，并最终设计出了“BrainWave-α”。BrainWave-α 是一款便携式创造力脑电检测设备，可与配套的 BrainWave-α PC 端软件结合使用。

### Design notes

The aim of this study is to explore the EEG features related to creativity in creative thinking activities based on EEG signals. Our goal is to achieve an automated assessment of creativity and to design a portable creativity EEG device with a small number of electrodes. Through the EEG experiment, we collected the EEG data of the subjects when they performed conventional imagination and creative imagination, and explored the EEG channels and frequency bands related to people's creative thinking through the difference test. The study found that in creative thinking, the energy value of the alpha band located on the F3, FP1 and FP2 electrodes can reflect the level of creativity of an individual. Further, through user research, competitive product analysis and Kano model, we determined the functional requirements of the product and finally designed the "BrainWave-α". The BrainWave-α is a portable creativity EEG detection device that can be used in combination with the accompanying BrainWave-α PC software.

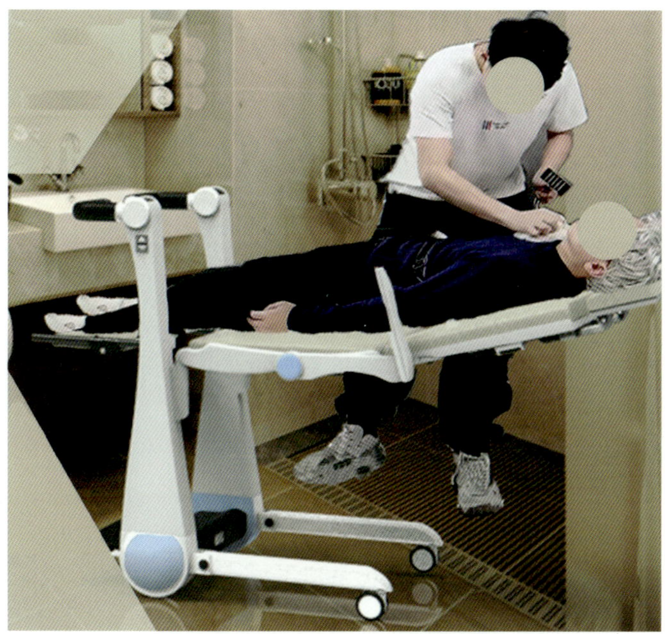

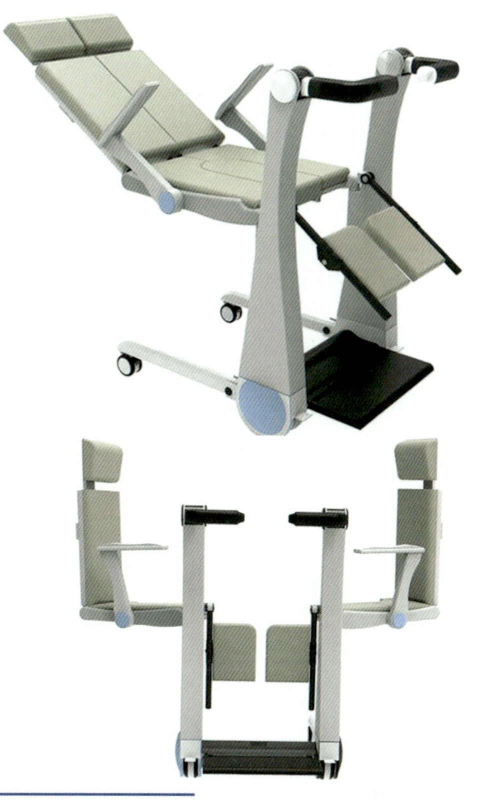





# 智能助老洗浴椅

## Intelligent Bath Chair Assisting the Elderly

作　　者：张　琪　陈墨桐
指导老师：王　坤
所在院校：内蒙古工业大学

### 设计说明

失能老人的洗浴问题是日常生活中一个较为棘手的挑战。在传统的护理过程中，护理人员需要多次转移老人，而且老人在洗澡时往往只能保持固定姿势，这大大降低了老人的洗浴体验。针对这一问题，我们设计了一种专门用于帮助下肢瘫痪老人舒适洗澡的设备。这款助老洗浴机器人具有可调节高度的座椅和180度开合的设计，这让护理人员能够轻松地将老人从床上转移到浴室，避免了传统的“背”和“抱”方式所带来的不便。此外，老人可以在这个设备上完成整个洗浴过程，减少了移动次数，也降低了护理人员的工作量。在洗浴中，老人可以根据自己的舒适度调整靠背和腿部支撑，靠背甚至可以调至水平位置，将设备转变为一张简易浴床，便于护理人员为老人清洁身体。

### Design notes

The problem of bathing the disabled elderly is a more difficult challenge in daily life. In the traditional nursing process, the nursing staff needs to transfer the elderly several times, and the elderly often can only maintain a fixed position when taking a bath, which greatly reduces the bathing experience of the elderly. In response to this problem, we designed a device specifically to help the elderly with lower limb paralysis taking a bath comfortably. With an adjustable height seat and a 180-degree opening and closing design, the robot allows caregivers to easily transfer the elderly from the bed to the bathroom, avoiding the inconvenience of traditional "back" and "hold" methods. In addition, the elderly can complete the entire bathing process on this device, reducing the number of moves and reducing the workload of caregivers. In the bath, the elderly can adjust the backrest and leg support according to their comfort, and the backrest can even be adjusted to the horizontal position, transforming the equipment into a simple bath bed, which is convenient for the nursing staff to clean the body for the elderly.

金
奖
银
奖
铜
奖
优秀奖

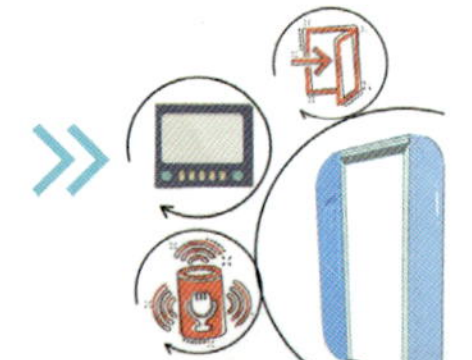

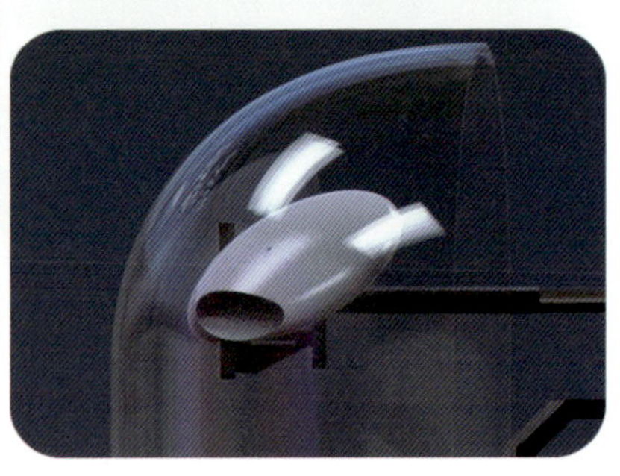

# 一站式胶囊——基于疫情防控的交互设计

## ONE-STOP CAPSULE —— Interaction Design Based on Epidemic Prevention and Control

作　　者：张怡帆　范国荣
指导老师：苗　秀　边　坤　郭媛媛
所在院校：内蒙古科技大学

### 设计说明

公共服务机器人技术催生了智能防疫机器人的出现。这些机器人解放了从事机械化工作的劳动力，避免了人与人之间的直接接触和交叉感染的风险，提高了检测效率，并大幅减轻了人们在通勤以及工作人员在疫情管控方面的压力和负担。

### Design notes

At present, the combination of epidemic prevention demand and the existing public service robot technology has spawned the emergence of intelligent epidemic prevention robots. These robots free up the labor force engaged in mechanized work, avoid direct contact between people and the risk of cross-infection, improve the efficiency of detection, and greatly reduce the stress and burden of people in commuting and of staff in epidemic control.

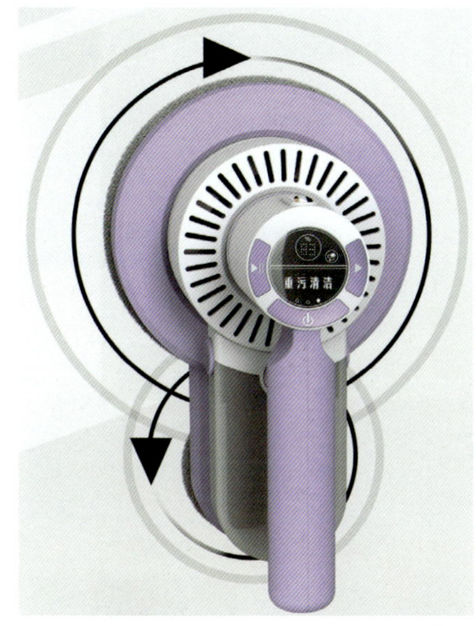


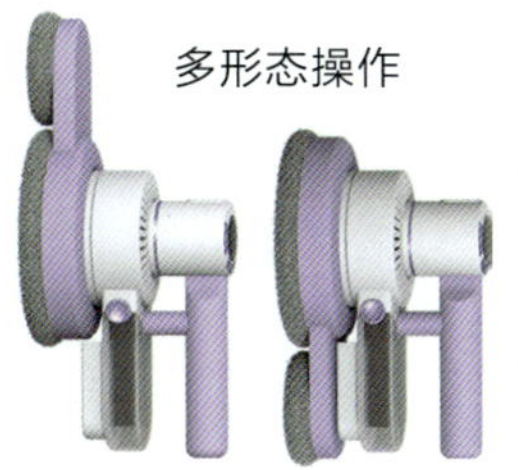


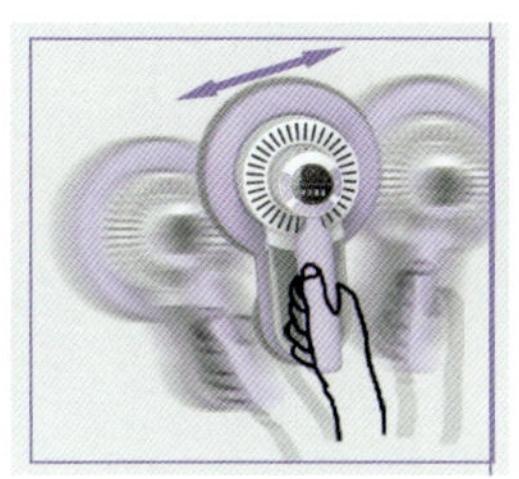

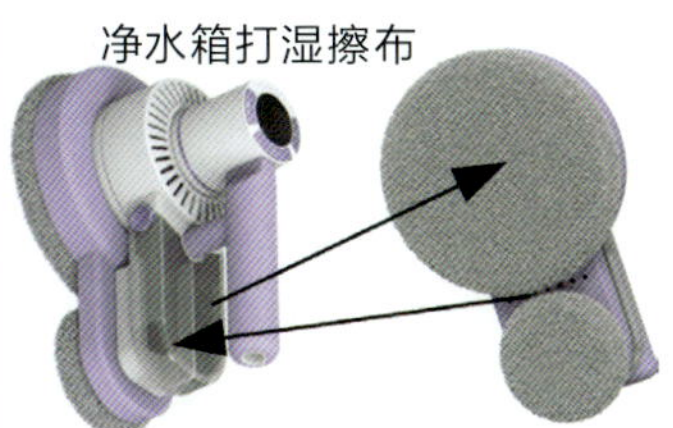


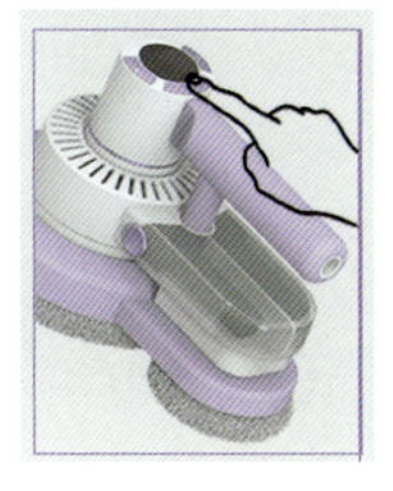

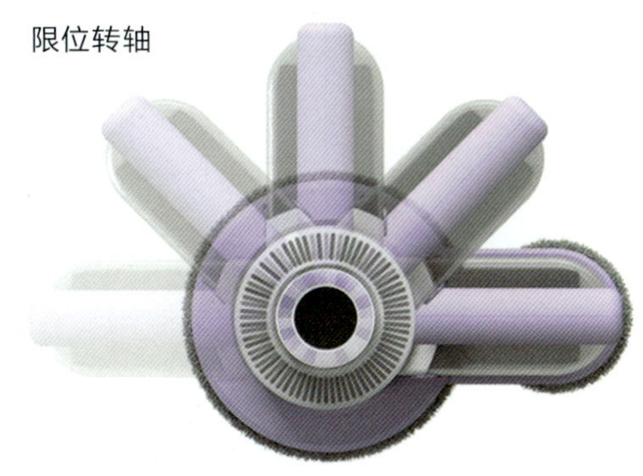


# 转——玻璃清洁器设计

## ROTARY ——
## Electric Glass Cleaner

作　　者：孙倩哲　郭小慧
指导老师：刘　日
所在院校：内蒙古工业大学

### 设计说明

本设计整体外观通过圆的大小变化来实现，形成了一大一小的双圆擦头。这种创新的操作方式旨在改善玻璃清洁工作，解决玻璃边角难以清洁和局部顽固污渍去除不彻底的问题。把手部分采用了限位转轴设计，确保在旋转时能够固定方向，同时以机械化方式切换不同的清洁模式，减少不必要的电力消耗。该设备采用单水箱设计，内部将净、污水分隔开。净水箱负责供应水分以保持擦布湿润，而污水箱则用于收集清洁过程中产生的污水，形成一个“净水擦拭、污水收集”的循环系统，从而保持机器的工作效果和效率。承载底座不仅可以用来存放玻璃清洁器和清洗擦布，还配备了有线连接以持续为机器供电，延长了单次使用时间。擦头具有吸附功能，可以减轻手持时的负担。它根据不同的清洁模式进行旋转震动，适应不同环境，满足用户的需求，实现了功能的最大化。

### Design notes

The overall appearance is designed by changing the size of the circle, forming a large and small double circular wiping head. This innovative method of operation is designed to improve glass cleaning and solve the problem of difficult to clean glass corners and incomplete removal of local stubborn stains. The handle part uses a limit shaft design to ensure that the direction can be fixed when rotating, while mechanizing to switch between different cleaning modes, reducing unnecessary power consumption. The equipment uses a single tank design, and the net sewage is separated inside. The clean water tank is responsible for supplying water to keep the wipe wet, while the sewage tank is used to collect the sewage generated during the cleaning process, forming a "clean water wiping, sewage collection" circulation system, so as to maintain the working effect and efficiency of the machine. Not only can the carrier base be used to store glass cleaners and cleaning wipes, it is also equipped with a wired connection to continuously power the machine, extending the single use time. The rubbing head has adsorption function, which can reduce the burden of the hand-held. It rotates and vibrates according to different cleaning modes, adapts to different environments, meets the needs of users, and realizes the maximum function.

# 基于模块化的可上下楼梯消杀配送机器人设计

## Sanitization and Delivery Robot Design for Going up and down Staiers



作　　者：柴有源　秦园园
指导老师：苗　秀　郭媛媛　边　坤
所在院校：内蒙古科技大学

### 设计说明

空间消毒是必不可少的，特别是在学校、居民楼等人流密集的区域。由于大多数低层建筑没有安装电梯，人工进行消杀既耗时又费力。因此，设计一款能够自主上下楼梯进行消杀的机器人显得尤为重要。本产品将麦克纳姆轮通过轮组的方式进行再设计，以解决上下楼梯的问题，并且通过内置的陀螺仪和升降装置，确保机器在上下楼梯时始终保持平衡，防止侧翻。此外，机器采用模块化设计，使得后部的消杀装置可以替换，同时具备环境消杀和货物配送两种功能。我们还对交互界面进行了设计，增加了一些交互表情以提高产品的趣味性和用户粘性。用户可以通过机器上的屏幕进行部分操作和数据监测。

### Design notes

During the epidemic, space disinfection is an essential measure, especially in crowded areas such as schools and residential buildings. Since most low-rise buildings do not have elevators installed, manual sterilization is time-consuming and laborious. Therefore, it is particularly important to design a robot that can automatically go up and down stairs to sanitize. This product redesigns the McNamum wheel by wheel grouping to solve the problem of going up and down stairs, and through the built-in gyroscope and lifting device, ensure that the machine is always balanced when going up and down stairs, preventing cartwheels. In addition, the machine adopts a modular design, so that the rear sanitization device can be replaced, and has both environmental sanitization and cargo distribution functions. We also designed the interactive interface and added some interactive expressions to improve the fun and user engagement of the product. Users can perform part of the operation and data monitoring through the screen on the machine.

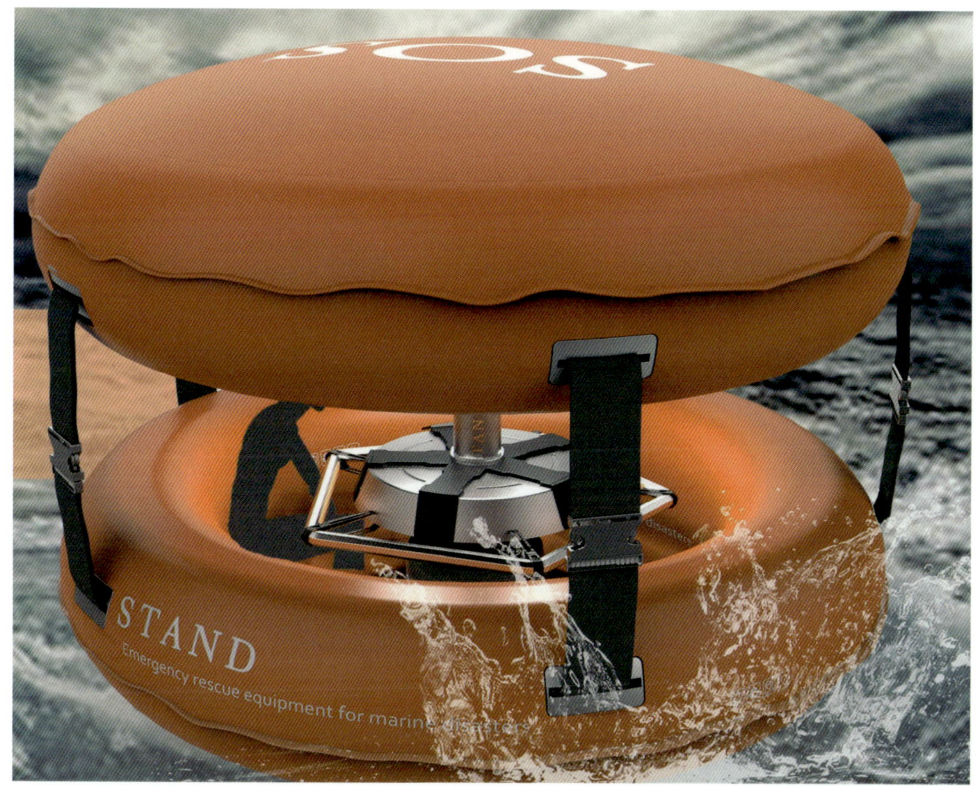

1. 发生事故后

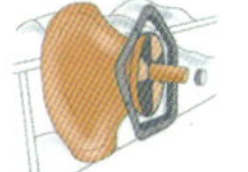
2. 将充气阀门用力拉出，救生装置迅速充气膨胀

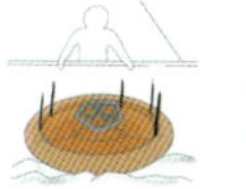
3. 遇难者将充好气的救生装置从船上扔下

4. 遇难者可以迅速爬进救生装置进行避难

5. 连接两个救生船形成一个半封闭空间

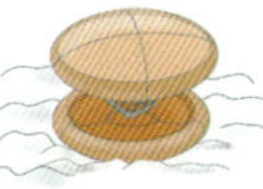
6. 人体重力为救生艇提供足够低的重心，从而使救生艇在海中不断翻滚也不会沉没



# 汉堡包救援船

## Hamburger Rescue Ship

作　　者：姚文博　姜志文
指导老师：石　钧
所在院校：内蒙古师范大学

### 设计说明

海难事故通常包括碰撞、浪损、触礁、搁浅、爆炸、沉没和失踪等类型。近年来，虽然从全球海滩事故发生的总体趋势来看有所减缓，但海滩事故的数量依然较高，形势依旧不容乐观。此外，现有的救援船普遍存在容易倾覆的风险，因此我们设计了这款专门针对水上救援的防倾覆救援船。

### Design notes

Marine accidents usually include collision, wave damage, running on rocks, stranding, explosion, sinking and missing. In recent years, although the overall trend of global beach accidents has slowed down, the number of beach accidents is still high, and the situation is still not optimistic. In addition, existing rescue boats are generally at risk of capsizing, so we designed this anti-capsizing rescue boat specifically for water rescue.

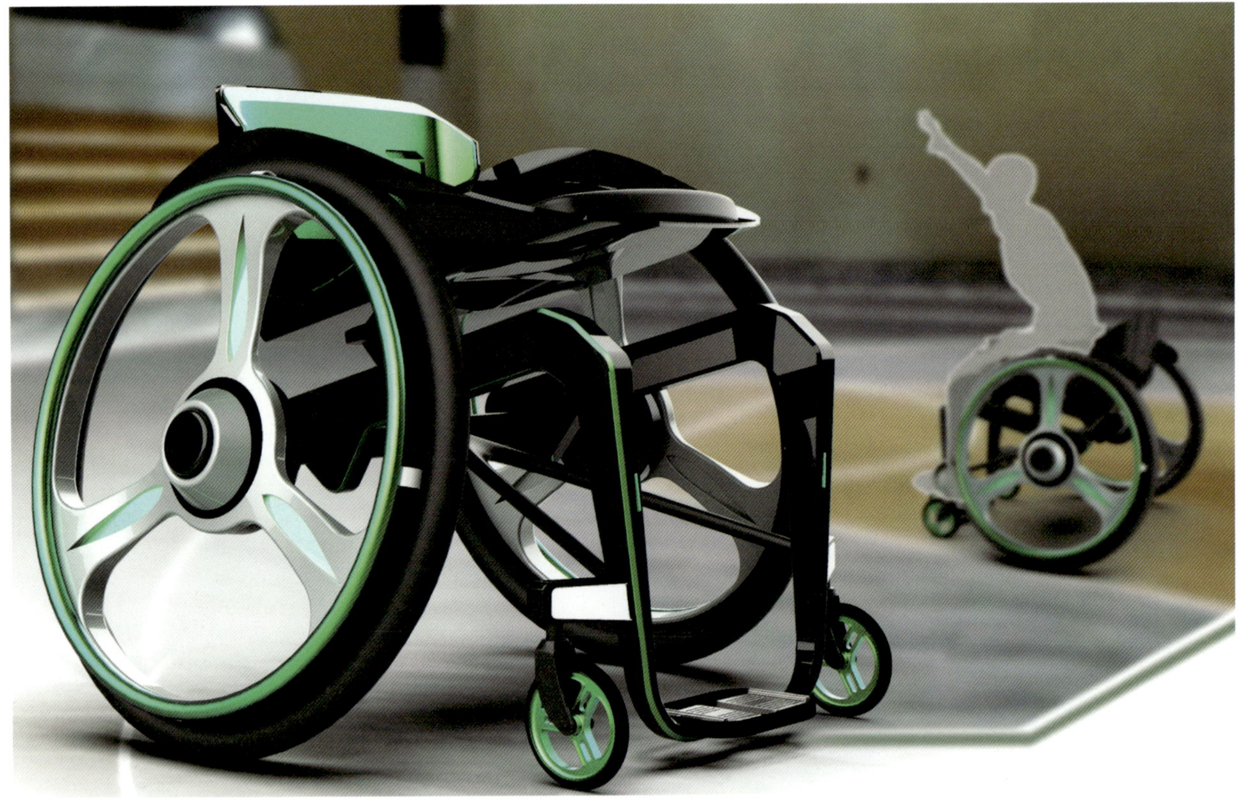



# 另一种“跳跃”——运动型轮椅设计

## ANOTHER "JUMP" —— Sport Wheelchair Design

作　　者：姜志文　姚文博
指导老师：石　钧
所在院校：内蒙古师范大学

### 设计说明

我们以运动型轮椅的结构和外观为基础进行了创新设计，通过借鉴螳螂的仿生结构，对轮椅的支架结构和轮胎轮毂进行了优化。同时，我们还采用了仿生绿色的理念对轮椅的材质进行了渲染。这些设计充分展现了轮椅的运动感，使得热爱运动的残障人士能够充分体验到轮椅带来的视觉效果。

### Design notes

We carried out an innovative design based on the structure and appearance of the sports wheelchair, and optimized the support structure and tire hub of the wheelchair by drawing on the bionic structure of the praying mantis. At the same time, we also adopted the concept of bionic green to render the material of the wheelchair. These designs fully demonstrate the sense of movement of the wheelchair, so that people with disabilities who love sports can fully experience the visual effects brought by the wheelchair.

# SAM——土壤高温杀虫灭菌绿色农机

# SAM — High Temperature Insecticidal Sterilization Green Agricultural Machinery for Soil

作　　者：倪茗瑄　马佳聪　冷守铭
指导老师：王成玥　陈　峰
所在院校：沈阳理工大学

## 设计说明

本设计基于可持续设计理念，在旋耕土壤的同时，利用电发热装置产生的高温对土壤中残留的化学农药、病毒以及害虫进行高温灭杀。该设备在日常还可以进行深松耕土工作，实现1+1>2的效果。它可以有效杀灭包括根结线虫在内的病菌和害虫，作业效率高。由于不使用高剧毒农药，因此避免了高剧毒农药带来的水源污染、土壤污染以及农产品中农药残留超标等问题，从根本上做到了环境保护。

## Design notes

This design is based on the concept of sustainable design. While rotating tilling the soil, it uses the high temperature generated by the electric heating device to sterilize the residual chemical pesticides, viruses and pests in the soil at high temperature. The equipment can also carry out deep-loosening tillage work in daily life to achieve the effect of 1+1 > 2. It can effectively kill germs and pests including root knot nematodes, and the operation efficiency is high. Because highly toxic pesticides are not used, problems such as water pollution, soil pollution and excessive pesticide residues in agricultural products caused by highly toxic pesticides are avoided, and these public hazards are fundamentally eliminated.

# 盈车嘉穗——海水稻排灌装备

## THE GRAIN HARVEST — Irrigation and Drainage Eguipment for Sea Rice



作　　者：王　威　陈欣雨　刘子圣
指导老师：高　崇　刘冀伟
所在院校：沈阳航空航天大学

### 设计说明

海水稻并不是用海水直接灌溉，而且盐碱地的含盐量必须低于1%，浓度过高会导致产量减少甚至颗粒无收。目前，许多水稻专家都在研究海水稻，以提高其耐盐碱性。以目前的技术来看，海水的含盐量约为3%，而海水稻的耐盐碱性只有1%，因此还不能直接使用海水进行灌溉。只有解决咸水灌溉问题，才能对农民和土地管理者有所帮助，否则大量土地将面临抛荒的风险。水稻的需水量应遵循“多排少补”的原则进行排灌，这就需要良好的排水系统，既能充分满足泡田洗碱的需求，也能做到灌溉、排水自如，为水稻高产提供必要的条件。为此，设计一款能根据当地海水盐浓度以及所需盐浓度智能调节的灌溉排水智能机械设备是十分必要的。

### Design notes

Sea rice is not directly irrigated with seawater, and the salt content of saline-alkali land must be less than 1%, which can lead to reduced yield or even no grain harvest. At present, many rice experts are studying sea rice to improve its saline-alkali tolerance. With current technology, the salt content of sea water is about 3%, and the salt and alkali tolerance of sea rice is only 1%, so it can not be directly used for irrigation. Only by solving the problem of saltwater irrigation can farmers and land managers be helped, or much land will be at risk of being abandoned. The water requirement of rice should follow the principle of "more discharge and less supplement" for irrigation and drainage, which requires a good drainage system, which can fully meet the needs of soaking the field and washing alkali, but also can achieve irrigation and drainage freely, providing the necessary conditions for high yield of rice. Therefore, it is necessary to design an intelligent irrigation and drainage mechanical equipment that can adjust intelligently according to the salt concentration of local seawater and the required salt concentration.

# 海底光缆庇护者

## Submarine Cable Shelter

作　　者：李嘉圆　郭彦君　于熙梦
指导老师：赵雨晴　高　崇
所在院校：沈阳航空航天大学

### 设计说明

海底光缆与陆地光缆最大的区别在于其“铠装保护”。海底光缆故障的处理通常包括三个步骤：首先，使用扩频时域反射仪定位大致的故障位置；其次，利用潜水机器人找到受损海底光缆的精确位置，切断故障部分，并将剩余两端拖回修理船进行修复；最后，用备用海底光缆连接受损光缆的两个断点，测试无误后再放回海底。本设计方案采用了VR技术支持，技术人员可以通过VR技术远程操控机器人内外的机械手，进行精准操作。VR控制器可以操纵机器人的手臂进行拾取和放置操作，这项功能可用于训练机器人执行任务等。

### Design notes

The biggest difference between submarine cable and land cable is its "armor protection". The fault handling of submarine cable usually includes three steps. First, the fault location is located by using spread spectrum time domain reflectometer. Then, the submersible robot is used to find the precise location of the damaged submarine cable, cut off the faulty part, and tow the remaining two ends back to the repair ship for repair. Finally, the two breakpoints of the damaged cable are connected with the backup submarine cable, and the test is correct before being returned to the sea floor. This design scheme uses VR technology support, technicians can remotely control the robotic arm inside and outside the robot through VR technology, perform accurate operation. The VR controller can manipulate the robot's arm for picking and placing operations, a feature that can be used to train the robot to perform tasks, among other things.

金奖 银奖 铜奖 优秀奖



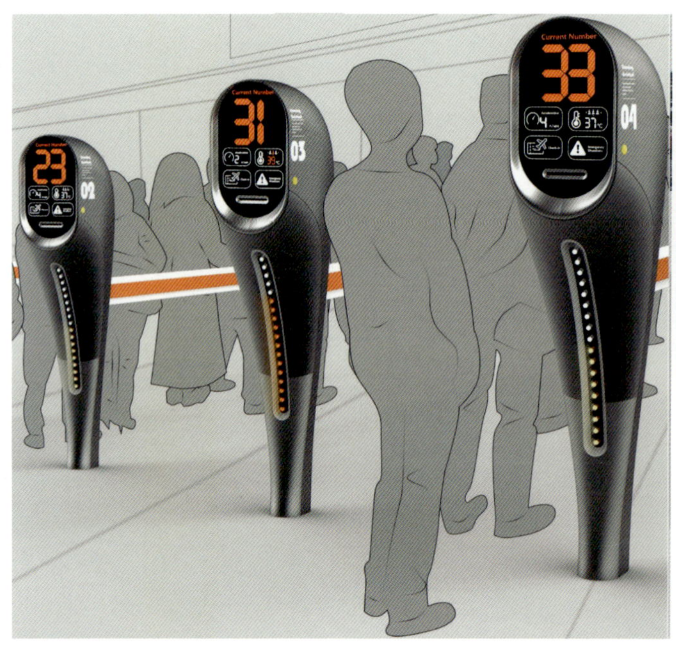


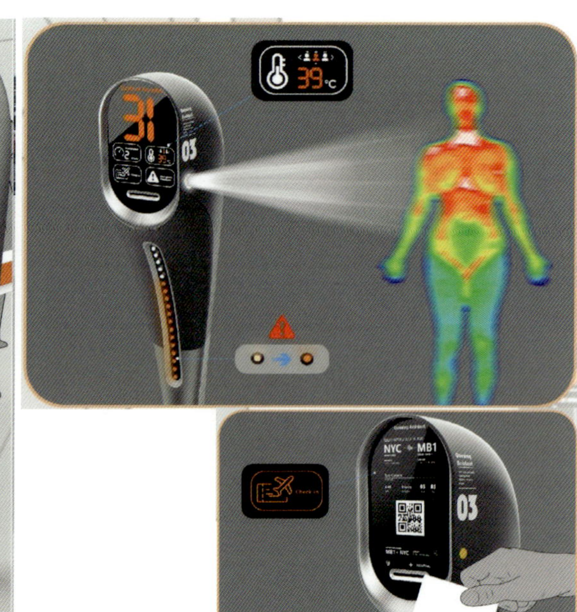


# 排队助手

## Queuing Assistant

作　　者：安童宇　崔　靖　普竟一　闫懋松
指导老师：许　坤　包海默　乔　松
所在院校：大连民族大学

### 设计说明

在过安检、购票或检票时，我们经常需要排队等候，而队伍往往很长且人员杂乱，使我们难以判断各队伍的人数，从而无法决定加入哪个队伍。排队助手能很好地解决这一问题。它通过内置的传感器记录排队人数，并将数字显示在电子屏幕上。同时，排队助手还能计算每条队伍的前进速度，帮助人们做出更明智的选择。除了以上功能，它还可通过外置的温度传感器检测人们的体温，当发现有体温异常的人时，它会发出警报提醒。此外，它还具备购票和取票功能，将取票机的功能集于一身。在紧急情况下，我们还可以通过它呼叫工作人员。

### Design notes

When going through security check, buying tickets or checking tickets, we often have to wait in lines, and the lines are often long and cluttered, making it difficult for us to judge the number of people in each line, and thus unable to decide which line should we join. The Queuing Assistant solves this problem well. It records the number of people waiting in line through a built-in sensor and displays the numbers on an electronic screen. At the same time, the Queuing Assistant can also calculate the forward speed of each queue to help people make more informed choices. In addition to the above functions, the Queuing Assistant can also detect people's body temperature through an external temperature sensor. When it detects someone with an abnormal body temperature, it sends out an alert. In addition, it also has ticket purchase and ticket collection functions, which integrates the functions of the ticket collection machine. In case of emergency, we can also call staff through it.

# 水下工作者

## Underwater Worker

作　　者：刘书铭　陈　淼　赵博文
　　　　　王　昊
指导老师：许　坤　包海默　乔　松
所在院校：大连民族大学

## 设计说明

水下工作者是一款采用最新技术的水下探测考察机器人。通过优化其内部的视觉识别系统、驱动方式、框架结构以及整体仿生外观等软硬件设计，全面提高了产品的性能。它能够有效解决深海探测、检测和考察过程中的智能性、准确性、工作效率和稳定性等问题。水下工作者为水下观测和考察提供了更加强大的功能，使执行水下任务变得更加方便快捷。操作人员只需在控制室内，就能清晰地了解水下情况，并实时绘制地形图，为后续任务的执行打下良好基础。此外，机器的分层安装设计方便了维修，而内部多层防护的结构则可以支持更长时间的任务执行。

## Design notes

Underwater Worker is an underwater exploration robot using the latest technology. By optimizing its internal visual recognition system, driving mode, frame structure and overall bionic appearance and other hardware and software design, the performance of the product is comprehensively improved. It can effectively solve the problems of intelligence, accuracy, working efficiency and stability in the process of deep-sea exploration, detection and investigation. Underwater workers provide more powerful functions for underwater observation and investigation, making the execution of underwater tasks more convenient and fast. Only in the control room, the operator can clearly understand the underwater situation and draw the topographic map in real time, laying a good foundation for the implementation of subsequent tasks. In addition, the layered installation design of the machine facilitates maintenance, while the multiple layers of protection of the internal structure can support longer task execution.

# 折叠卡车

## Folding truck

作　　者：闫懋松　普竟一　崔　靖
安童宇
指导老师：乔　松　许　坤
所在院校：大连民族大学

### 设计说明

在当今飞速发展的社会中，交通运输扮演着重要的角色。厢货车在交通运输中具有不可替代的作用。随着经济的飞速发展，道路上的厢货车数量日益增多，超车现象频繁发生。由于车身过长，货车司机在行驶过程中会面临许多视野盲区，这给小型车辆带来了很多安全隐患。这款折叠卡车可以根据货物的数量调整箱体的体积，通过三种状态的变化来降低重心和减小体积，从而降低事故风险。

### Design notes

In today's rapidly developing society, transportation plays an important role. Van has an irreplaceable role in transportation. With the rapid development of the economy, the number of vans on the road is increasing. Because the body is too long, the truck driver will face many blind areas of vision in the process of driving, and the phenomenon of overtaking occurs frequently, which brings a lot of safety risks to small vehicles. This folding truck came into being, it can adjust the volume of the box according to the number of goods, through three state changes to reduce the center of gravity and reduce the volume, thereby reducing the risk of accidents.

# 自动化多功能叉车

## Automated Multifunctional Forklift

作　　者：唐元昊　苏英奇　陈　熠
指导老师：赵轶博
所在院校：沈阳大学

### 设计说明

在日益复杂的物流环境中，随着智能化技术的进步以及社会发展新的需求和因素的影响，叉车在物流系统中扮演着重要的角色。因此，叉车的外部形态、作业模式和使用效率都需要不断地发展和突破。鉴于市场上现有叉车的基础功能和优化空间还有所不足，我们提出了这款新产品。本产品具有绿色环保、创新引领的特点，其造型新颖，且符合未来工业设计发展趋势。

### Design notes

In the increasingly complex logistics environment, with the progress of intelligent technology and the influence of new needs and factors of social development, forklifts play an important role in the logistics system. Therefore, the external form, operation mode and use efficiency of forklifts need to be continuously developed and broken through. In view of the shortcomings of basic functions and optimization space of existing forklifts on the market, we proposed this new product. This product has the characteristics of green environmental protection and innovation, its novel shape, and in line with the development trend of industrial design in the coming years.

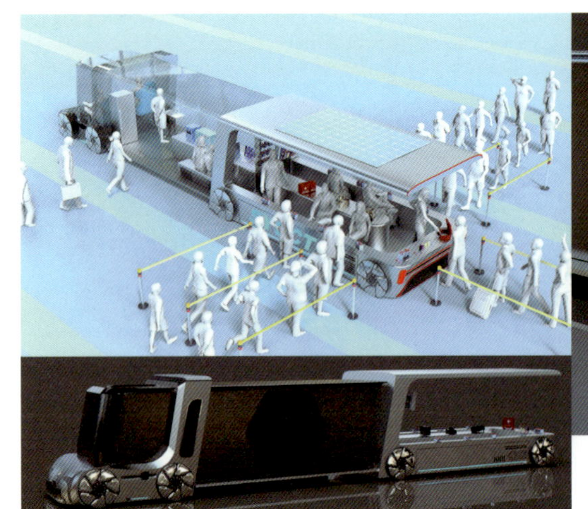

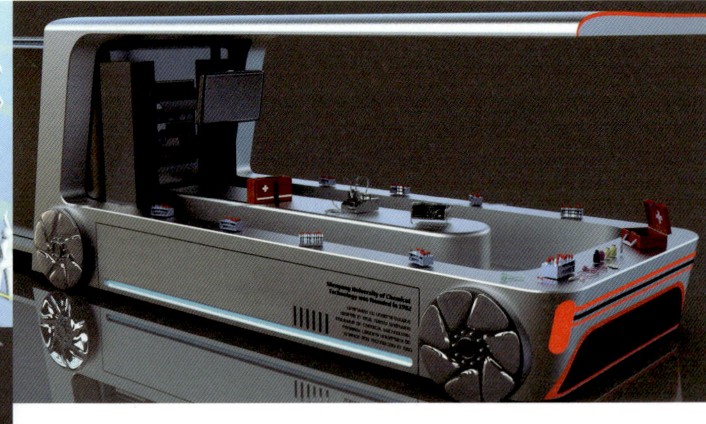

# 分区化核酸检测医疗车

## Partitioned Nucleic Acid Detection Medical Vehicle

作　　者：王可艺
指导老师：佟　建
所在院校：沈阳化工大学



### 设计说明

本产品是一款分区化核酸检测医疗车，主要针对核酸检测时排队密集、秩序混乱以及感染风险高的问题而设计。该车整体造型具有未来科技感，在医疗区采用了流线型的 U 型设计，这不仅美观，还提高了实用性和功能性。U 型设计可以让工作人员分散工作，使排队检测的人员呈扇形排列，从而增加间隔距离。该检测车采用模块化伸缩设计，前半部分是驾驶舱，车厢中部为功能区。分离后，可以通过侧面车门进入，向前通往驾驶舱，向后则通向 U 型医疗区。车内设施齐全，包括消毒仪、全自动核酸检测仪器、冰箱、电脑、洗手台等，旨在提供舒适的医疗环境并提高工作效率。此外，这款多功能模块化医疗车节能环保，其主要能源为电能，可通过太阳能电池板和充电桩进行充电。

### Design notes

This product is a zoned nucleic acid detection medical vehicle, which is mainly designed to solve the problems of dense queue, chaotic order and high infection risk during nucleic acid detection. The overall shape has a sense of future technology, and the streamlined U-shaped design is adopted in the medical area, which is not only beautiful in appearance, but also improves practicality and functionality. The U-shaped design allows the staff to work separately, so that the person in line for detection are fanned out, thereby increasing the spacing distance. The test vehicle adopts modular telescopic design, the front half is the cockpit, the middle of the carriage is the functional area. After separation, it can be accessed through side doors, leading forward to the cockpit and backward to the U-shaped medical area. There are complete facilities in the car, including disinfection inst-rument, automatic nucleic acid testing instrument, refrigerator, computer, wash basin, aiming to provide a comfortable medical environment and improve work efficiency. In addition, this multi-functional modular medical vehicle is energy efficient and environmentally friendly, and its main energy source is electricity, which can be charged by solar panels and charging piles.

# 指尖制作人
# ——视障人士编曲套件

## FINGERTIP PRODUCER —— Arranger Kit for the Visually Impaired

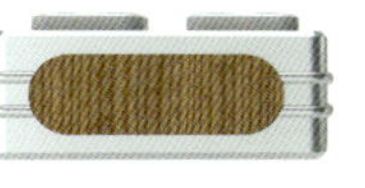
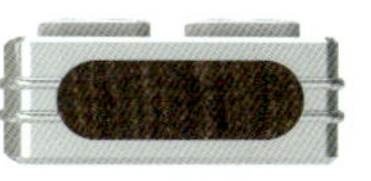

作　　者：董佳宁　王　威　吴家昊
指导老师：卜立言　高　崇
所在院校：沈阳航空航天大学

### 设计说明

指尖制作人是一款专为视障人士设计的辅助编曲工具包，旨在帮助他们实现音乐梦想。在过去，视障人士需要依赖阅读屏幕的软件来完成工作，而现在我们通过实体套件进行音乐编排，使他们能够直接在指尖上制作音乐。这款产品简化了工作流程，提高了工作效率，减少了视障人士在工作中遇到的障碍，帮助他们获得更好的工作机会，确保他们有一份体面的工作，从而促进视障者就业。

### Design notes

FINGERTIP PRODUCER is an assistive arrangement kit designed for the visually impaired to help them achieve their musical dreams. Whereas in the past, visually impaired people relied on software that read screens to get things done, now we have music orchestration through physical suites that enable them to make music directly at their fingertips. This product simplifies the work process, improves work efficiency, reduces the obstacles that visually impaired people encounter at work, helps them get better job opportunities, ensures that they have a decent job, and thus promotes the employment of visually impaired people.

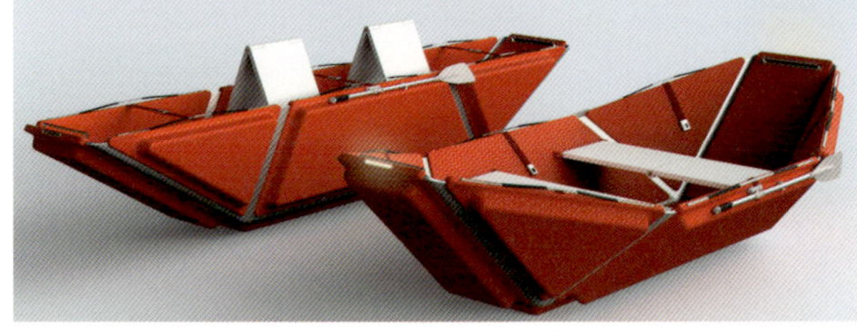

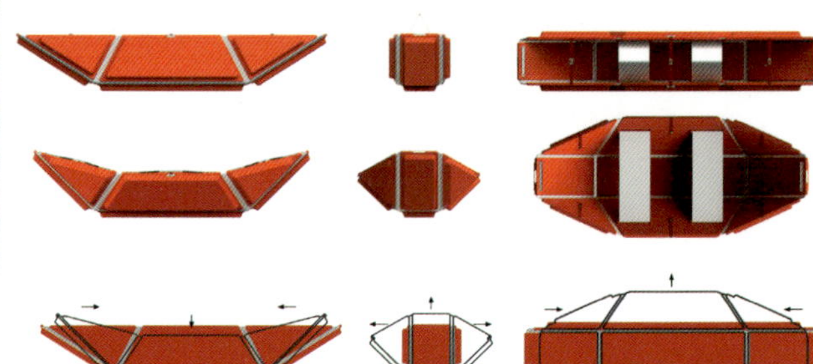



# 折纸救生船

## Origami Lifeboat

作　　者：陈奕凝
指导老师：曹伟智
所在院校：鲁迅美术学院

### 设计说明

针对轮船储备救生船数量不足的实际问题，我们设计了以节省空间为目的的折纸救生船，以增加更多的救生设备储备。该设计的灵感来源于折纸船，通过其折叠结构，实现了对救生船的重新设计。折纸救生船结构简洁，整体由九个加厚的浮力板和软连接的反光材料组成。这种救生船操作简便，只需依次解开三条绑带即可完成展开。座位设计为可折叠板，并配备有定位锁，只需伸直座位并使用定位锁，就能固定救生船的开合角度。此外，折纸救生船还配有发光信号灯、船桨和安全把手，确保在紧急情况下能够快速且安全地使用。

### Design notes

In view of the shortage of lifeboats, we designed an origami lifeboat to save space, so as to add more life-saving equipment. The design is inspired by origami boats, and through its folding structure, the redesign of the lifeboat is realized. The structure of the origami lifeboat is simple, consisting of nine thickened buoyancy plates and soft connected reflective materials. The lifeboat is easy to operate and can be deployed by untying three straps in turn. The seat is designed as a folding plate and is equipped with a positioning lock, which can be fixed to the opening and closing angle of the lifeboat by simply extending the seat and using the positioning lock. In addition, the origami lifeboats are equipped with light signals, paddles and safety handles to ensure that they can be used quickly and safely in an emergency.

# 运货者

## Carrier

作　　者：冯德泽　高　涵　杜皓宇
　　　　　张子骁
指导老师：包海默　乔　松　许　坤
所在院校：大连民族大学

### 设计说明

在小尺度作业空间使用传统叉车进行搬运工作时，由于叉车是单向操作，操作员需要不断转头来装卸货物，这不仅不安全，还降低了工作效率。运货者是一款新型的双向履带叉车，它与传统叉车不同，其前后双向叉子的设计使得叉车能够两边同时进行操作，从而提升了工作效率。而履带设计则允许货物从车身中间穿过，这样叉车在不需要转头的情况下就能完成装卸货物，非常方便，大大减少了工作的复杂性和危险性。加高的车身可将货物运送并存放在车内，临时作为仓库使用。车身内的货物和两端的叉子在装载时有助于平衡叉车的重心，并防止因货物过重而导致的翻车情况，从而增强了叉车的安全性。这款叉车的多功能性和专业性使其特别适合于超市、仓库等需要频繁装卸货物的场所。

### Design notes

When using a traditional forklift truck for handling work in a small working space, because the forklift is one-way operation, the operator needs to constantly turn the head to load and unload the cargo, which is not only unsafe, but also reduces the work efficiency. The carrier is a new type of bidirectional crawler forklift, which is different from the traditional forklift, and its front and rear bidirectional fork design enables the forklift to operate on both sides at the same time, thereby improving the work efficiency. The track design allows the goods to pass through the middle of the forklift body, so that the forklift can complete the loading and unloading of goods without the need to turn the head, which is very convenient and greatly reduces the complexity and danger of the work. The increased body allows goods to be transported and stored in the forklift, temporarily used as a warehouse. The cargo in the body and the forks at both ends help balance the center of gravity of the forklift during loading and prevent rollover caused by excessive cargo, thus enhancing the safety of the forklift. The versatility and professionalism of this forklift make it particularly suitable for supermarkets, warehouses and other places that require frequent loading and unloading of goods.

# 货物支柱

## Cargo Prop

作　　者：丁继豪　张卓亚　王　冲
姜嘉钰
指导老师：许　坤　盖甄迪
所在院校：大连民族大学



### 设计说明

随着科技的进步和国内外物流业的迅猛发展，世界各地的货物运输企业不断发展壮大，业务范围也在逐步扩大。同时，货物的运输种类也在不断丰富。目前，一辆货车很难同时完成不同种类货物的运输任务，往往因为货物的各种限制导致运输空间的大量浪费，而且在运输过程中，货物极易发生危险，造成不必要的损失。货物支柱是一款可以固定车厢内货物的设备，其升降和翻折结构使其能够适应多种类型的货物固定需求。该设备能够实现对有堆叠限制的货物、需要分类的货物、贵重及异形货物的运输。在不使用时，设备可以存放在车厢内的板凹槽中，大大增强了货物在运输途中的安全性。同时，它提高了货物运输效率，使用起来方便灵活，成为现代物流行业不可或缺的运输设备。

### Design notes

With the progress of science and technology and the rapid development of logistics industry at home and abroad, cargo transportation enterprises around the world continue to grow and expand, and their business scope is gradually expanding. At the same time, the types of transportation of goods are also constantly enriched. At present, it is difficult for a truck to complete the transportation of different kinds of goods at the same time, often because of various restrictions on goods lead to a large amount of waste of transportation space, and in the process of transportation, goods are prone to danger, causing unnecessary losses. The cargo prop is an eguipment that can hold the cargo in the carriage, and its lifting and folding structure enables it to adapt to the needs of many types of cargo fixation. The equipment enables the transportation of goods that are stacking limited need to be sorted, valuable and irregular. When not in use, the equipment can be stored in the plate groove in the carriage, which greatly enhances the safety of the goods in transit. At the same time, it improves the efficiency of cargo transportation, with the character of convenient and flexible, and has become an indispensable transportation equipment in the modern logistics industry.

# CAT BLOCK ——积木猫箱

## CAT BLOCK——Building Block Cat box

作　　者：徐旻一
指导老师：曹伟智　刘勃峥
所在院校：鲁迅美术学院

### 设计说明

“CAT BLOCK”积木猫箱是一款专为宠物猫设计的便携式出行猫箱，它不仅方便猫咪出行，还具有积木拼接的特性，让主人可以随时随地为猫咪搭建一个“宠物乐园”。这款猫箱创新性地优化了休息区、玩耍区、进食区等不同功能区域，采用集成化设计将各个区域融为一体，同时实现了空间的伸缩自由调节。在家中，主人可以根据空间限制搭建出猫窝、猫爬架等不同形态的猫咪活动区域。出行时，可以将部分模块拆卸下来，简单拼装组成便携式猫箱。这样既解决了宠物猫对陌生物品产生的排斥心理和应激反应问题，也加强了宠物主人与宠物之间的互动，增进了双方的感情。

### Design notes

"CAT BLOCK" building block cat box is a portable travel cat box designed for pet cats. It is not only convenient for cats to travel, but also has the characteristics of building blocks, so that owners can build a "pet paradise" for cats anytime and anywhere. This cat box innovatively optimizes different functional areas such as the rest area, the play area, and the eating area, and uses an integrated design to integrate each area, while realizing the free adjustment of space expansion. At home, the owner can build different forms of cat activity areas such as catteries and cat climbing frames according to space restrictions. When traveling, some modules can be removed and simply assembled to form a portable cat box. This not only solves the psychological rejection and stress response problems caused by pet cats to strange objects, but also strengthens the interaction between pet owners and pets, and enhances the feelings of both sides.

金奖 银奖 铜奖 **优秀奖**

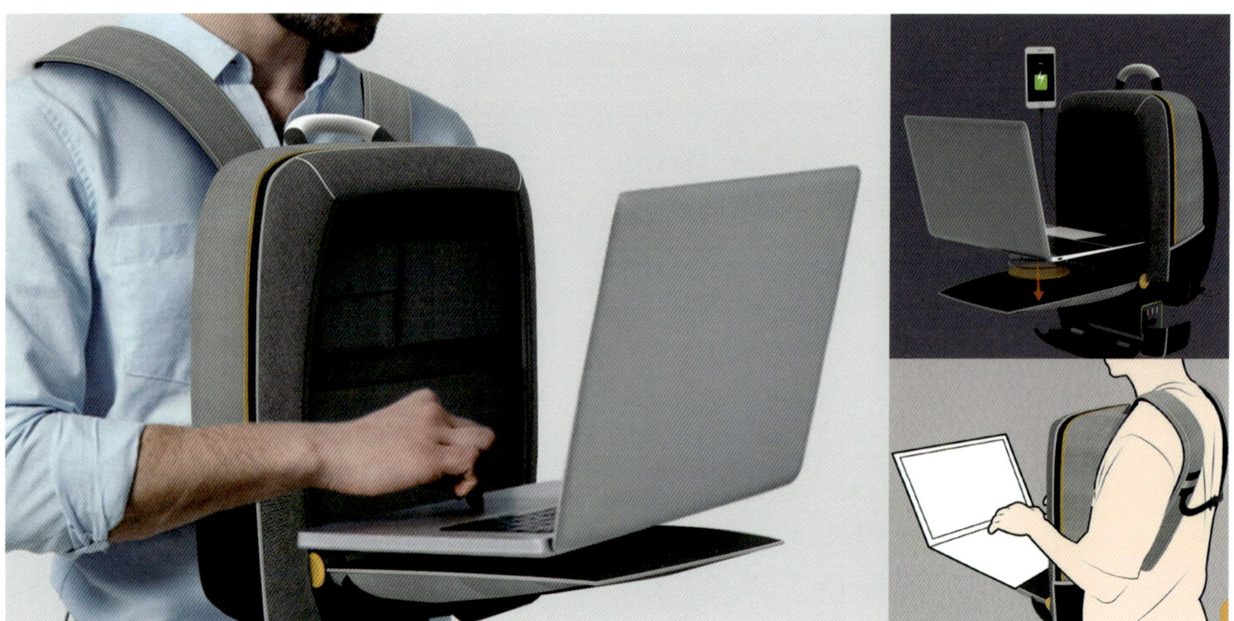

# 随行办公桌

## Backpack Desk

作　　者：丁继豪　张卓亚　姜嘉钰
王　冲
指导老师：许　坤　乔　松　包海默
所在院校：大连民族大学



### 设计说明

随着现代社会的迅速发展，人们的工作节奏也变得越来越紧凑。一些上班族经常需要出差，并可能临时接到公司的任务。在户外时，他们往往不能及时找到合适的办公场所，无法实现随时随地办公。随行办公桌应运而生，这是一款可以随时变形为办公桌的背包。用户只需将背包背到前面，放下背包的前硬板，便可将其变形为一个小型办公桌，方便在桌上进行简单的办公活动。这一设计解决了用户急需处理工作事务却找不到合适办公环境的尴尬局面，方便快捷且使用简单。作为正常的背包使用时，内置的收纳袋可以完美地收纳电脑、平板和书籍。背包内的电池盒还可以为办公设备提供电源补给，避免出现在外使用时电量不足的问题，让用户能够放心地在外办公，减轻使用者的焦虑。

### Design notes

With the rapid development of modern society, people's working rhythm has become more and more tensional. Some office workers often need to travel and may receive temporary assignments from the company. In outdoors, they usually can not find a suitable office space in time, and can not achieve office at anytime and anywhere. The accompanying desk came into being, a backpack that can transform into a desk at any time. The user may simply put the backpack to the front, and put down the front hard board of the backpack, and transforms it into a small desk for simple office activities on the desk. This design solves the awkward situation that users urgently need to deal with work affairs but can not find a suitable office environment, and is convenient, fast and simple to use. When used as a normal backpack, the built-in storage bag is perfect for storing computers, tablets and books. The battery box in the backpack can also provide power supplies for office equipment, avoid the problem of insufficient power when used outside, so that users can rest assured that they can work outside and reduce their anxiety.

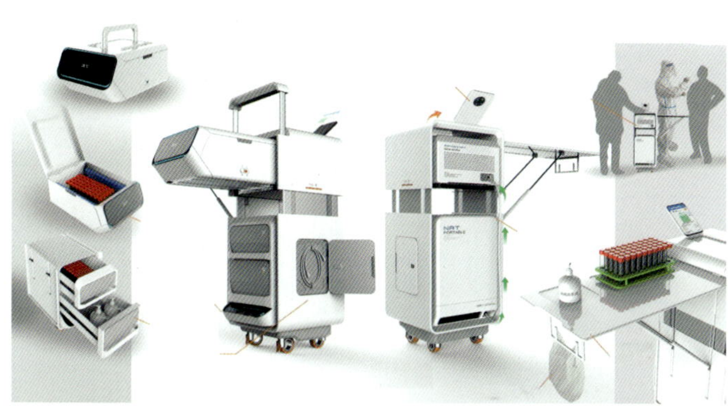

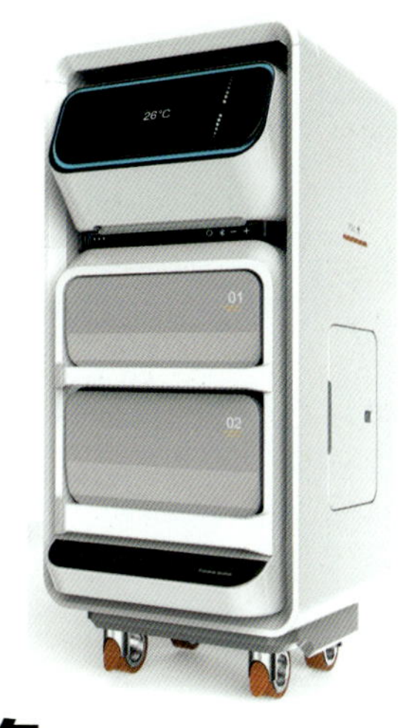

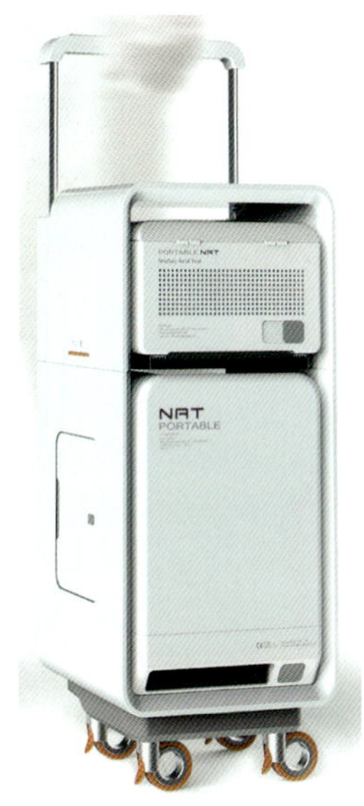

# PORTABLE NAT
# ——便携式核酸检测箱

## PORTABLE NAT —— Portable Nucleic Acid Detection Box

作　　者：张梓滢
指导老师：曹伟智　刘勃峥
所在院校：鲁迅美术学院

## 设计说明

核酸检测遇到突发情况时，医务人员的转移十分不便。这款便携式核酸检测医疗箱为医务人员在户外进行核酸检测提供了便利，无需再搬动桌子或转移运输工具等繁琐事项，实现了必要工具的整合。产品采用了模块化设计，侧面设有折叠桌板，主体包括可单独转移的恒温送检箱、音响、检测用具存储箱和移动电源等模块，移动电源可在紧急情况下为恒温送检箱和医务人员提供电源。功能的集成是本次设计的核心和亮点，外观设计采用拉杆箱结构，内部支架可升降调整高度。内部还加入了音响，便于广播通知；顶部则配备了智能扫码摄像头，免去了志愿者人力扫描。该设计确保一名医务人员即可完成整个检测工作，这不仅为医务人员提供了便利，还提高了检测效率并降低了检测成本。

## Design notes

Under the background of nucleic acid testing, the transfer of medical personnel has become very inconvenient in the event of an emergency. This portable nucleic acid testing kit provides convenience for medical personnel to conduct nucleic acid testing outdoors, without the need to move tables or transport machine and other trifles, achieving the integration of necessary tools. The product adopts a modular design, with a folding table on the side, and the main body includes a separate removable constant temperature detection box, sound system, detection equipment storage box and mobile power supply modules, which can provide power for the constant temperature detection box and medical personnel in emergency situations. The integration of functions is the core and highlight of this design. The appearance design adopts the structure of the draw-bar box, and the internal bracket can be lifted and lowered to adjust the height. The interior of it is also added a sound system, which is easy to broadcast announcements; the top is equipped with a smart scan code camera, which eliminates the need for human volunteers to scan, ensuring that a medical worker can complete the entire inspection. This not only provides convenience for medical personnel, but also improves the efficiency and reduces the cost of testing.

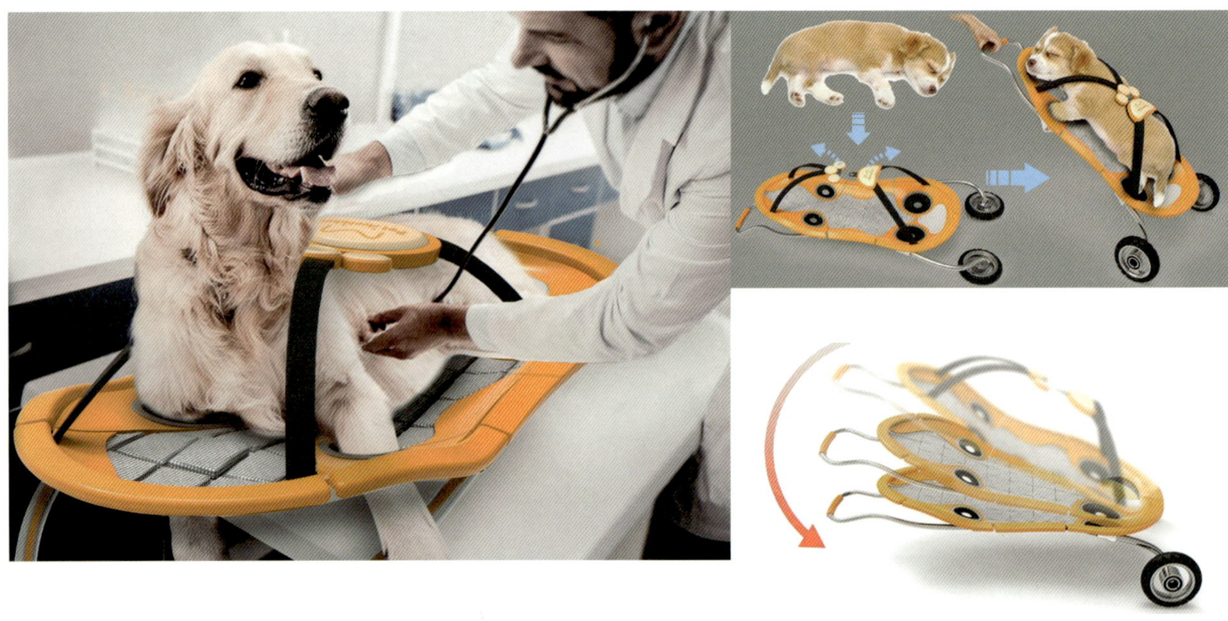

金奖 银奖 铜奖 **优秀奖**

212

# 宠物担架

## Pet Stretcher

作　　者：伍柯屹　邵一迪　莘佳琪
　　　　　陈　彤　侯　亮
指导老师：许　坤　包海默　乔　松
　　　　　盖甄迪
所在院校：大连民族大学

### 设计说明

当宠物受伤后，人们通常会选择抱起它们前往宠物医院就诊，但这样做很容易造成二次伤害。目前市场上缺乏专为宠物设计的担架，而宠物担架这一产品恰好解决了这一问题。宠物担架有两种使用方式：首先，在运输过程中，将宠物放置在担架上并扣好背带，即可由一人拉着担架将宠物运送至医院；其次，在救治过程中，只需将轮胎顺时针旋转90°，再调整把手至适当角度，便可将担架平放在地上，方便对宠物进行急救。宠物担架在四肢位置采用了扭扭杯式的可伸缩结构，以适应不同尺寸宠物的四肢安置，并牢牢固定宠物四肢，防止二次伤害。担架中间部分采用柔软亲肤材料，可实现全方位拉伸，以适应不同体型的宠物。

### Design notes

When pets are injured, people usually choose to pick them up and go to the pet hospital for treatment, but it is easy to cause secondary injuries. At present, there is a lack of stretchers designed for pets on the market, and this pet stretcher just solves the problem. There are two ways to use the pet stretcher: first, in the transportation process, the pet is placed on the stretcher and the strap is fastened, so that one person can pull the stretcher to transport the pet to the hospital; secondly, in the treatment process, just rotate the wheel 90° clockwise, and then adjust the handle to the appropriate angle, you can put the stretcher flat on the ground, convenient for pet's first aid. The stretcher adopts a stretchable structure of the twist cup in the position of the limbs to adapt for the limbs of pets of different sizes, and firmly fix the limbs of pets to prevent secondary injuries. The middle part of the stretcher is made of soft skin-friendly materials, which can be stretched in all aspects in order to adapt to different types of pets.

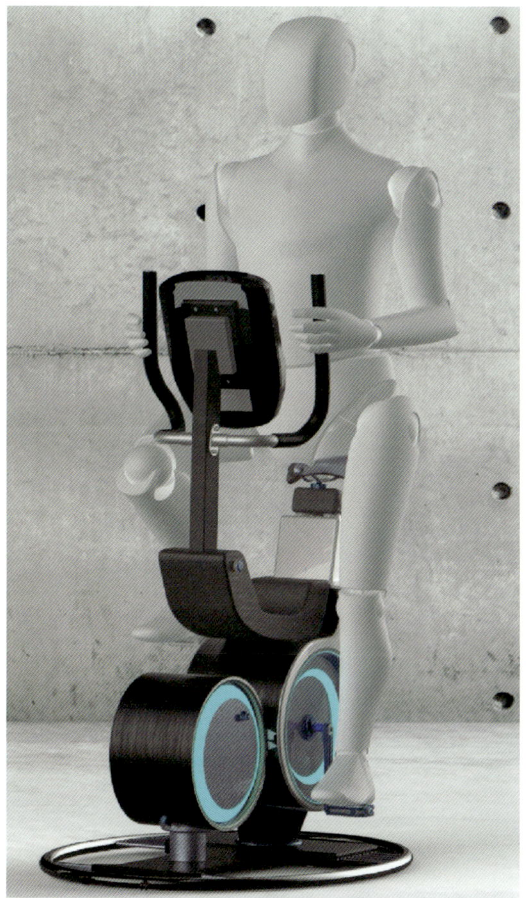

# 智能化上下肢康复训练器

## Intelligent Upper and Lower Limb Rehabilitation Trainer

作　　者：赵越久　王新荣　王建明
张小燕　张　飘
指导老师：刘华伟
所在院校：北华大学

### 设计说明

上下肢康复训练器整体采用立式结构设计，设计主题为轻便简洁。该机器具备智能化功能，在使用者进行康复时，会根据当前的恢复情况自动调整速度。训练器支持主动和被动双模式运动，还可与手机连接，实现播放音乐的功能，适用于手脚运动不灵敏的患者。

### Design notes

The upper and lower limb rehabilitation trainer is designed with vertical structure and with the theme of lightness and simplicity. The machine has an intelligent function, when the user is recovering, the system will automatically adjust the speed according to the current recovery situation. The trainer supports active and passive dual-mode movement, and can also be connected with the mobile phone to achieve the function of playing music, which is suitable for patients with insensitive hand and foot movement.

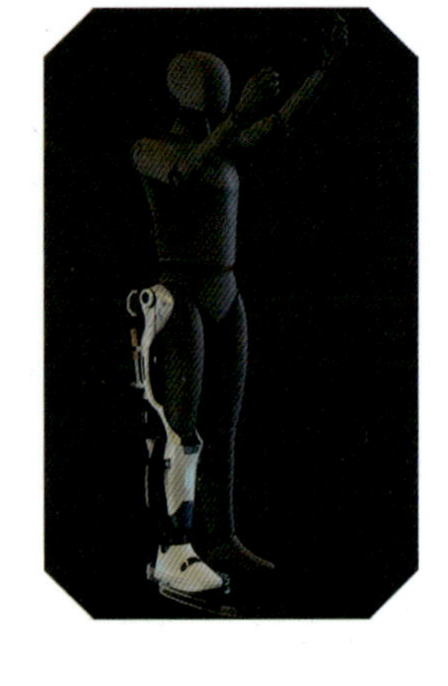

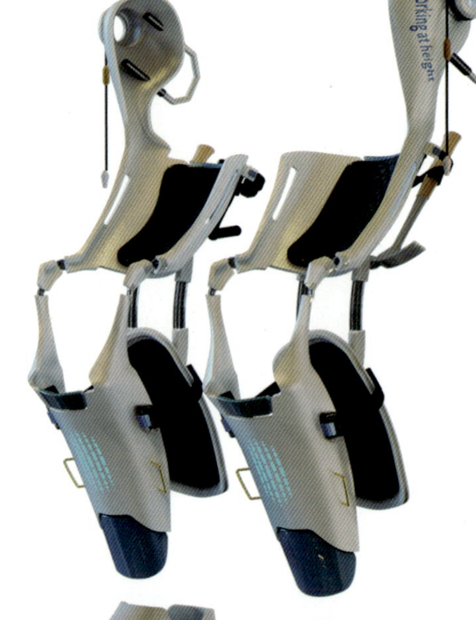

# 高空作业辅助外骨骼

## Assists Exoskeletons for High-Altitude Operations

作　　者：刘昊天　刘欣欣　张　晗
范琼月
指导老师：李　莹　霍思光
所在院校：吉林师范大学

### 设计说明

如今，随着科技的进步，外骨骼技术正逐步融入各行各业。从高空作业者的需求出发，他们目前使用的装备存在显著的稳定性和便携性问题。为此，我们设计了这款外骨骼，它采用半包裹式结构，旨在提高施工者活动的便捷性。该外骨骼的优势包括提升安全性、设备便携、便于施工、增强支撑力、减轻施工人员腿部压力。

### Design notes

Nowadays, with the progress of science and technology, exoskeleton technology is gradually integrated into all walks of life. From the needs of high-altitude operators, the equipment they currently use has significant stability and portability problems. To this end, we designed this exoskeleton, which uses a semi-wrapped structure to improve the mobility of the builder. The exoskeleton's advantages includes increase safety, portability, ease of construction, enhanced support, and reduced stress on the builder's legs.

# 智能化图书馆

## Intelligent Library

作　　者：丁柯疆　林　娜　王　敏
指导老师：谢亚蒙　王鹰飞
所在院校：长春建筑学院

### 设计说明

随着科学技术的进步和人工智能技术的普及，本设计旨在将人工智能技术引入建筑领域，用以管理图书，达到节约管理成本和提高管理效率的目的。设计的关键在于考虑引入人工智能对建筑空间的影响。当新兴的工业设计与传统的建筑理念相结合时，将催生出哪些新的空间设计方法？这正是我们团队探讨的问题。该设计选址于一个相对较小的地块上，创意源自一个长方形体量，并通过内部退台处理形成。从外观上看，它四平八稳、平和而沉着，外立面采用多孔铝板包裹，隐约透露出室内空间，寓意着中国古代读书人内敛而不露锋芒。进入室内，可以看到层层递进的退台，隐喻着深藏不露的学识。设计这个建筑的初衷是创造一个能够逐步引导学生进入并激发学习欲望的空间。

### Design notes

With the progress of science and technology and the popularization of artificial intelligence technology, this design aims to introduce artificial intelligence technology into the field of architecture to manage books, achieving the purpose of saving management costs and improving management efficiency. The key of the design is to consider the impact of the introduction of artificial intelligence on the building space. When the new industrial design is combined with the traditional architectural concept, what new spatial design methods will be spawned? This is exactly what our team is asking. The design is located on a relatively small plot, and the idea is derived from a rectangular volume, which is formed by internal retreat treatment. From the appearance, it is smooth, peaceful and calm, the facade is wrapped in porous aluminum plates, which vaguely reveals the interior space, implying the talent of ancient Chinese scholars who were restrained but not revealing their edge. Entering the room, you can see the successive steps of the stage, a metaphor for the hidden knowledge. The original intention of the building was to create a space that would gradually lead students into and stimulate their desire to learn.

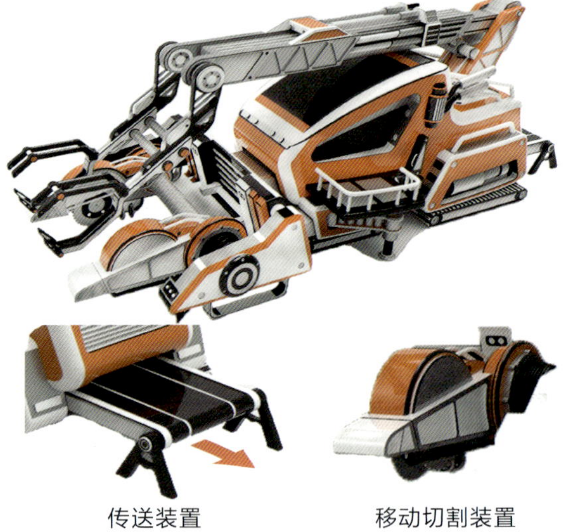

传送装置　　移动切割装置

# 切割式采冰运输机

## Cutting Ice Conveyor

作　　者：刘宸汛
指导老师：何丽鹏
所在院校：长春工业大学

### 设计说明

大型冰块是重要的自然资源，不仅提供固态水源，还为制作冰雕、构建冰上运动场等提供原材料。为了便于采集和运输冰块，同时突出产品造型，我们设计了一款综合性的大型冰上作业机械设备。该设备集成了双向切割冰块、旋转式双抓手采冰以及可伸缩的多单元连接传送带运输功能，旨在提高工作效率。

### Design notes

Large ice blocks are an important natural resource, not only providing solid water source, but also providing raw materials for making ice sculptures and constructing ice sports arenas. In order to facilitate the collection and transportation of ice, while highlighting the product shape, we designed a comprehensive large-scale ice operation machinery and equipment. The device integrates bidirectional ice cottecting, rotating double-gripper ice harvesting, and retractable multi-unit connected conveyor to increase efficiency.

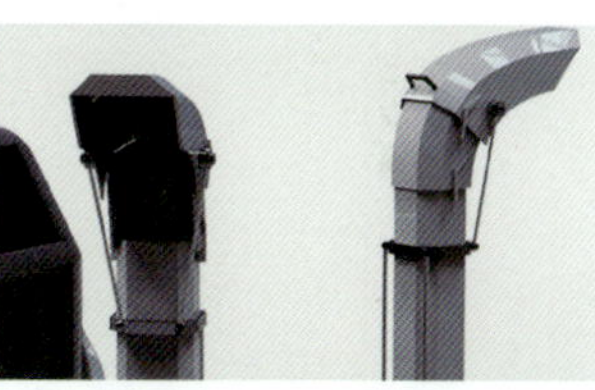

# 除雪车设计

## Snow Thrower Design

作　　者：吴俊彤
指导老师：端文新
所在院校：长春工程学院

### 设计说明

本设计是一辆封闭式驾驶舱除雪机，能够为操作者提供一个温暖的工作环境。前端采用旋转绞龙将积雪粉碎并收集，随后通过抛雪筒内的叶轮转动，将积雪从两侧的抛雪筒弯管抛出，以达到清理积雪的目的。本机器配备有多区域玻璃窗，以拓宽作业视野。抓握型履带设计使其能适应各种地形的积雪清理工作。此外，其独特的外观设计也为道路增添了一项设计亮点。

### Design notes

The design is a snow thrower with a closed cockpit that provides a warm working environment for the operator. The front end uses a breaking rotator to crush and collect the snow, and then rotates through the impeller in the snow throwing cylinder to throw the snow from the elbow of the snow throwing cylinder on both sides to achieve the purpose of snow clearing. The machine is equipped with multi-area glazing window to broaden the view of the work. The grip track design makes it suitable for snow clearing in various terrains. In addition, its unique exterior design also adds a highlight to the road.

金
奖
银
奖
铜
奖
优秀奖

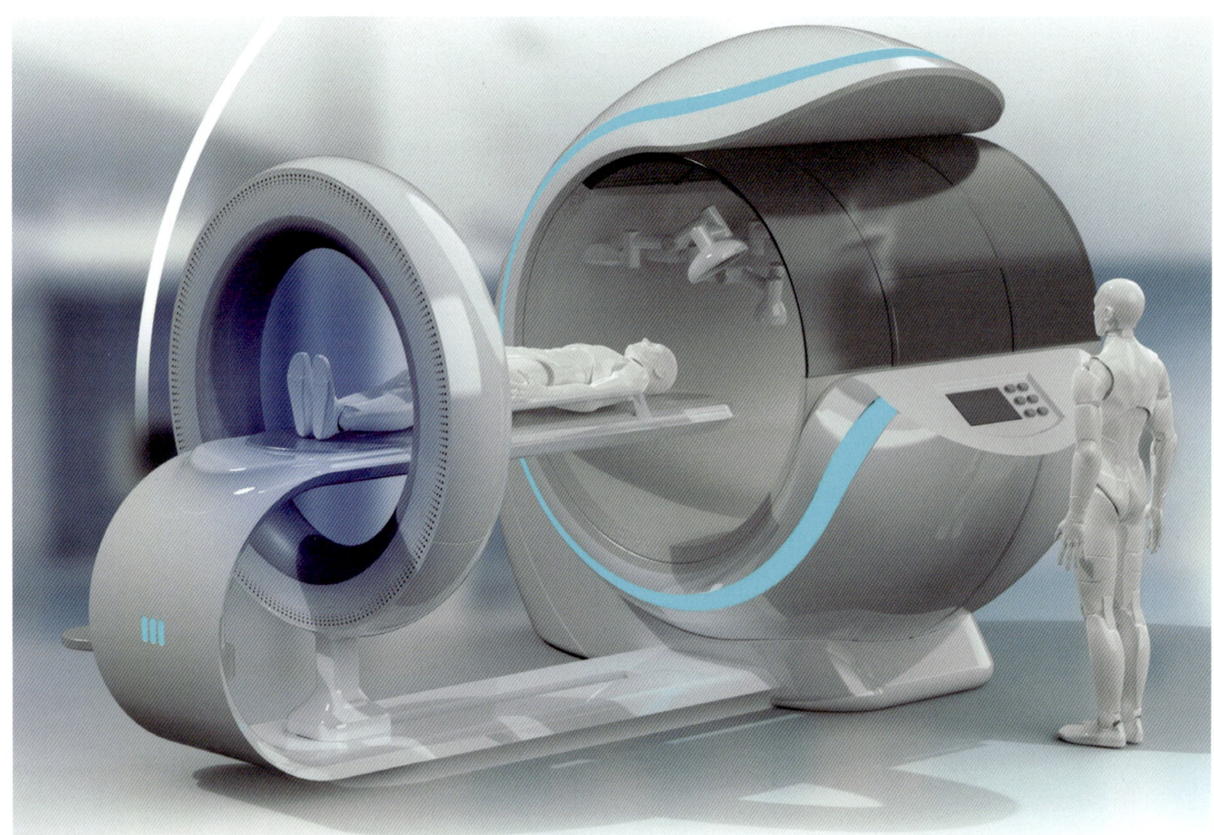



# 紫外线光疗仪概念设计

## Conseptual Design of Ultraviolet Phototherapy Instrument

作　　者：于东立　许润泽　张　依
　　　　　伍　泽　杜文慧
指导老师：刘华伟
所在院校：北华大学

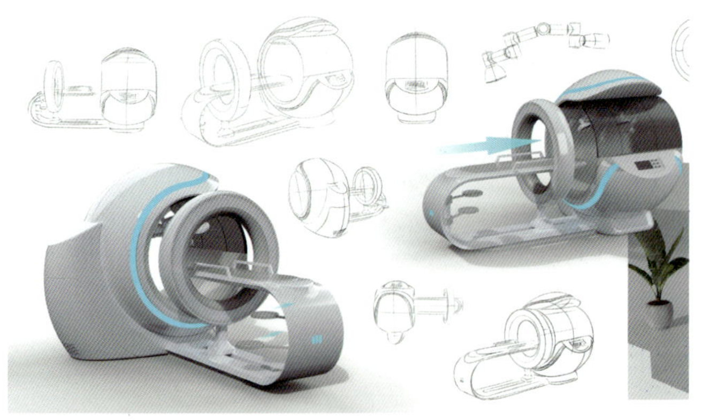

### 设计说明

紫外线光疗是治疗蕈样真菌病，尤其是早期蕈样真菌病的有效方法之一。患者平躺后，通过扫描系统对其进行扫描以记录病变位置，然后将信息上传至控制平台。随后，紫外线光疗灯根据扫描信息对病患部位进行精确照射，以避免对其他身体部位造成损伤。

### Design notes

Ultraviolet phototherapy is one of the effective methods to treat MF, especially to early MF. After lying flat, the patient is scanned through a scanning system to record the location of the lesion, and the information is then uploaded to the control platform. The UV phototherapy light then illuminates the patient's diseased location precisely based on the scan information to avoid damage to other body parts.

# 模块化防疫充气帐篷

## The Modular Epidemic Prevention Inflatable Tent

作　　者：沈　桐
指导老师：孙晓航
所在院校：吉林大学艺术学院

### 设计说明

模块化防疫充气帐篷是专为核酸检测工作设计的，其灵感来源于蜂巢结构，由多个六角柱形单元组成。这些模块可以自由组合和拆分，以实现合理的功能分区，并能够适应不同的场景和环境需求。此外，这种快速充气帐篷可以显著减少人力和时间的消耗。

### Design notes

The modular epidemic prevention inflatable tent is an efficient solution specifically designed for nucleic acid detection work, inspired by the honeycomb structure and consisting of multiple hexagonal columnar units. These modules can be freely combined and split to achieve a reasonable functional partition, and can adapt to different scenarios and environmental requirements. In addition, this rapidly inflatable tent can significantly reduce the consumption of manpower and time.

APP 界面设计

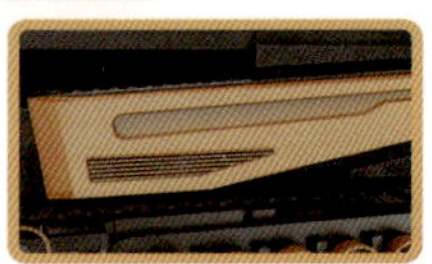

# 水稻收割打捆一体机

## Rice Harvesting Baler



作　　者：王潇晗　于婷婷　王妙冉
指导老师：李　博
所在院校：东北林业大学

### 设计说明

这款农机是专为山地丘陵地区设计的小型水稻收割打捆一体机。它采用履带式车轮，有效降低重心，确保在陡峭的丘陵地形中稳定作业。机身设计运用了视觉动力学原理，既显得稳重又保持灵活性。作为一款小型水稻收割机，它功率适中且采用电力驱动，实现了节能环保。驾驶室的设计遵循人机工程学原则，减少驾驶员操作时的疲劳感。在软件方面，本机采用共享模式，分为管理端和用户端。村民可以通过手机应用程序（APP）的用户端来租赁共享农机，而村委会则可以通过 APP 的管理端进行设备管理。这一模式解决了因农机价格高昂、使用率不高而导致的资源浪费问题，有助于节约社会资源。

### Design notes

This agricultural machine is a small rice harvesting and bundling machine designed for mountainous and hilly areas. It uses tracked wheels to effectively lower the center of gravity and ensure stable operation in steep, hilly terrain. The design of the body uses the principle of visual dynamics to maintain stability and flexibility. As a small rice harvester, it has moderate power and electric drive, which realizes energy saving and environmental protection. The design of the cab follows ergonomic principles to reduce the fatigue of the driver during operation. In terms of software, the machine adopts the sharing mode, which is divided into management side and client side. Villagers can rent and share agricultural machinery through the user side of the mobile application (APP), and the village committee can manage the equipment through the management side of the APP. This model solves the problem of resource waste caused by the high price of agricultural machinery and low utilization rate, and helps to save social resources.

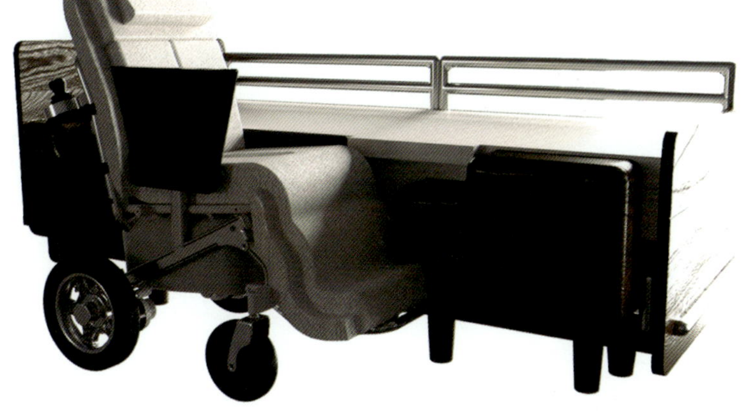



# 重度失能老人居家护理轮椅床概念设计

# Conceptual Design of Wheelchair Bed for Home Care of Severely Disabled Elderly

作　　者：李方旭
指导老师：周　岩
所在院校：哈尔滨工业大学

## 设计说明

随着人口老龄化问题的加剧，失能老人在社会中的比例迅速上升，引发了一系列社会问题。在这种背景下，我计划通过分析国内外轮椅床的情况，以实际需求为指导，设计一款适合重度失能老人居家护理的轮椅床。我将运用 SET 因素和使用方式分析等方法，进行设计需求和审美理论的探讨，确定用户的审美偏好和情感需求，从而界定产品开发的范围，并分析市场现状。接下来，我会设定设计目标，开展市场调研，明确设计的目的和技术实用性。基于这些信息，我将提出设计方案，包括外观造型、材料选择和配色方案的决策。最终，我将借助计算机辅助工具完成产品设计及其流程展示，并为失能老人设计简洁的交互界面，以满足他们的心理和生理需求。

## Design notes

With the aggravation of the aging of the population, the proportion of disabled elderly people in the society has increased rapidly, which has caused a series of social problems. In this context, I plan to design a wheelchair bed suitable for home care for severely disabled elderly people by analyzing the situation of wheelchair beds at home and abroad, guided by actual needs. I will use the SET series of factors and the method of application analysis to discuss the design needs and aesthetic theory, determine the user's aesthetic preference and emotional needs, so as to define the scope of product development and analyze the market status. Next, I will set design goals, carry out market survey, and clarify the purpose and technical practicability of the design. Based on this information, I will propose the design, including the appearance, material selection and color scheme decisions. Finally, I will use computer-aided tools to complete product design and process display, and design a simple interactive interface for disabled elderly people to meet their psychological and physical needs.

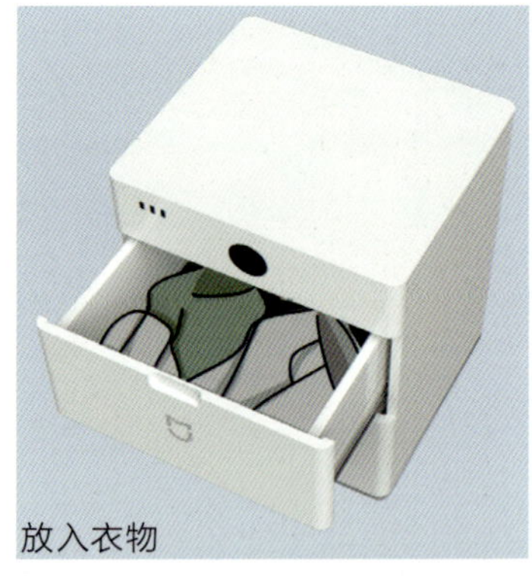
放入衣物

加热衣物

# 暖衣床头柜

## Cloth-Heating Bedside Cupboard



作　　者：李　治　黄佳文　王添运
王昱夫　李文希
指导老师：朱　磊　周　岩
所在院校：哈尔滨工业大学

### 设计说明

本产品主要面向长期居住在南方地区，尤其是南北交界地带的用户。这些区域冬季气温偏低，且缺乏足够的取暖设施，居民早晨起床时常常需要面对冰冷的衣物。本产品在床头柜中集成了加热衣物的功能，并通过手机 APP 进行控制，提供智能化的加热体验。产品在加热过程中能够保持静音，确保不会干扰用户的睡眠。产品配备了3个 USB 接口，即使有多个设备需要充电也无需担心，只需将衣物放入抽屉并关闭，按照设定好的时间到时便会自动加热，确保用户能穿上温暖的衣物。此外，用户还可以打破常规随时手动启动加热功能，不受预设时间的限制。

### Design notes

This product is mainly for users who live in the southern region for a long time, especially the North-South border zone. In these areas, where winter temperatures are low and adequate heating facilities are lacking, residents often have to face cold clothes when they get up in the morning. This product integrates the function of heating clothes in the nightstand, and is controlled by the mobile phone APP to provide intelligent heating experience. The product can be kept silent during heating to ensure that the user's sleep will not be disturbed. Equipped with 3 USB ports, even if multiple devices need to be charged, there is no need to worry. Simply put the clothes in the drawer and close them, and when the set time is up, they automatically heat up, ensuring that the user can wear warm clothes. In addition, users can also break the routine and manually start the heating function at any time, regardless of the preset time limit.

# 无人机土壤采样

## Soil Sampling By UAV

作　　者：朱　莹　张　跃　赵睿思
鲁新蕊　孙　萌
指导老师：赵爱丽　曲庆峰
所在院校：黑龙江八一农垦大学

### 设计说明

复杂多变的野外环境为土壤采样技术人员带来了许多挑战。为了解决这一问题，我们设计了一款集无人机与取样机于一体的土壤勘测设备。该设备主要通过无人机对环境进行勘察和定位，然后由取样机完成土壤采样工作。取样机采用螺旋钻头、收集转盘和试管等结构，能够对不同位置的土壤进行有效采集，直到试管装满，即标志着一次采样任务的完成。这一过程极大地简化了土壤采集工作，解决了传统人工采集土壤时面临的困难。

### Design notes

The complex and varied field environment presents many challenges for soil sampling technicians. In order to solve this problem, we designed a soil survey equipment that integrates UAV and sampler. The equipment mainly surveys and locates the environment through the UAV, and then the soil sampling is completed by the sampler. The sampler adopts the structure of screw drill, collection turntable and test tube, which can effectively collect soil at different locations until the test tube is filled, which marks the completion of a sampling task. This process greatly simplifies soil collection and solves the difficulties faced by traditional manual soil collection.

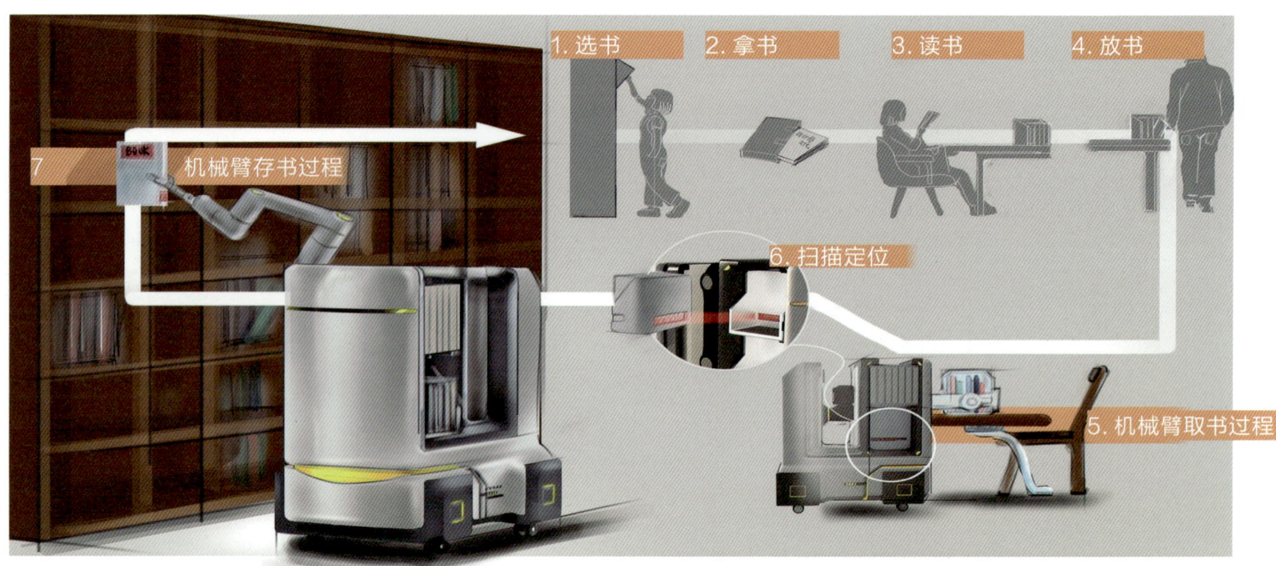


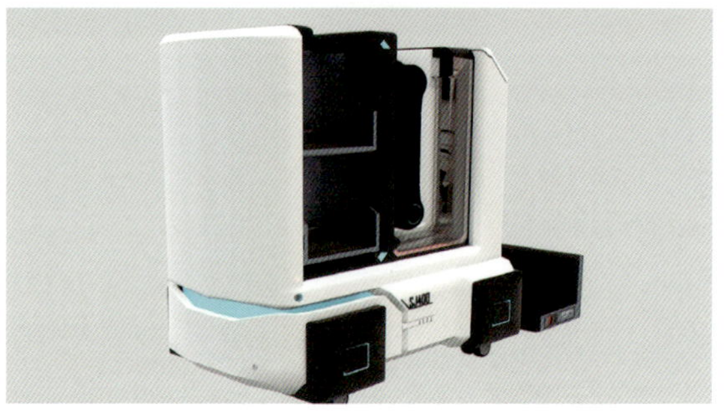

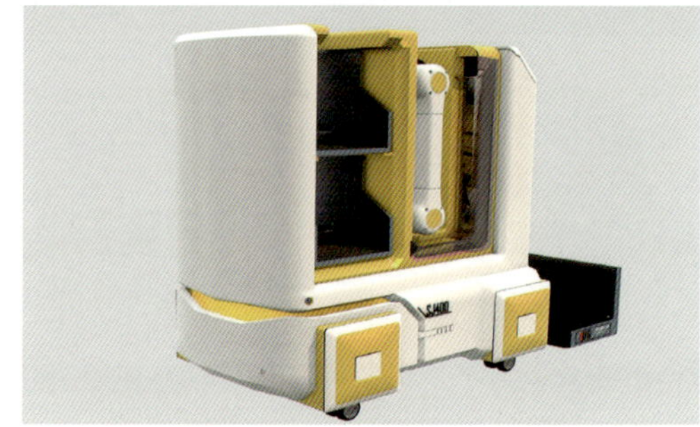



# 图书馆自动还书装置

## Library Automatic Book Return Device

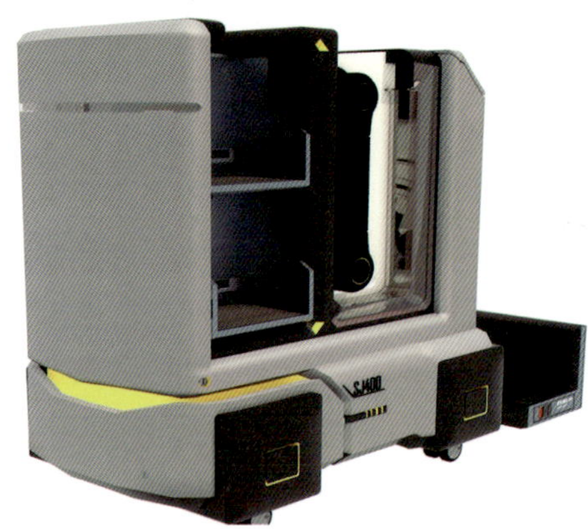

作　　者：苏永进　朱　莹　吕相衡
指导老师：张　颖　张义峰
所在院校：黑龙江八一农垦大学

### 设计说明

设计这款图书馆自动还书装置的目的是为了构建适应现代化图书馆的新型阅读方式。该设备利用了现有稳定的机械臂工作机制，以实现对读者借阅流程的服务以及满足图书馆日常运营的需求。自动还书过程需要与其他设备的协同配合才能顺利进行。

### Design notes

The purpose of this automatic book return device is to build a new reading mode that adapts to the modern library. The device makes use of the existing stable robotic arm working mechanism to the robotic service to the reader borrowing process and meets the needs of the daily operation of the library. The automatic book return process requires coordination with other devices to proceed smoothly.

# 冬日之歌——基于声热交互的冰溜子收集与利用装置

## SONG OF WINTER —— Icicle Collection and Utilization Device Based on Sound and Heat Interaction

作　　者：吴胜龙　孟欣宇　赖　燚
　　　　　汤　斯
指导老师：邱信贤　唐岳兴
所在院校：哈尔滨工业大学

### 设计说明

在冬季，高纬度地区常因地理位置特殊而遭受极寒天气，冰冻期持续时间较长。屋檐下滴落的水珠容易形成冰锥，这些冰锥可能会造成人身伤害、物品损坏或建筑本体受损，并在融化后影响人们的行走安全。为此，我们设计了一款装置，其核心是通过微型传感器来监测风速、云层和温度数据，并通过促进冰锥的融化来解决上述问题。同时，该装置还利用计算机技术分析融水的声音强度，并将其转换为听觉旋律和视觉投影，以此与过往人群互动。此外，装置收集的融水也可以进行回收再利用。我们希望这一装置不仅能解决实际问题，还能通过多学科交叉融合和新技术运用，实现与自然的互动，将自然界的信息转化为听觉和视觉的体验，从而找到一种自然、可持续、人机交互的未来派解决问题的新方法。

### Design notes

In winter, high latitudes often suffer from extremely cold weather due to their special geographical location, and the freezing period lasts longer. Dripping water under the eaves can easily form ice cones, which may cause personal injury, damage to goods or the building itself, and affect people's walking safety after melting. To this end, we designed a device whose core is to monitor wind speed, clouds and temperature data through tiny sensors, and to solve these problems by facilitating the melting of the icicle. The device also uses computer technology to analyze the sound intensity of meltwater and convert it into auditory melodies and visual projections to interact with passing crowds. In addition, the meltwater collected by the device can also be recycled. We hope that this device will not only solve practical problems, but also interact with nature through multidisciplinary integration and the use of new technologies, transforming information from nature into auditory and visual experiences, thereby to discover a natural, sustainable, human-computer interactive futuristic problem-solving approach.

金
奖
银
奖
铜
奖
优秀奖

# Uciclo

## Uciclo



作　　者：张峻玮　刘雨妍　貊祖运
叶绿满山　周雨薇　吴　桐
指导老师：韩　挺
所在院校：上海交通大学

### 设计说明

在这个追求效率至上的共享时代，仍有一群人坚持寻求私人定制的出行方式。因此，我们引入了既相互对立又相辅相成的4个主题——春、夏、秋、冬，以此作为不同出行模式的象征，并提供与之相应的个性化出行服务体验。这四个主题代表单独的车辆，它们既可以独立存在，也可以组合成一个整体。车内的部件设计为模块化，可以拆卸并循环利用。每个主题都配备了与其季节特征相符的色彩、材料和表面处理工艺以及人机交互界面，广泛采用全息投影和手势控制操作。整车设计融合了中国元素和现代主义风格，旨在为用户提供一个在共享环境下仍能享受个性化出行的独特体验。

### Design notes

In this era of sharing where efficiency is paramount, there is still a group of people who insist on seeking a personal customized way to travel. Therefore, we introduced four opposing and complementary themes – spring, summer, autumn and winter – as symbols of different travel patterns and to provide personalized travel experiences accordingly. The four themes represent individual vehicles that can either stand on their own or be combined into a whole. The components inside the car are designed to be modular and can be disassembled and recycled. Each theme is equipped with colors, materials and CMF in line with its seasonal characteristics, HMI, as well as a human-computer interaction interface with extensive holographic projection and gesture control operations. The vehicle design incorporates Chinese elements and modernist style, aiming to provide users with a unique experience that can still enjoy personalized travel in a shared environment.

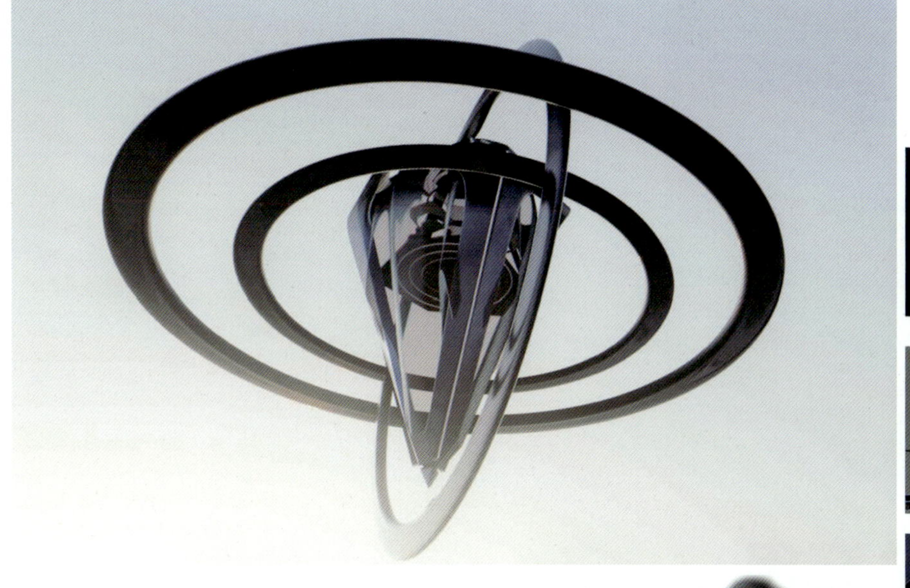

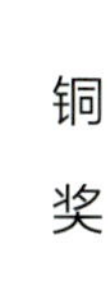

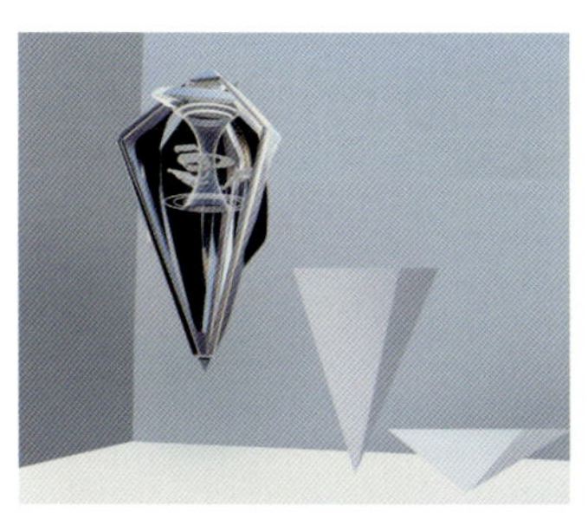

# 星环——未来飞行器概念设计

## STAR-RING——Future Aircraft Concept Design

作　　者：王新乔　貊祖运　周雨薇
指导老师：傅　炯
所在院校：上海交通大学设计学院

### 设计说明

本作品立足于开源节流的可持续理念，着眼于未来新兴人类的生活和出行模式，并对超越一级文明水平后的资源利用进行深入思考。主要设计概念“STAR-RING”灵感源自于恒星系的结构。其中，能源系统具备自主导航至地心进行充电的能力，而主舱体则设计为能够搭载乘客，实现通勤、游览和休闲等多功能一体化的出行体验。

### Design notes

This work is based on the sustainable concept of increase income and reduce expenditure, focuses on the life and travel mode of emerging human beings in the future, and deeply thinks about the utilization of resources after surpassing the level of first-class civilization. The main design concept "STAR-RING" is inspired by the structure of the stellar system. Among them, the energy system has the ability to autonomously navigate to the earth's core for charging, while the main cabin is designed to carry passengers, achieving a multi-functional travel experience such as commuting, sightseeing and leisure.

# 城市应急充电救援车概念设计

## Conseptual Design of City Emergency Charging Rescue Vehicle

作　　者：盛锦灏
指导老师：蔡萌亚
所在院校：上海工程技术大学



### 设计说明

车辆在行驶过程中因突发状况而中断行程一直是常见的交通问题之一。随着新能源时代的到来，众多汽车企业纷纷设计并投产新能源汽车，电能逐渐打破了传统化石能源的垄断地位。随着新能源汽车市场份额的逐步增加，其配套的充电设施也受到了广泛关注。尤其是新能源汽车在行驶过程中可能因能源不足而中断行程的问题尚未得到足够的重视。未来的应急救援车辆必须能够满足新能源汽车的应急充电需求。针对当前情况，我们提出了一个面向未来的解决方案：为应对新能源汽车在正常行驶中可能遇到的电能不足、电池故障、充电系统故障或车载交互系统故障等突发事件，设计一款新型的、具备应急救援功能的社会交通服务概念车。这款概念车旨在提供有效的应急救援解决方案，能够解决故障新能源汽车的充电和自动检修等问题。

### Design notes

It is always one of the common traffic problems that the vehicle interrupts its journey due to unexpected conditions. With the arrival of the new energy era, many automobile companies have designed and put into production of new energy vehicles, and electric energy has gradually broken the monopoly status of traditional fossil energy. With the gradually increased market share of new energy vehicles, their supporting charging facilities have also received widespread attention. In particular, the problem that new energy vehicles may interrupt their journey due to lack of energy during driving has not received enough attention. Future emergency rescue vehicles must be able to meet the emergency charging needs of new energy vehicles. In view of the current situation, we propose a solution for the near future: in order to cope with the unexpected events that new energy vehicles may encounter during normal driving, such as insufficient power, battery failure, charging system failure or on-board interaction system failure, a new social transportation service concept car with emergency rescue function is designed. The concept car is designed to provide an effective emergency rescue solution that can solve problems such as charging and automatic maintenance of broken down new energy vehicles.

# 基于老龄化社区的无人运输智能车及服务系统设计

# Design of Unmanned Transportation Intelligent Vehicle and Service System Based on Aging Community

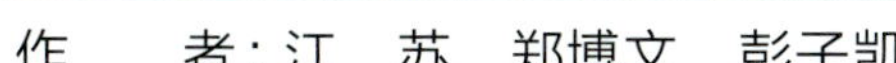

作　　者：江　苏　郑博文　彭子凯
徐佳怡　容嘉穗

指导老师：李雪楠

所在院校：华东理工大学

## 设计说明

本项目设计了一款专为应对老龄化社区挑战的无人配送交通工具及其服务系统。它旨在解决日益严峻的老龄化问题以及与之相关的社区服务设施发展不均衡的现状。项目以人文关怀为核心理念，在视觉、信息和情感三个方面进行了适老化设计。无人运输车的形态和交互设计也体现了这一理念。我们希望通过这些适老化的设计元素，能够减少老年人在使用这类产品时的抵触情绪，增强他们的归属感，并帮助他们更好地融入现代化信息社会。

## Design notes

This project designed an unmanned delivery vehicle and its service system to meet the challenges of aging communities. It aims to address the increasingly serious problem of aging in the related uneven development of community service facilities. The project takes humanistic care as the core concept, and carries out age-appropriate design in three aspects: vision, information and emotion. The form and interaction design of unmanned vehicles also reflect this concept. We hope that these age-appropriate design elements can reduce the resistance of the elderly when using such products, enhance their sense of belonging, and help them better integrate into the modern information society.

# 浚洙号——高速应急排水工程车

# JUNZHU ——High Speed Emergency Drainage Engineering Vehicle

作　　者：林煊力　朱　力　刘　雨　张一领
程宇航
指导老师：高　瞩
所在院校：上海工程技术大学

## 设计说明

针对特大暴雨灾害，我们设计了一款高速应急排水车，它具有快速移动救援的能力和适应多种地形的应急排水优势，有助于减轻城市内涝带来的影响。

## Design notes

In response to heavy rainstorm disasters, we designed a high-speed emergency drainage vehicle, which has emergency drainage advantages which possess the ability to move quickly and adapt to a variety of terrain, helping to reduce the impact of urban waterlogging.

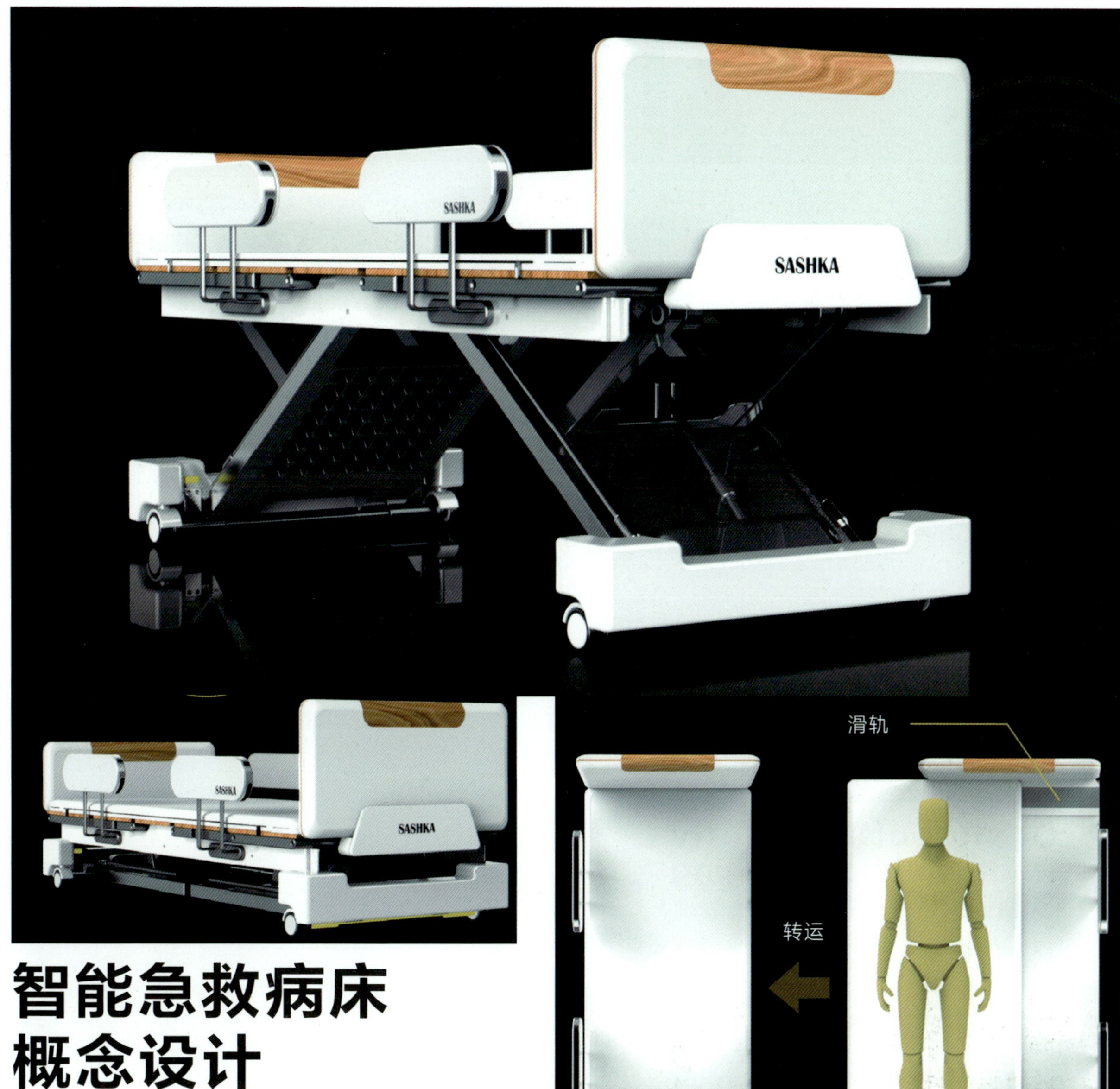


# 智能急救病床概念设计

## Conceptual Design of Intellgent Emergency Bed

作　　者：李子坤
指导老师：徐江华
所在院校：广州美术学院

### 设计说明

这款急救病床的设计包括可调节的底座脚轮和液压支撑柱，使得病床的角度和高度可以根据不同的使用需求进行调整。此外，床体还配备了滑轨系统，以便于在转运病人时减轻医护人员的工作负担，并减少对病人造成的潜在二次伤害。

### Design notes

The design of this emergency bed includes adjustable base casters and hydraulic support columns, so that the angle and height of the bed can be adjusted according to the needs of different uses. In addition, the bed is equipped with a sliding rail system to reduce the workload of medical staff during patient transport and reduce potential secondary injuries to patients.

# 蜻蜓光伏充电站设计

## Design of Dragonfly Photovoltaic Charging Station

作　　者：陈雯琦
指导老师：杨景裕
所在院校：上海杉达学院

### 设计说明

本产品的灵感源自于将光伏充电桩与斜列式停车位相结合，主要应用于高速公路的户外充电站。整体设计采用“三角形”元素组合，旨在营造一种无限延伸的速度感。产品名称中包含“蜻蜓”，寓意产品设计的核心理念——提供轻盈且高效的使用体验。提升使用效率的三大方向包括，一是维护光伏面板的清洁并调整其角度以提升光能储存效率；二是通过斜列式的停车位设计实现便捷停车；三是在充电桩内置收纳圆盘，引导用户更加高效地进行日常管理。

### Design notes

This product is inspired by the combination of photovoltaic charging piles and diagonal parking spaces, which are mainly used in outdoor charging stations on highways. The overall design uses a combination of "triangular" elements to create a sense of speed that extends indefinitely. The name of the product contains the word "Dragonfly," which symbolizes the core concept of product design —— to provide a light and efficient user experience. Three directions to improve the efficiency of use, include: first, maintain the clean photovoltaic panel and adjust its angle to improve the efficiency of light energy storage. The second is to realize convenient parking through the design of diagonal parking space. The third is the built-in storage disc in the charging pile to guide users to conduct daily management more efficiently.

# ING 时间画——松江布文化衍生产品设计

## ING TIME PAINTING——Derivative Product Design Based on Songjiang Cloth

作　　者：蔡　萌
指导老师：王思萍
所在院校：东华大学

### 设计说明

松江布的传统文化反映了松江府地区以织布为生的织女们的生活方式和情感记忆。这一缓慢而真挚的生活美学在现代依然值得传承和延续。ING 时间画是一款概念交互时钟，它从松江布图案的再设计、产品交互体验以及松江布文化内涵的传达三个角度进行创新。这款时钟利用钟面上展示的抽象化松江布代表性图案来实时记录家庭环境中的声音和温度变化，促使用户感知并反思近期的家庭氛围，从而映射出松江布文化中关于织造生活和沉淀时间的设计主题。此外，该时钟配套的小程序允许用户选择不同的松江布图案显示在钟面上，了解松江布的文化故事，在线报名参与由松江布展馆组织的线下体验活动，以及购买相关的文化创意商品等。

### Design notes

The traditional culture of Songjiang cloth reflects the life style and emotional memory of the weaving girls in Songjiang Prefecture. This slow and sincere aesthetic of life is still worth inheriting and continuing in modern times. ING Time Painting is a conceptual interactive clock, which innovates from three perspectives: the redesign of Songjiang cloth patterns, the interactive experience of products and the transmission of the cultural connotation of Songjiang cloth. The clock uses the abstract representative patterns of Songjiang cloth displayed on the clock to record the sound and temperature changes in the home environment in real time, prompting the user to perceive and reflect on the recent family atmosphere, thus reflecting the design themes of the Songjiang cloth culture about weaving life and settling time. In addition, the small program accompanying the clock allows users to select different Songjiang cloth patterns to be displayed on the clock face, learn about the cultural story of Songjiang cloth, register online to participate in offline experience activities organized by the Songjiang Cloth Exhibition Hall, and buy related cultural and creative goods.

金奖

银奖

铜奖

优秀奖

# FOR YOU——老年电动代步车

## FOR YOU——Elderly Electric Scooter

作　　者：施天宇
指导老师：于　炜
所在院校：华东理工大学

### 设计说明

这是一款专为老年人设计的电动代步车。由于腿脚不便和体力下降，老年人的出行能力受到限制。这款老年电动代步车以其操作简便、轻巧便携和低速行驶的安全性为特点，成为老年人理想的出行工具。本产品创新性地将代步车与拉杆购物车融合在一起，实现了一键分离功能，方便老年人在出行时携带和购买物品。其外观设计简洁时尚，摒弃了市面上传统代步车的陈旧感。该代步工具的车型尺寸、制动系统、拐角型车把手和皮质软坐垫等配置都考虑到了老年人的生理需求。本产品的目标是简化老年人的出行过程，扩展他们的生活范围，并丰富他们的社交生活。

### Design notes

This is an electric mobility scooter designed for the elderly. The ability of going around of the elderly is limited due to leg inconvenience and decreased physical strength. This electric mobility scooter for the elderly is characterized by its easy operation, lightweight, portable and the safety of low-speed driving, which make it an ideal travel tool for the elderly. This product innovatively integrates the mobility scooter with the trolley cart to realize the one-click separation function, which is convenient for the elderly to carry and buy items when traveling. The appearance design is simple and fashionable, abandoning the old sense of the traditional scooter on the market. The size of the car, the braking system, the corner handlebars and the leather cushion are all designed to take into account the physiological needs of the elderly. The goal of this product is to simplify the travel process for the elderly, expand their range of life, and enrich their social life.

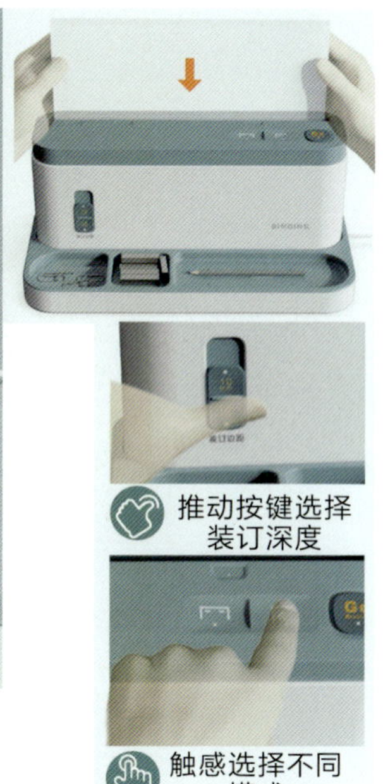


# 公共使用多功能装订机

## Multifunctional Binding Machine In Opening Office

作　　者：胡　喆
指导老师：陈明阳　莫逸凭
所在院校：上海第二工业大学

### 设计说明

许多现代化办公环境采用开放式工作空间的设计，并配备共用设备以降低成本和提高资源利用率。这样的设计旨在追求更高效、节能的工作方式。通过调研发现，在这种环境中，文具使用存在一些问题，例如订书机频繁被借用和文具损耗等现象。针对装订文件时遇到的问题进行深入研究后，我发现用户在装订前常需重复整理文件，这大大降低了装订过程中的效率。此外，更换订书钉不便，以及使用回形针时消耗率高且容易导致文件损坏等问题也十分常见。为了提升用户的装订体验，我的设计对现有的个人装订模式进行了改进，使其更方便用户使用，减少了打印、装订和签名行为之间的距离，从而有效提高了办公效率。

### Design notes

Many modern office environments are designed with open work spaces and are equipped with shared equipment to reduce costs and improve resource utilization efficiency. This design aims to pursue a more efficient and energy conservation way of working. Through investigation, it is found that in this environment, there are some problems in the use of stationery, such as frequent borrowing of stapler and stationery loss. After an in-depth study of the problems encountered in the binding of documents, I found that users often need to organize documents repeatedly before binding, which greatly reduces the efficiency of the binding process. In addition, problems such as the inconvenience of replacing staples, and the high consumption rate and easy to cause file damage when using paper clips are also very common. In order to enhance the user's binding experience, my design has improved the existing personal binding mode, making it more user-friendly and reducing the distance between printing, binding and signature activities, thereby it could effectively improving office efficiency.

# 新型可除尘公交车

## New Type of Dust-Proof Bus

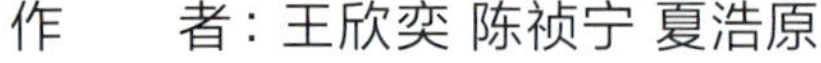

作　　者：王欣奕 陈祯宁 夏浩原
指导老师：William Volcoff Matthew Rhoades
所在院校：上海视觉艺术学院

### 设计说明

这是一辆将来可以净化空气的公共汽车。它可以改善跑步时的空气质量，从而清洁城市。

### Design notes

This is a bus that can clean the air in the future. It can improve the air quality while jogging, so as to clean the city.

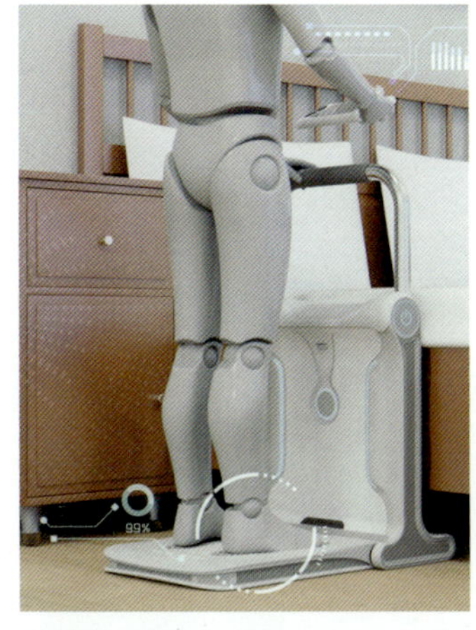

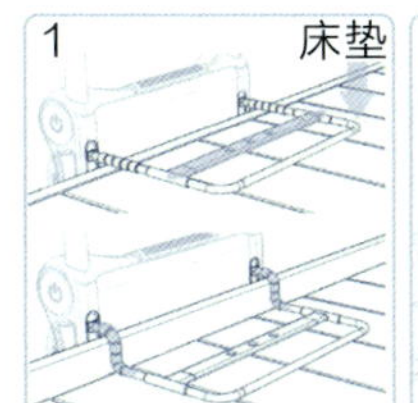

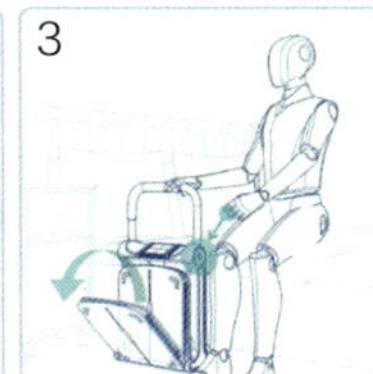


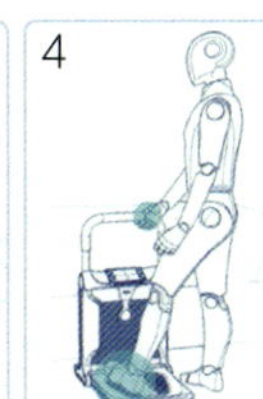


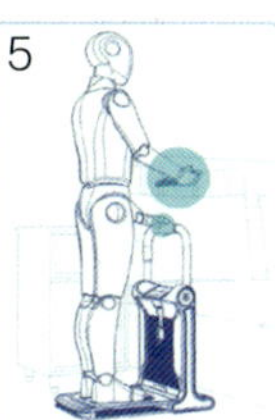


# 糖尿病高龄患者的智能健康监测仪设计

# Design of Intelligent Health Monitor for Elderly Patients with Diabetes

作　　者：张怡卿
指导老师：倪敏娜
所在院校：东华大学

## 设计说明

糖尿病足智能监测仪是一款结合物联网技术，专门用于监测足底温度和压力异常的医疗设备，其智能化和便捷性特点特别适合高龄糖尿病患者在家中进行健康管理。然而，市面上现有的产品主要关注技术和定制化，往往忽略了老年患者的特殊需求。为此，我们通过对高龄患者的居住环境和行为习惯进行深入研究分析后，完成了该产品及其交互界面和服务系统的设计。该产品集成了多种智能交互功能，能够准确收集足部数据，并直观地反馈监测结果与及时提供疾病风险预警。在人机工程学方面，我们根据老年人的身体尺寸和患者的足部特征进行分析，以确定产品设计的细节和使用流程的最优化。此外，围绕监测仪构建的数字模块和服务架构能够在多种场景下提供高效的智能监测和医疗协同服务，从而有助于预防疾病的发生，并推动适老化智慧健康领域的发展。

## Design notes

Diabetic Foot Intelligent Monitor is a medical device that combines the Internet of Things technology and is specially used to monitor the abnormal temperature and pressure of the sole, and its intelligent and convenient characteristics are especially suitable for elderly diabetic patients to carry out health management at home. However, existing products on the market focus on technology and customization, often ignoring the special needs of elderly patients. To this end, we completed the design of products, and its interactive interfaces and service systems through in-depth research and analysis of the living environment and behavioral habits of elderly patients. The product integrates a variety of intelligent interactive functions, can accurately collect foot data, and provide intuitive monitoring results feedback and disease risk warning in. In terms of ergonomics, we carry out analyses based on the body size of the elderly and the foot characteristics of the patient to determine the details of the product design and the optimization of the use process. In addition, the digital module and service architecture built around the monitor can provide efficient intelligent monitoring and medical coordination services in a variety of scenarios, thereby helping to prevent the occurrence of diseases and promote the development in the field of smart health of aging.

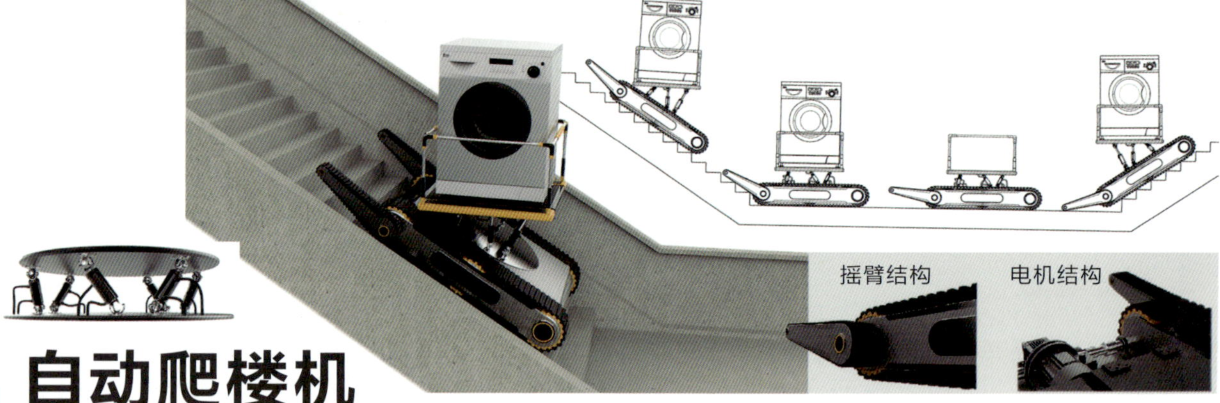


# 自动爬楼机

## Automatic Stair Climbing Machine

作　　者：刘　杰
指导老师：张　磊
所在院校：南通大学

## 设计说明

通过采用六轴位移平台的载物结构，并结合电动履带驱动方式，该机器在上下楼梯时能够始终保持载物平台的平衡。这一设计解决了传统搬运重物上下楼时货物不稳定以及耗费时间和人力的问题。机身前部配备了红外距离传感器和摄像头，这些设备使得机器能够自动识别路线和障碍物，从而实现自动化运作。

## Design notes

By using a six-axis displacement platform load structure, combined with electric track drive, the machine can always maintain the balance of the load platform when going up and down the stairs. This design solves the problem of unstable cargo and time-consuming and manpower when carrying heavy loads up and down stairs. The front of the body is equipped with infrared distance sensors and cameras, which allow the machine to automatically identify routes and obstacles, thus enabling its automatic operation.

# GAIA——荸荠智能运输装备

## GAIA —— Water - Chestnut Intelligent Transportation Equipment

作　　者：陈尚仪　余唯一
指导老师：孙宁娜　张　凯
所在院校：江苏大学

### 设计说明

整体功能造型与当前5G 背景下的智能动态趋势相契合，适用于农业机械和交通运输工具，使之内容丰富且能有效解决采收替换箱体劳动力需求大、效率低下的问题。机器的功能与主机紧密配合，配备全息屏以实时显示工作状态和基本信息。在配色方面，机器以银灰、白色为主色调，辅以黑灰色，刀架顶部的荧光绿色反光面起到警示作用。叉车的底座前方设有前叉刀，上方设有后叉刀。前叉刀主要负责拿取空箱体和协助其他叉车放置已装载完成的箱体，后叉刀则用于拿取箱体。双头刀架设计便于更换箱体和拿取物品，有效利用空间并增加机器的层次感。智能全息屏和感应跟踪系统的加入基于对未来趋势的预测，通过独特的视觉动效符合未来科技风格。前后刀架的距离可调节，为确保重心稳定，机体部分将增加重量。搬运和替换箱体的任务需要叉车间的协同配合以提高效率。

### Design notes

The overall functional modeling is in line with the intelligent dynamic trend under the current 5G background, and is suitable for agricultural machinery and transportation vehicles, making it content rich and effectively solving the problem of large labor demand and low efficiency of the harvesting and replacement box. The functions of the machine are closely coordinated with the main machine, and the holographic screen is equipped to display the working status and basic information in real time. In terms of color matching, the machine is mainly silver gray and white, supplemented by black gray, and the fluorescent green reflective surface on the top of the tool holder plays a warning role. A front fork knife is arranged in front of the base of the forklift truck, and a rear fork knife is arranged above it. The front fork knife is mainly responsible for taking the empty box and assisting other forklifts to place the loaded box.The rear fork knife is used to take the box. The double head tool holder is designed for easy replacement of boxes and access to goods, making efficient use of space and increase the layer of the machine. The addition of the intelligent holographic screen and the induction of the tracking system is based on the prediction of future trends, and it conforms to the future technology style through unique visual dynamic effects. The distance between the front and rear tool holders can be adjusted, and the body part will increase in weight to ensure a stable center of gravity. The task of moving and replacing boxes requires collaboration between forklifts to increase efficiency.

# 高压电缆除冰预警设计

## High Voltage Cable Deicing Machine

作　　者：陈　榕　巢文博　孙　超
　　　　　孟忠涛
指导老师：张　凯　孙宁娜
所在院校：江苏大学

1. 工作人员将机器安装在高压电缆上

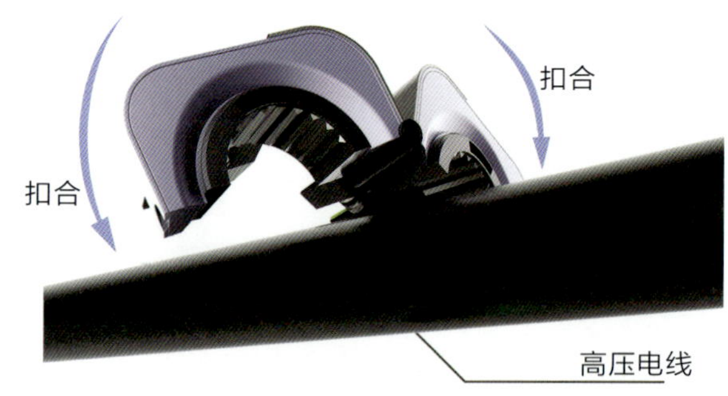


2. 将连接组件的装置与除冰机器装接上

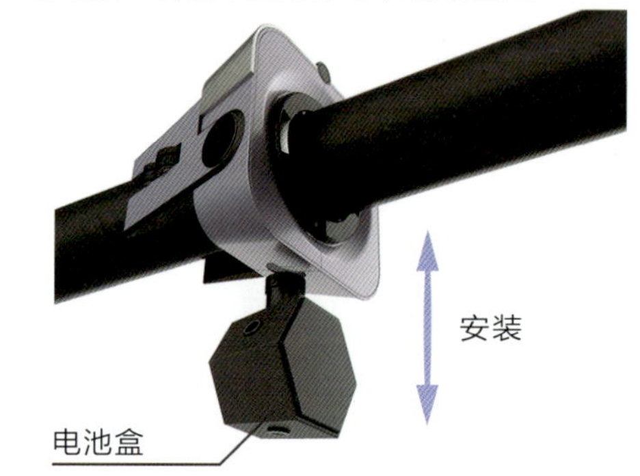


3. 在电池盒提供动力下，机器沿着电缆前进除冰

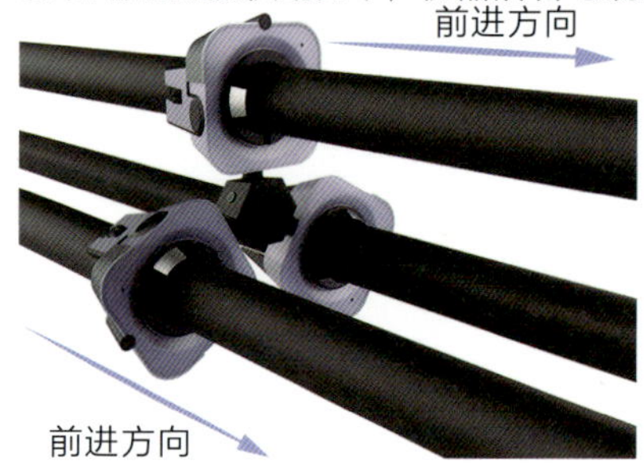


### 设计说明

这是一款专为高压电缆设计的除冰机器，主要应用于偏远山区和公路上的高压电缆，用以解决极端寒冷天气引起的电缆损坏问题。该机器首先利用前端的除冰刀去除大块积冰，然后通过中间的旋转滚刀刮除电缆表面的冰层，从而解决了传统高空手工除冰效率低下且危险的问题。

### Design notes

This is a deicing machine designed for high-voltage cables, mainly used in remote mountainous areas and high-voltage cables on highways to solve the problem of cable damage caused by extreme cold weather. The machine first uses the ice cutter at the front to remove large pieces of ice accumulation, and then scrapes the ice on the surface of the cable through the rotating hob in the middle, thus solving the problem of inefficient and dangerous traditional high-altitude manual deicing.

# 仓库消防员团队

# Team of Warehouse Firefighters

作　　者：王瑞杰　李　泽　邵一鸣
指导老师：苏　旭
所在院校：徐州工程学院

## 设计说明

仓库火灾通常被分为 A、B、C、D 四种类型，每种类型都需采用相应的灭火方法。在实际情况中，尤其是码头仓库，很难迅速确定火灾的具体类型，夜间发生火灾时，往往无法及时发出警报并采取正确的应对措施。为此，设计了一套配套的灭火机器人系统来解决这一问题。这个“消防员团队”由 A、B、C 和 D 四种类型的灭火机器人、一个中央控制台以及烟雾报警器组成。它能对火灾进行早期预警，其中的探测器能及时监测到火情并确定火灾类型，然后将火灾的具体位置和类型信息发送至中央控制台。中央控制台依据预先设置的路线图指派适当的灭火机器人前往起火点执行灭火任务。

## Design notes

Warehouse fires are usually classified into A, B, C, D four types, each type needs to adopt the corresponding fire extinguishing method. In the actual situation, especially in the dock warehouse, it is difficult to quickly determine the specific type of fire, and when the fire occurs at night, it is often impossible to issue the alarm in time and take the correct response measures. Therefore, a set of fire-fighting robot system is designed to solve this problem. This "firefighter team" consists of four types of fire-fighting robots, A, B, C and D, a central console and smoke alarms. It provides early warning of fires, in which the detector can detect the fire in time and determine the type of fire, and then send the specific location and type of fire information to the central control panel. According to the pre-set roadmap, the central control panel assigns the appropriate fire extinguishing robot to the fire location to carry out the fire extinguishing task.

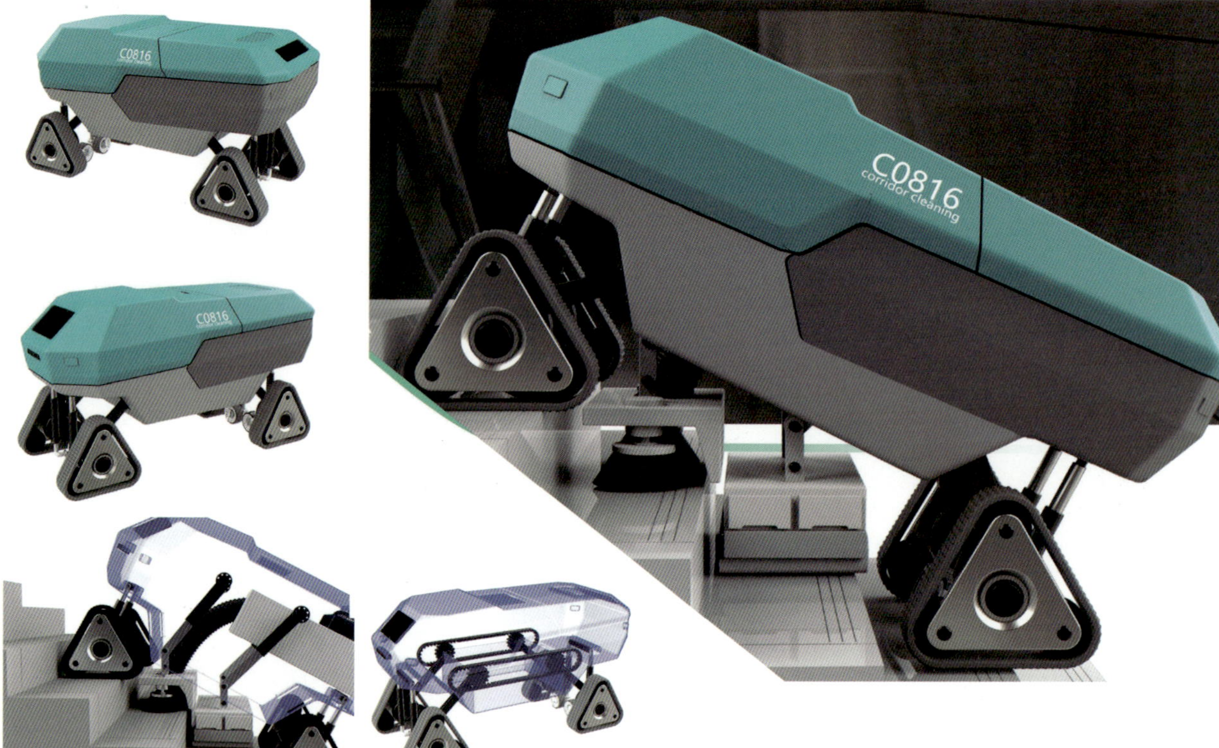


# 楼道智能清洁设备



## Corridor Intelligent Cleaning Equipment

作　　者：陈　昕
指导老师：陆建华
所在院校：南通大学

### 设计说明

当前，智能清洁技术正快速发展，各种清洁设备已被广泛应用于道路等公共环境中。与手工清扫相比，智能清洁产品不但能提供较为标准化的清洁效果，还能显著提高工作效率，并大幅减少人力投入。然而，现有的智能清洁设备在应用于公共空间时还存在局限性，大多数设备只能在相对平坦的地面上工作，在需要上下台阶的楼梯公共空间中，尚未有成熟的解决方案。针对楼梯间清洁所面临的挑战，本设计旨在探索并开发一种新型的智能清洁设备，以适应楼梯等公共空间的清洁需求。

### Design notes

At present, intelligent cleaning technology is developing rapidly, and a variety of cleaning equipment has been widely deployed in the environment such as roads, public spaces and families. Compared with manual cleaning, smart cleaning products can not only provide a more standardized cleaning effect, but also significantly improve work efficiency and significantly reduce human input. However, the existing smart cleaning equipment still has limitations when applied to public spaces, most of the equipment can only work on relatively flat ground. In the need for stairs up and down steps in the public space, there is no mature solution. In view of the challenges of cleaning betweem stairs, this design aims to explore and develop a new type of intelligent cleaning equipment to adapt to the cleaning needs of public spaces such as stairs.

# 全自动上线电缆涂覆机器人

# Automatic On-Line Cable Coating Robot

作　　者：钱青云　朱丹丹
指导老师：戚玥尔　沈瀚文　张梦盈
所在院校：浙江工业大学之江学院

## 设计说明

传统的绝缘化改造方案存在诸多不足，如覆盖率低、施工难度大、需要较长的停电时间、安全风险高以及人工效率低下。而本产品能够在带电状态下操作，能够自动上线并通过远程操控完成高压裸导线的涂覆作业。本产品可节省能源总费用的35%，将施工周期缩短至原来的1/180，勘测费用减少至原来的33%，并将人员投入减少至原来的1/10。

## Design notes

There are many shortcomings in the traditional insulation renovation scheme, such as low coverage, difficult operation, long power failure time, high safety risk and low labor efficiency. The product can be operated in a live state, can be automatically on-line and complete the coating of high voltage bare wire through remote control. Using this product can reduce the price by 35%, shorten the construction period to 1/180 of the original, reduce the survey cost by 67%, and reduce the personnel input to 1/10 of the original.

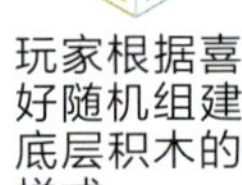
玩家根据喜好随机组建底层积木的样式

玩家通过投掷积木，将积木朝上的颜色堆叠到相应的位置上，如果没有对应颜色的话，意味着游戏结束

玩家需要将手中的积木与周围积木的顶面或侧面进行颜色对应，颜色相同被认为搭建成功，否则为搭建失败

玩家在游戏过程中可以借助一些特殊模块来减少对手搭建的可能性，以此来获得胜利

随着搭建的进行，当一名玩家无法再继续搭建时，游戏结束，另一名玩家获胜

距离模块

惊喜模块

# ADVENTURE——儿童空间数智玩具

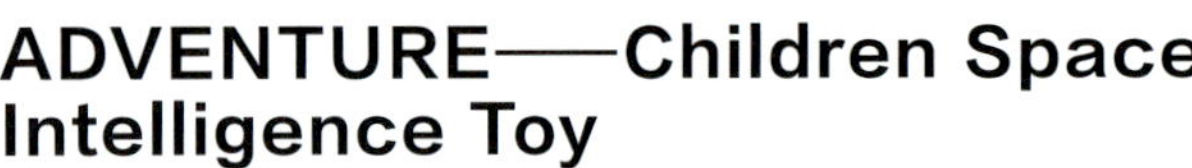

## ADVENTURE——Children Space Intelligence Toy

作　　者：王鑫峪　孙恺睿　徐庆龙　钟梓锐
指导老师：梁玲琳　沈　嘉　胡旭冲
所在院校：浙江理工大学

## 设计说明

Adventure 是一款注重空间感知能力的交互式积木玩具。它在传统的“井”字棋游戏基础上，融入了多感官交互反馈和立体化玩法的创新，通过引人入胜的空间关系引导儿童娱乐和学习。这款积木由普通模块和功能模块组成，其中功能模块采用 Arduino 技术设计。孩子们可以通过控制空间距离来堆叠积木，并通过这种方式获得相应的交互反馈，这不仅增强了他们对空间距离的感知能力，也促进了与其他伙伴之间的互动交流。

## Design notes

Adventure is an interactive building block toy with space perception. On the basis of the original "well" word chess game, multi-sensory interactive feedback and three-dimensional game innovation are added to guide children to have fun with interesting spatial relationships. The building blocks are divided into ordinary modules and functional modules. The functional modules are designed by Arduino, and the stacking of building blocks is realized by children's control of space distance, and corresponding interactive feedback is generated, so as to realize the perception of space distance and the interaction with partners.

打开套盒取出玩具　选取目标颜色，撕开包装　倒入魔法药水 1

# CCCOLOR——学龄前儿童色彩教育玩具设计

## CCCOLOR —— Design of Educational Colour Toys for Preschool Children

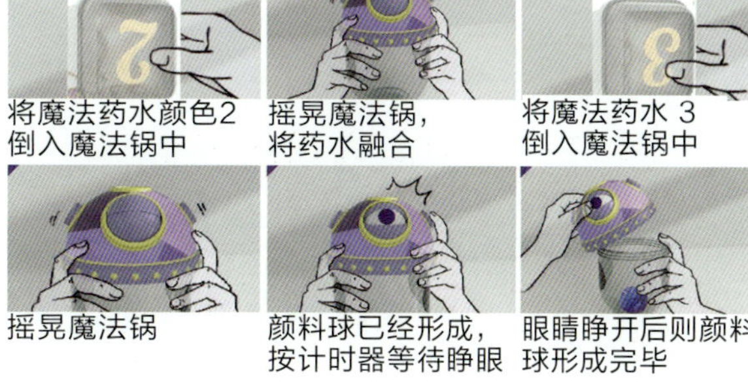

将魔法药水颜色2倒入魔法锅中　摇晃魔法锅，将药水融合　将魔法药水 3 倒入魔法锅中

摇晃魔法锅　颜料球已经形成，按计时器等待睁眼　眼睛睁开后则颜料球形成完毕

作　　者：董宇舒　孙世龙　杨洪旭
指导老师：梁琳玲
所在院校：浙江理工大学

### 设计说明

CCCOLOR 以“学龄前儿童色彩教育”为核心主题，致力于探索多样化的儿童绘画方法、创新的立体色彩材料以及开放式的绘画空间。我们采用“海藻酸钠颜料球”来激发儿童的天性，旨在突破传统的、平面的、静态的色彩教育模式，打造一套新颖的、立体的、动态的色彩教学体系。本玩具套件的视觉主题为“色彩魔法”，并以“see see Color”作为设计理念，期望通过结合儿童的视觉和触觉体验，营造一次丰富的色彩感官之旅。

### Design notes

With "Color Education for Preschool Children" as its core theme, CCCOLOR is committed to exploring diversified children's painting methods, innovative three-dimensional color materials and open painting space. We use "sodium alginate paint ball" to stimulate the nature of children, aiming to break through the traditional, flat and static color education mode, and create a novel, three-dimensional and dynamic color teaching system. The visual theme of this toy kit is "Color Magic" , and the design concept is "See See Color" , hoping to create a rich color sensory journey by combining children's visual and tactile experience.

# "汛息" X-DRONE 模块化应急救援无人机

## X-DRONE Modular Emergency Rescue Drones

作　　者：涂裕哲　龙志宇　李婷婷
指导老师：刘　星
所在院校：杭州电子科技大学

### 设计说明

为了满足洪涝灾害救援中救援人员和被困人员的需求，X-Drone 应急模块化救援智能无人机采用模块重组技术，能够搭载生命探测、定位、救援救生指引等多种功能模块。该无人机适用于水域和空域的救援任务，能够灵活地协助救援人员进行高效救援行动。

### Design notes

In order to meet the needs of rescue workers and trapped people in flood disaster, X-Drone emergency modular rescue intelligent UAV adopts module reorganization technology, and can carry a variety of functional modules such as life detection, positioning, rescue and life-saving guidance. The UAV is suitable for rescue missions in water and airspace, and can flexibly assist rescue workers in efficient rescue operations.

# 智能机场跟随行李车设计

## Intelligent Airports Follow Luggage Cart Design

作　　者：周　涵　梅怡茵
指导老师：殷晓晨
所在院校：合肥工业大学

### 设计说明

飞机作为一种重要的交通工具，因其快捷和便利而被许多人选择用于长途旅行。在机场，我们常看到乘客携带沉重的手提箱和行李，显得疲惫且行动吃力。针对这一问题，开展了一项关于机场行李问题的设计研究。在该设计研究中，我们对用户的行为进行了分析，并关注了行李箱在旅途中的搬运问题，提出了智能跟随行李车的概念设计，旨在解决用户在机场托运、拖拉、举起和等待行李箱时遇到的困难。根据设计规划，用户可以在候机厅、到达厅以及公共交通站点等公共区域通过指引找到并使用这种行李车。此外，在产品交互方面，它不仅具备智能控制、查询、使用、跟随等软件交互功能，还细致考虑了人机交互的细节设计，如方便用户存取行李。本设计旨在提升机场的出行体验，帮助人们更加便捷、智慧地出行。

### Design notes

As an important means of transportation, airplane is chosen by many people for long-distance travel because of its speed and convenience. At airports, we often see passengers looking tired and struggling to move heavy suitcases and luggage. To solve this problem, a design study on airport baggage was carried out. In this design study, we analyzed the user's behavior, and paid attention to the handling of luggage during the journey, and proposed the conceptual design of intelligent luggage cart, which aims to solve the difficulties encountered by users when they are checking in, dragging, lifting and waiting for luggage at the airport. According to the design plan, users can find and use the luggage cart in public areas such as departure halls, arrival halls and public transport stations through the guidance. In addition, in terms of product interaction, it not only has intelligent control, query, use, follow and other software interaction functions, but also carefully considers the details of human-computer interaction design, such as convenient user access to luggage. This design aims to enhance the travel experience of the airport and help people travel more conveniently and intelligently.

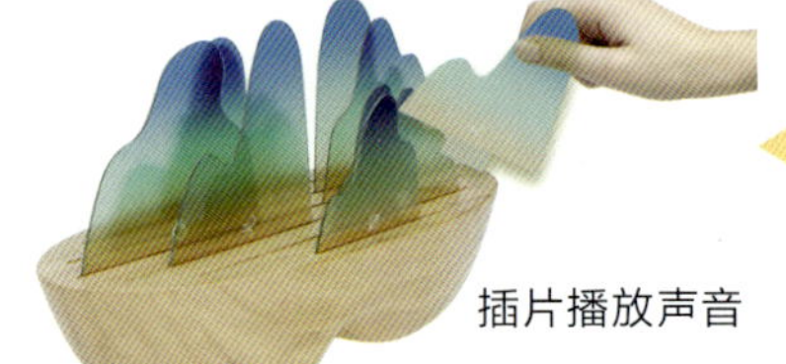


# 千里江山——创意交互音响

## QIANLI JIANGSHAN ——Creative Interactive Audio

作　　者：韩诗祺
指导老师：李　颖
所在院校：安徽工程大学

### 设计说明

千里江山——创意交互音响，是一款灵感来源于千里江山图的创新产品，融合了当下流行的“白噪音”概念。它将声音视觉化为江山，通过听觉和视觉的结合，创造了独特的体验。每一片山体代表一种白噪音声音，如“水流声”“篝火声”，共有八种不同的声音供用户选择和组合，定制个性化的音乐体验。通过左右滑动，用户可调整音量。该产品采用了创意技术，当插片底部的特殊音片与插孔接触时，即可实现开关效果。整体设计以简约流畅的线条为主，展现了国风的典雅与质朴，上下部分的设计呼应了千里江山的主题。开关和扬声器设计尽量简洁隐形，确保了良好的视觉效果。这样，该音响不仅可以作为音乐播放设备，还能作为艺术品进行展示，是一次视觉和听觉创新结合的新享受。

### Design notes

QianLi Jiangshan – White Noise audio is an innovative product inspired by the Chinese painting of Qianli Jiangshan, integrating the current popular concept of "white noise". It visualizes sound and creates a unique experience through the combination of hearing and sight. Each mountain represents a white noise sound, such as "water" , "campfire" , with total of eight different sounds for users to choose and combine, customizing the personalized music experience. By swiping left and right, the user can adjust the volume. The product uses creative technology that allows the switch effect to be achieved when a special sound disc at the bottom contact with the jack. The overall design is based on simple and smooth lines, showing the elegance and simplicity of the national style, and the design of the upper and lower parts echoes the theme of Qianli Jiangshan. The switch and speaker design is as simple and invisible as possible to ensure a good visual effect. In this way, the stereo can be displayed not only as a music playback device, but also as a work of art, which is a new enjoyment combining visual and auditory innovation.

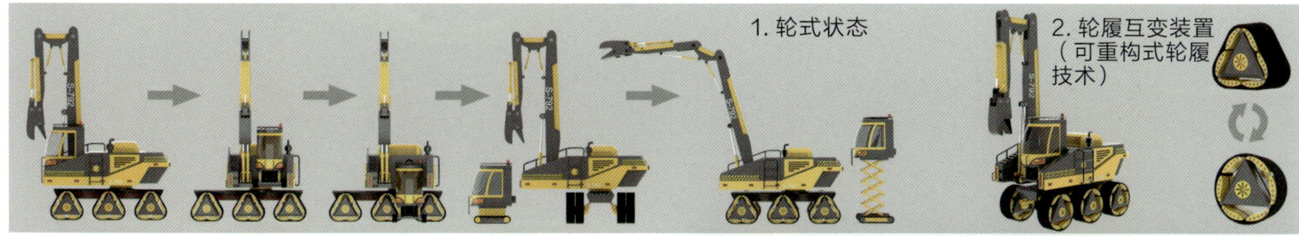


1. 在高危环境下工程车通过旋转90°，将升降装置下降至地面，使驾驶室与主机分离，再将驾驶室驾驶至安全工作区后，开始进行作业

2. 道路上行驶时，切换为轮式驾驶，可提高驾驶速度，方便工程转场

# 高危环境下的分离式工程作业车概念设计

# Concept Design of Separate Engineering Work Vehicle in High Risk Environment

作　　者：苏育斌　陈青青
指导老师：鲁江伟
所在院校：泉州师范学院

## 设计说明

此款工程车的设计特点在于其驾驶室与主机之间采用可分离结构，这种设计在确保驾驶室人员安全的同时，为驾驶员提供了更开阔清晰的视野。此外，该工程车配备了六个轮履互换装置，使其驱动能力能够适应各种地形，既灵活又稳定，这不仅提升了操作人员的舒适性，也有效提高了工作效率。设计理念上，它旨在高危工作环境中实现更安全的操作。用途方面，该工程车能够在挖掘、破拆和灾害响应等场合下，帮助操作人员在特殊且危险的工况中安全作业，同时解决了传统工程车转场时对拖车的依赖问题。其主要功能包括驾驶室与主机的分离、轮履互换以及装置的互换。

## Design notes

The design of this engineering vehicle is characterized by the use of a separable structure between the cab and the main engine, which ensures the safety of the driver while providing a wider and clearer vision for the driver. In addition, the engineering vehicle is equipped with six wheel and track interchanges, making its adapt to various terrain, both flexible and stable, which not only improves operator's comfort, but also effectively improves work efficiency. Conceptually, it is designed to enable safer operations in high-risk work environments. In terms of use, the engineering vehicle can help operators work safely in special and dangerous working conditions in excavation, demolition and disaster response occasions, and solve the problem of traditional engineering vehicles depending on trailers when transfer. Its main functions include the separation of cab and main engine, wheel and shoe interchange and device interchange.

# 极限营救计划——可移动式山地越野跑救援装置概念设计

## EXTREME RESCUE PLAN——Concept Design of Mobile Mountain Trail Running Rescue Device

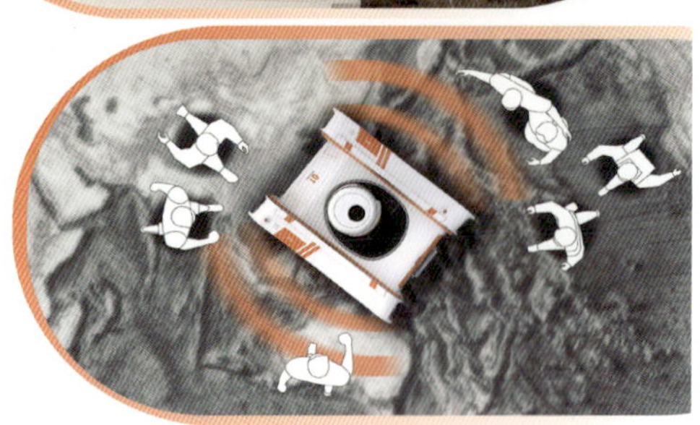

作　　者：邓依婷
指导老师：周　涛
所在院校：厦门大学

### 设计说明

本设计项目旨在为山地越野跑比赛提供一套专门的救援装置体系。该体系主要针对比赛中选手可能面临的突发性天气和其他紧急状况，包括救援车、救援手环以及固定救援包等多种载体。通过结合机动的主动救援模式和被动的自救模式，这套系统旨在最大程度上适应越野跑复杂多变的环境，并为参赛选手提供及时有效的救援。

### Design notes

This design project aims to provide a special rescue device system for mountain trail running competition. The system is mainly for unexpected weather and other emergency situations that competitors may face in the competition, including a variety of carriers such as rescue vehicles, rescue bracelets and fixed rescue kits. By combining mobile active rescue mode and passive self-rescue mode, the system is designed to adapt to the complex and changeable environment of trail running to the maximum extent and provide timely and effective rescue for competitors.

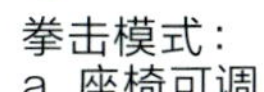

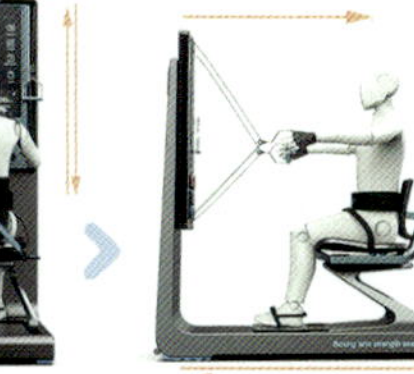

拳击模式：
a. 座椅可调

b. 屏幕可调

拉力模式：
a. 座椅可调

b. 人机工学靠背



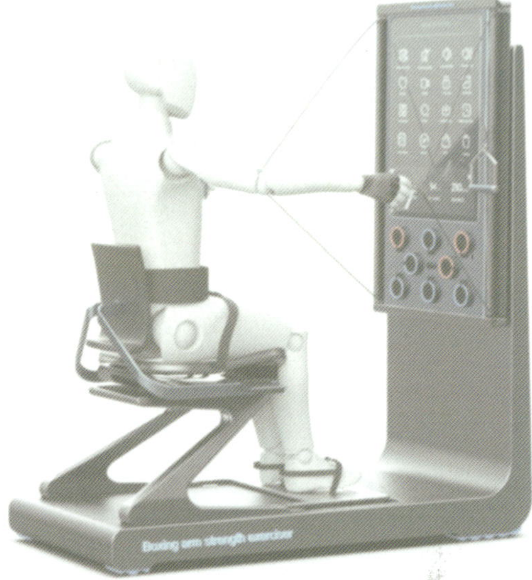



# 下肢残障者拳击手臂力量训练器

## Boxing Arm Strength Training Device for People with Lower Limb Disabilities

作　　者：戴嘉明
指导老师：刘晓宏
所在院校：福州大学

### 设计说明

下肢残障者拳击手臂力量训练器为下肢残疾群体的运动流程提供了规范，涵盖了从前期的拉力和拳击双模式运动到后期运动数据的反馈。有效的运动不仅能促进残疾群体身体多个系统的健康发展，还是增强体质的重要方式。下肢残障者拳击手臂力量训练器的设计通过简单的结构转换，使下肢残障者能够进行不同强度的拳击训练和不同角度的拉力训练。结合虚拟现实技术，该训练器提供渐进式训练内容，并通过音乐、画面、文字和语音提示等手段激励使用者，提高运动规范性。这有助于残疾群体正常参与健身活动，并享受多样化的运动乐趣。

### Design notes

The boxing arm strength trainer for lower limb disabled people provides a standard for the movement flow of lower limb disabled people, covering from the dual mode movement in early pull and boxing to the data feedback of the later movement. Effective exercise can not only promote the healthy development of multiple systems of the body of the disabled group, but also be an important way to enhance physical fitness. Lower limb disabled boxing arm strength trainer is designed to enable lower limb disabled people to perform different intensity boxing training and different angles of tension training through simple structural transformation. Combined with virtual reality technology, the trainer provides progressive training program, and inspires users by means of music, pictures, text and voice prompts to improve movement standardization. This helps disabled groups to participate in fitness activities normally and enjoy a variety of sports.

金奖 银奖 铜奖 优秀奖

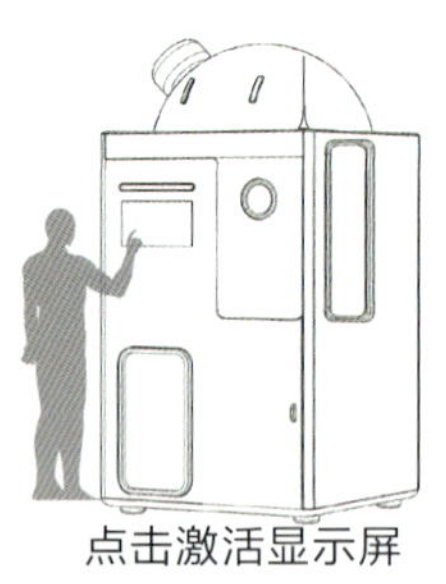
点击激活显示屏

扫脸登录

根据屏幕提示投放瓶子

确认积分后浏览商城

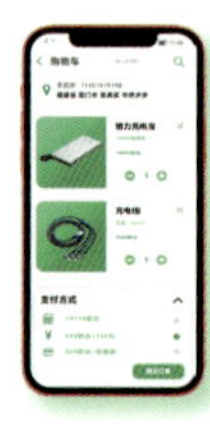
APP 界面设计

# 瓶小 C——与碳账户绑定的饮料瓶回收服务

# PINGXIAO C—— Beverage Bottle Recycling Service Bound to Carbon Account

作　　者：许鸿钧　庞浩歌　张雅聪
甘宇芃　陈　亮
指导老师：林敬亭
所在院校：华侨大学

## 设计说明

随着社会对节能减排这一环保话题关注度的提高，政府正在逐步推行碳积分政策以减少企业的碳排放。在个人层面，多家银行已经建立了碳账户系统，旨在鼓励居民采取低碳生活方式。本服务通过积分兑换机制激励市民将饮料瓶回收至服务网点。服务提供方（平台）在处理这些饮料瓶并将其卖给工厂的同时，所获得的碳配额还可以在碳交易市场上进行转售。

## Design notes

With the increasing attention to the environmental protection topic of energy conservation and carbon emission reduction in the society, the government is gradually implementing the carbon credits policy to reduce the carbon emissions from enterprises. At the individual level, several banks have set up carbon account systems aimed at encouraging residents to adopt a low-carbon lifestyle. The service encourages the public to return beverage bottles to the service outlets through the point redemption mechanism. While the service provider (platform) handles these beverage bottles and sells them to the factory, the carbon credits obtained can also be resold on the carbon trading market.

金式

木式

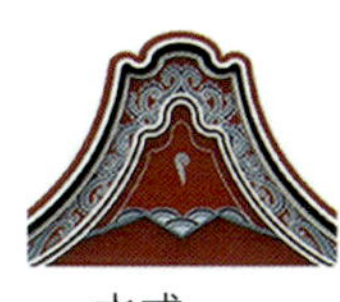

水式

火式

土式



# 厝错碟叠——闽粤五行山墙蘸料碟

# WUXING GABLES—— Minyue Five Elements Mountain Wall Dipping Plate

作　　者：吴　奕
指导老师：吴正仲　管幸生
所在院校：福建工程学院

## 设计说明

五行山墙是一种以“金、木、水、火、土”五种元素为样式进行装饰和构造的山墙，这种风格在中国福建闽南地区、广东潮汕地区、台湾地区的传统民居中十分流行。本作品巧妙地将闽粤地区的特色建筑结构与蘸料碟融合设计，方碟的两侧模仿了山墙的造型，而方碟中间则凸起一个三角形的“屋顶”。使用此碟时，可将食物上多余的蘸料在“屋顶”处刮去，随后蘸料会沿着“屋檐”流回碟中，仿佛雨水顺着屋檐淅沥流淌，既实用又富有地方特色。

## Design notes

Wuxing gable is a gable which is decorated and constructed with five elements of "gold, wood, water, fire and earth". This style is very popular in traditional houses in southern Fujian Province, Chaoshan region of Guangdong Province and Taiwan Province of China. This work cleverly integrates the characteristic architectural structure of Fujian and Guangdong regions with the dipping dish. The two sides of the square dish imitate the shape of the gable wall, and the middle of the square dish has a triangular "roof" . When using this dish, you can scrape the excess dip on the food on the "roof" , and then the dip will flow back to the dish along the "eaves" , just like rain flowing along the eaves, which is practical and rich in local characteristics.

# 应急救援叉车模组

## Emergency Rescue Forklift Module



作　　者：孙嘉棋　葛　秋
指导老师：杨　梅
所在院校：山东科技大学

### 设计说明

在当前社会中，自然灾害和人为灾害频繁发生，为了更好地进行救灾和救援工作，应急救援叉车模组可以迅速解决各种问题。这些车组适用于交通事故处理、自然灾害后的房屋搭建及救援、临时设施的搭建等方面，能够快速高效地完成任务，同时节省人力物力。全车采用模块化设计，前置车组能够进行360°旋转，并配备顶部环绕高清摄像头以及后视镜摄像头，确保施工安全。前置车组还配有可拆卸的叉车臂，安装后可用作叉车以搬运重物。履带式设计使其能够适应多种路面行驶。后置运输搬运车组则负责运输，并通过机械臂灵活操作搬运货物。分体式设计允许单独操作各个部分，而液压臂可以轻松抬起重物。后置四驱系统弥补了前车动力不足的问题，并实现了差速转向功能。

### Design notes

In the current society, natural and man-made disasters occur frequently. In order to better carry out disaster relief and rescue work, the special vehicle group can quickly solve various problems. These vehicles are suitable for a variety of purposes such as traffic accident handling, house construction, rescue after natural disasters and the construction of temporary facilities. They can complete tasks quickly and efficiently, while saving manpower and material resources. The whole vehicle adopts a modular design, the front car can rotate 360°, and is equipped with a top surround HD camera and a rearview mirror camera to ensure construction safety. The front unit is also equipped with a detachable forklift arms that can be used as a forklift to lift heavy loads when the arms are installed. The track design allows it to adapt to a variety of road surfaces. The rear-mounted transport vehicle group is for transportation and the robotic arm can operate the cargo quickly. The split design allows individual section to be operated separately, while hydraulic arms can lift heavy loads with ease. The rear all-wheel-drive system makes up for the lack of power in the front car and realizes the differential steering function.

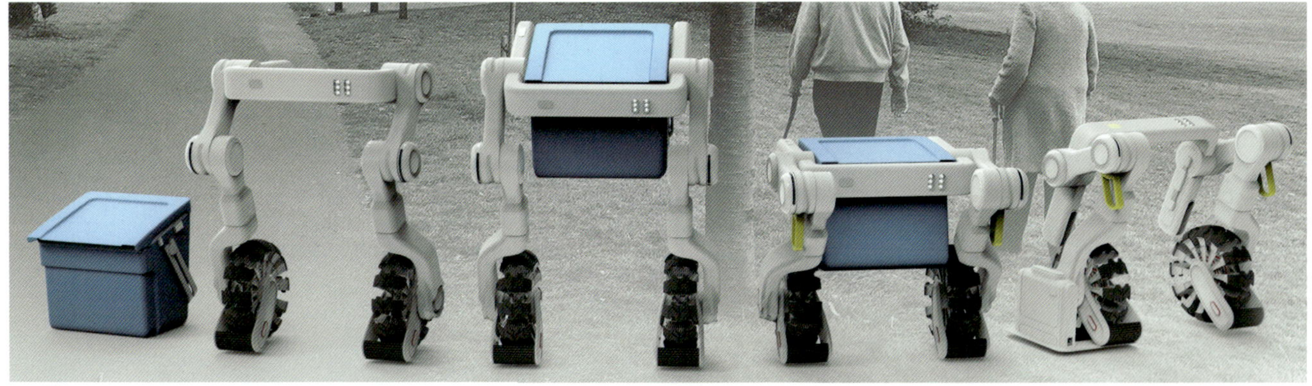

模式转换流程图

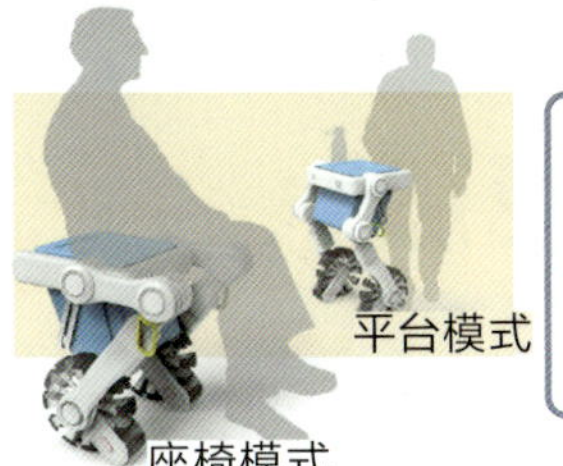


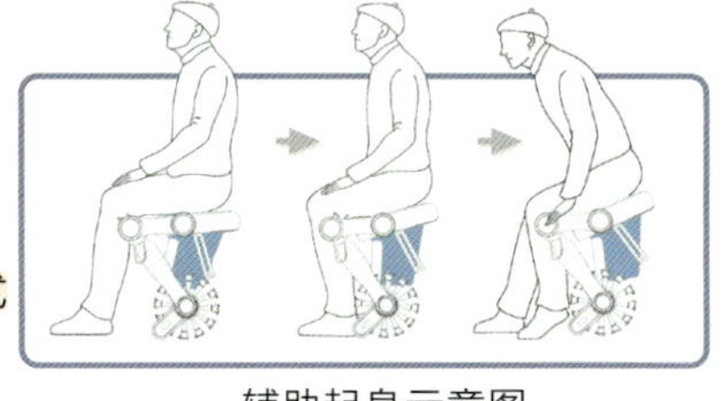

辅助起身示意图

# ALTER——老年人日常出行智能辅助伴侣

## ALTER —— an Intelligent Elderly-Assised Robot for Daily Trip

作　　者：刘亦鸣　王麒栋　蒋佶成
　　　　　张　翮　刘　涛
指导老师：郑　枫　杨　芳
所在院校：齐鲁工业大学

### 设计说明

Alter 是一款专为老年人设计的智能辅助伴侣，旨在帮助他们更轻松地进行日常出行。它拥有智能跟随购物模式、助行器模式、座椅模式和平台模式四种模式。考虑到当前城市中老年人的活动范围通常较小，且出行场景相对单一，他们往往选择步行作为主要的出行方式。基于两轮自平衡技术和智能跟随技术，我们设计了这款设备，以适应老年人在不同场景下的需求，使他们可以根据具体情况选择不同的 Alter 模式。此外，Alter 配备了完整的交互系统，能够全面提高用户的使用体验，真正实现轻松出行。与传统的助老产品不同，Alter 提供了一种全新的一站式出行辅助解决方案，它解决了不同场景下可能遇到的出行问题，能显著提升老年人的生活质量。

### Design notes

Alter is a smart assisted companion designed for the elderly to help them make their daily trips easier. It has four modes: smart shopping mode, walker mode, seat mode and platform mode. Considering that the activity range of the elderly in the current city is usually small, and the travel scene is relatively simple, they often choose walking as the main way of travel. Based on two-wheel self-balancing technology and intelligent following technology, we designed this device to adapt to the needs of the elderly in different scenarios, so that they can choose different Alter modes according to the specific situation. In addition, Alter is equipped with a complete interactive system, which can comprehensively improve the user's experience and truly realize easy travel. Unlike traditional elderly assistance products, Alter provides a new one-stop mobility assistance solution that solves mobility problems that may be encountered in different scenarios and significantly improves the quality of life of the elderly.

# 陆空两栖地震救援飞行汽车

## Amphibious Earthquake Rescue Flying Vehicle

作　　者：王　静　肖　慧　程　艳
　　　　　李永海
指导老师：李淑江
所在院校：青岛科技大学



### 设计说明

为了应对地震后道路塌陷导致救援车辆无法及时抵达受灾区域的问题，设计了一种集无人机与汽车特性于一体的交通工具。该设备拥有飞行和陆地行驶两种模式，并能通过轮胎翻转实现模式切换。它能够进行短距离飞行，在遇到道路不通的情况下，可从陆地模式转换为飞行模式，迅速越过障碍物，进入灾区。此外，它结合了无人机探测功能，能够快速侦测灾区的生命迹象，以便在72小时黄金救援时间内采取行动。该交通工具主要运用了无人机技术和人工智能技术，并整合了现代智能操纵系统，能确保迅速前往灾区执行救援任务。本无人机救援的功能包括搜索定位、远距离视频传输、投送食物及医疗设备、无人机测绘以及无人机消防等。为了保证高效性能，该交通工具采用了高升阻比车身结构技术增加阻力比，以适应飞行模式的需求，采用低空飞行智能驾驶技术以提高安全性。

### Design notes

In order to deal with the problem that rescue vehicles cannot reach the affected area in time due to road collapse after the earthquake, a vehicle that integrates the characteristics of drones and cars is designed. The vehicle has both flight and land driving modes, and can switch modes by flipping the tire. It can fly short distances quickly cross obstacles and enter the disaster area. In the case of road impassability, by switch the land mode to flight mode. In addition, combined with the drone detection function, the vehicle can quickly detect signs of life in disaster areas so that the action can be taken within 72 hours of rescue. It mainly uses drone technology and artificial intelligence technology, and integrates modern intelligent control systems to ensure the rescue missions can be carried out quickly in disaster areas. The functions of drone rescue include search and positioning, long-distance video transmission, delivery of food and medical equipment, drone mapping, and drone firefighting. In order to ensure efficient performance, the vehicle adopts the body structure technology of high lift-drag ratio to increase the drag ratio to meet the needs of flight mode. And its smart driving mode at low altitudes can improve safety.

# 高山滑雪机

## Alpine Ski Machines

作　　者：赵泽名　谭启轲　朱俊衡
张天宇
指导老师：包春燕
所在院校：山东科技大学

### 设计说明

高山滑雪机通过机械装置结构、使用方式和功能实现形式的创新设计，使得高山滑雪活动能够突破自然环境的限制，并为使用者提供优质的体验。该设计主要包含一个机械主体，它能够模拟滑雪时身体的扭转和腿部的活动，同时感应上肢动作，可为人体的整体活动检测提供支持，同时搭载一个头戴式 VR 设备，用于模拟滑雪场景。此外，还配备了一个屏幕，用于设置高山滑雪场的环境参数。

### Design notes

The innovative design of the form is achieved through the structure, use and function of the mechanical device, allows alpine skiing to break through the limitations of the natural environment and provides an excellent experience for the user. The design consists of a mechanical body that simulates body twists and leg movements while skiing, a pair of snowball fights that sense upper limb movements while providing support for overall movement detection of the human body, and a VR headset that simulates ski scenes. In addition, it is equipped with a screen for setting the environmental parameters of the alpine ski resort.

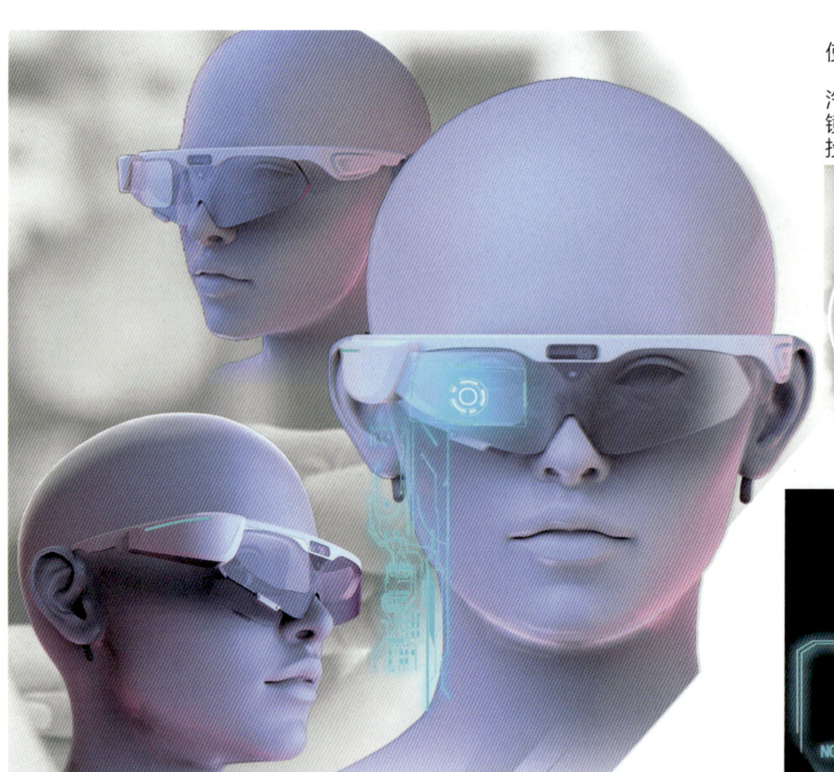

使用场景——出行

汽车鸣笛检测、方位检测
镜腿震动提醒
投影镜片显示具体方位

# 聋哑人智能眼镜设计

## Smart Glasses for Deaf-Mute



作　　者：张艺千　崔金媛　王丽娟
　　　　　贠铭鸿
指导老师：李峻峰
所在院校：潍坊学院

### 设计说明

这是一款专为聋哑人设计的概念眼镜，旨在帮助他们进行日常社交和出行。虽然聋哑人无法接收听觉信息，但他们可以接收视觉和触觉信息。因此，这款眼镜能够将外界的声音信息转换为视觉和触觉信号，使聋哑人能够通过这种方式实现无障碍的社交互动和安全出行。同时，当他们需要向外界传达信息时，这款眼镜可以通过摄像头识别手语动作，并利用智能计算技术将手语实时转换成语音。

### Design notes

This is a concept eyewear designed for deaf-mute people to help them with their daily socializing and mobility. Although deaf-mute people cannot receive auditory information, they can receive visual and tactile information. Therefore, the glasses are able to convert sound information from the outside world into visual and tactile signals, enabling deaf-mute people to achieve barrier-free social interaction and safe travel in this way. At the same time, when they need to convey information to the outside world, the glasses can recognize sign language movements through the camera, and use intelligent computing technology to convert sign language into speech in real time.

外展内收

前屈后伸

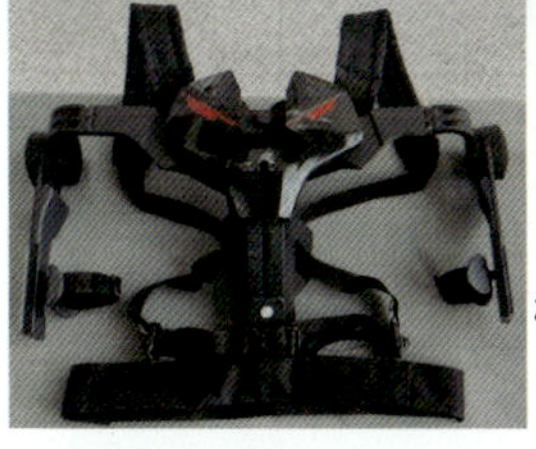

模型展示

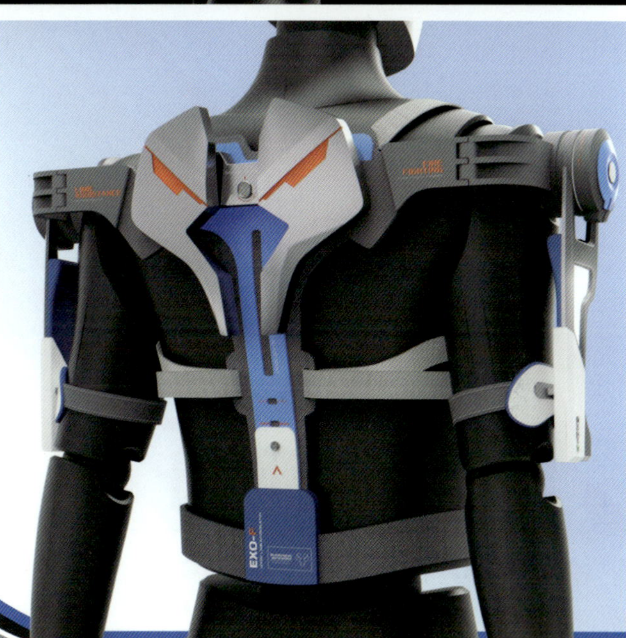

# 消防员搬运工具的人机一体化上肢外骨骼设计

## Design of Man-Machine Integrated Upper Limb Exoskeleton for Firefighten's Handling Tools

作　　者：王晓慧　戴煜雅　张永乐
尹雪冰
指导老师：赵　川
所在院校：青岛大学

### 设计说明

沉重的救援设备常常限制了救援人员的工作效率。将动力外骨骼助力机器人应用于消防救援和抗震救灾等领域，一方面可以减少救援人员的体力消耗；另一方面，它能够加快救援速度并提高救援效率，从而尽可能减少灾难造成的伤害。这款上肢外骨骼产品通过辅助消防员的上肢和肩部运动，为搬运消防工具等任务提供支持，帮助救援人员把握救援的“黄金时间”。

### Design notes

Heavy rescue equipment often limits the effectiveness of rescue workers. The application of powered exoskeleton assisted robot in fire rescue and earthquake relief can reduce the physical consumption of rescue workers on the one hand; on the other hand, it can speed up the rescue and improve the efficiency of the rescue, so as to minimize the damage caused by the disaster. This upper limb exoskeleton product helps rescue workers grasp the "golden time" of rescue by assisting firefighters with upper limb and shoulder movement, providing support for tasks such as carrying fire tools.

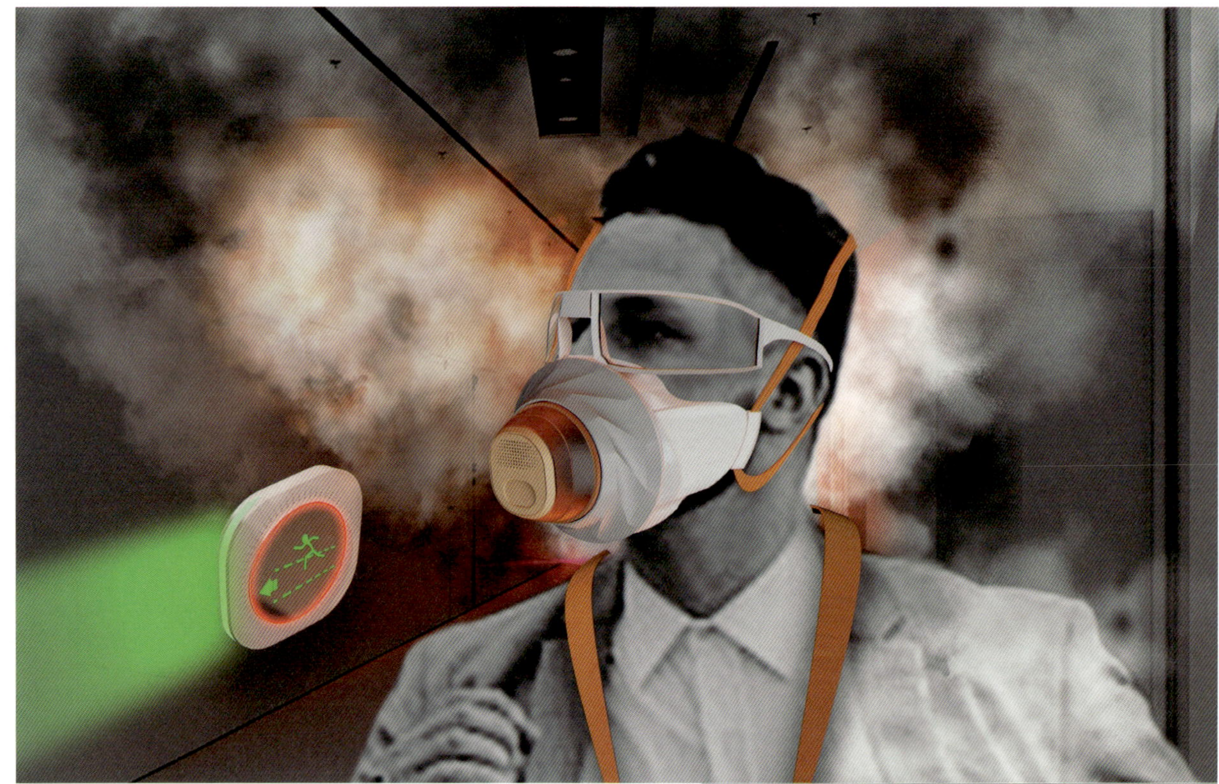

# 高层建筑消防救援装备设计

## Design of Fire Rescue Equipment for High-Rise Buildings

作　　者：路　源　赵雅艺　黄丽玲
　　　　　周诗雨
指导老师：徐　静
所在院校：郑州轻工业大学

### 设计说明

这套高层建筑消防救援装备是一款使用方便、效果显著且安全性高的产品，以最大程度地保护生命安全。其首要特点是配备呼吸面罩，从根本上解决因窒息导致的死亡问题；其次是护目镜，它能够保护眼睛，减少有害气体对器官的伤害。智能工牌在平时用于身份信息认证等基本功能，在火灾发生时能立即显示逃生路线的俯瞰图，并与公共装置的指引灯带配合，实现更加高效和安全的救援。

### Design notes

The main purpose of this high-rise building fire rescue equipment design is an easy to use, effective and high safety products, in order to maximize the protection of life safety. Its primary function feature is equipped with a breathing mask, which fundamentally solves the problem of death caused by asphyxia. The second is goggles, which can protect the eyes and reduce the damage of harmful gases to hurt the organs. In normal times, the smart work card is used for basic functions such as identity information authentication, and can immediately display the overlooking map of the escape route in the event of a fire, and cooperate with the guidance light belt of the public device to achieve a more efficient and safe rescue.

# 火柴蚊香

## Match Mosqutio Yepellent incense

作　　者：贾　优
指导老师：孙许方
所在院校：中原工学院

### 设计说明

在炎热的夏日生活中，许多人仍然选择使用蚊香。但经常遇到的问题是，点蚊香时打火机可能需要点火很长时间才能点燃蚊香，这样长时间使用打火机有时会导致其损坏。为此，设计了一种新型蚊香，其点火原理与火柴相似，在盒子顶部有一处可划火的地方，一划即可点燃蚊香。此外，即便是使用打火机，也能轻松地点燃这种蚊香，让人们能更加快捷方便地使用。

### Design notes

In hot summer, many people still choose to use mosquito repellent incense. However, the problem often encountered is that the lighter may take a long time to ignite when lighting the mosquito coil, so that the use of the lighter for a long time can sometimes lead to its damage. To this end, a new type of mosquito repellent incense is designed, whose ignition principle is similar to that of a match. There is a place on the top of the box where the fire can be struck, and the mosquito repellent incense can be lit with a stroke. In addition, even if you use a lighter, you can easily light this mosquito repellent incense, so that people can use it more quickly and easily.

金

奖

银

奖

铜

奖

优秀奖

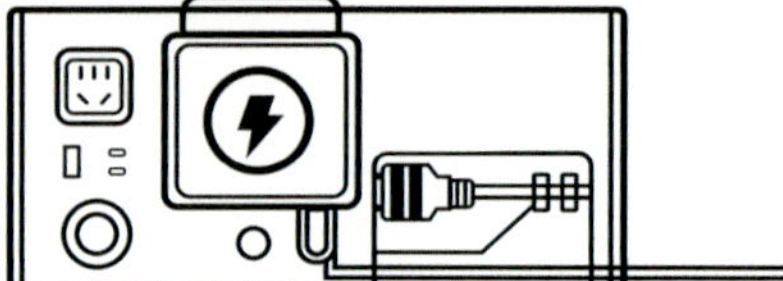

# MOBILE UNIVERSE ——户外移动电源

## MOBILE UNIVERSE —— Outdoor Mobile Power Supply

作　　者：许洲杰　何家宝　王中楷
　　　　　刘　欢
指导老师：张　婷　崔蒙蒙
所在院校：河南工业大学

### 设计说明

Mobile Universe 通过其“可移动式插孔”技术扩展了电子产品的使用距离和范围，解决了传统电源线长度不足的问题。这使用户在使用电源的同时可以自由活动，提高了用电的便利性和舒适度。此外，它还能解决多种电器同时用电的需求，避免了因多个电器集中使用造成的空间拥挤问题。如此，用户可以方便地与周围的人共享电源。

### Design notes

Mobile Universe expands the range of electronic products through its "mobile jack" technology, solving the problem of insufficient length of traditional power cords. This allows the user to move freely when using the power supply, improving the convenience and comfort of electricity. In addition, it can also meet the needs of a variety of electrical appliances used at the same time, avoiding space congestion caused by the centralized use of multiple electrical appliances. Users can easily share power with those around them.

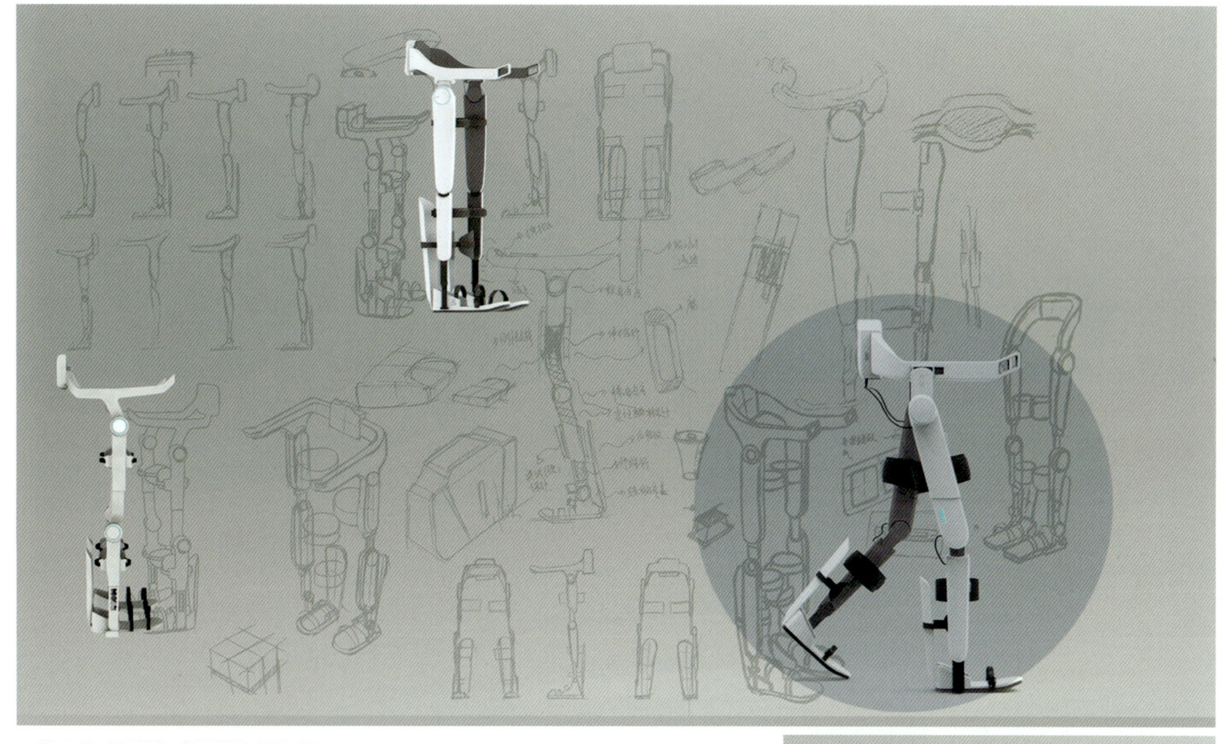

APP 界面设计图



# 基于自然人机交互的步态矫正外骨骼设计

# Design of Gait Correction Exoskeleton Based on Natural Human-Computer Interaction

作　　者：陈春雷　李柄兴　丁嘉昊
　　　　　常美玲
指导老师：徐　静
所在院校：郑州轻工业大学

## 设计说明

本设计能够提升用户的交互体验和使用意愿，满足患者独立行走和康复的基本需求。同时，我们建立了一个沟通平台，让患者、家庭和医护人员之间能够进行有效沟通，使患者能够更直观、详细地了解自身的康复训练和恢复情况，从而帮助医护人员更好地掌握患者的康复训练数据。

## Design notes

Through the natural human-computer interaction theory, we can enhance the user's interactive experience and use intention. This can meet the basic needs of patients to walk independently and recover. At the same time, we have established a communication platform, so that patients, families and medical staff can effectively communicate, and patients can understand their own rehabilitation training and recovery situation more intuitively and in detail, so as to help medical staff better grasp the rehabilitation training data of patients.

# 敦煌客制化键盘

## Dunhuang Customer -Based Keyboard

作　　者：胡晓琪　胡王毅文　夏　钰
指导老师：苏　晨
所在院校：湖北工业大学

### 设计说明

这款定制键盘的键帽搭配灵活，色彩和图案的选择具有很大的随机性。此设计灵感来源于我国文化的瑰宝——敦煌壁画，它采用了鲜艳的色彩搭配，并巧妙运用奇特的物体形象作为细节装饰。键盘机身的设计采用了类似人体工程学的形状，以增加使用时手指的舒适度。在键盘布局方面，取消了不常用的按键，并进行紧凑排列，从而缩小了键盘的整体尺寸，节省了空间成本，增加了键盘的精致感。

### Design notes

The keycaps of this custom keyboard are flexible, and the choice of colors and patterns has a great deal of randomness. Inspired by the Dunhuang murals, a treasure of Chinese culture, the design uses bright color combinations and clever use the imagery of strange objects as details. The body of the keyboard is designed with an ergonomically similar shape to increase the comfort of the fingers when using it. In terms of keyboard layout, we abandoned the infrequently used keys, and compact the arrangement, thus reducing the overall size of the keyboard, saving space costs, but also increasing the exquisite sense of the keyboard.

# 净海-21

## SEA CLEAN-21

作　　者：吴昭仪　杨圣林　刘凯铸
　　　　　高春虎　胡王毅文
指导老师：苏　晨　王海强　许德骅
　　　　　彭　魏
所在院校：湖北工业大学

### 设计说明

净海-21为海上溢油紧急处理船，适用于溢油的初期和中期阶段，也是控制溢油扩散的有效手段之一。当船体发生溢油事故时，船员需要根据海况和溢油情况，将储存在船体的围油栏装备放入水中，按照指示自动将溢油区域控制在可控范围内。同时，船员应找到溢油点进行封堵和修复。

### Design notes

The emergency treatment of oil spill at sea is applicable to the early and middle stage of oil spill, and is also one of the effective means to control the spread of oil spill. When an oil spill accident occurs in the hull, the crew needs to put the oil containment equipment stored in the hull into the water according to the sea conditions and the oil spill situation, and automatically control the oil spill area within the controllable range according to the instructions. At the same time, the crew should find the oil spill point to plug and repair.

金
奖
银
奖
铜
奖
优秀奖

# 崛起之神

## RISING GOD

作　　者：周俊宏　聂绪龙　尹　昊
王民航　陈嘉芃
指导老师：龚怡慧
所在院校：湖北工程学院

## 设计说明

“崛起之神”是一款独特的滑板车，与市面上的其他滑板车不同，它将滑板与滑板车的功能融合在一起，把手可折叠，确保无锐角以避免对儿童造成伤害。产品外观充满科技感，彰显了未来产品的设计特点。除了具备传统滑板的功能，它还增加了电动力驱动功能。在用户使用分析方面，该产品的主要使用者是儿童，因此产品的重量设计得较为轻便，使用了碳纤维板等轻量化材料。

## Design notes

"Rising God" is a unique scooter designed to accompany children as they grow. Unlike other scooters on the market, it combines the functionality of a skateboard and a scooter, and is designed to fold the handle and ensure that there is no acute angle to avoid injury to children. The appearance of the product is full of scientific and technological sense, highlighting the design characteristics of future products. In addition to the functions of traditional skateboards, it also adds an electric drive function. In terms of user usage analysis, the main users of the product are children, so the weight of the product is designed to be relatively light, using lightweight materials such as carbon fiber plates.

# 上肢体障碍者器械辅助健康管理系统

## Instrument-Assisted Health Management System for People with Upper Limb Disabilities

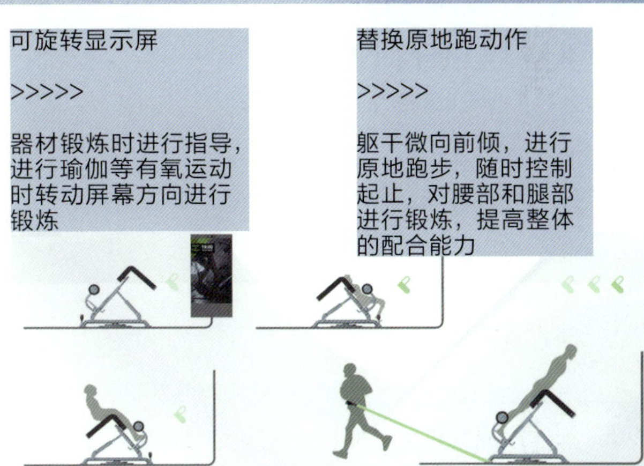


作　　者：董佳洁　郑冬秀　周子豪　高浩原

指导老师：李江泳

所在院校：湘潭大学

### 设计说明

对于肢体残疾人来说，在进行健身康复训练时，使用现有的常规肢体健身设备往往会给他们带来极大的不便。上肢体障碍者器械辅助健康管理系统是一款专为独臂或无臂等手臂肢体缺陷者设计的器械辅助健康管理系统，它可以锻炼腰部、腿部和肩部，同时具备娱乐功能，是一种改良的健身器材。

### Design notes

For people with physical disabilities, the use of existing conventional physical fitness equipment often brings them great inconvenience when carrying out fitness rehabilitation training. It is an instrument-assisted health management system designed for people with Cupper limb disabilities, such as with one arm or no arms. It can exercise the waist, legs and shoulders, while also providing entertainment. It is a modified fitness equipment.

# 山居秋暝——家具设计

## MOUNTAIN DWELLING AUTUMN NIGHT —— Furniture Design

作　　者：范君珂　孙　萌　徐晴晴
王艺菲　张　航
指导老师：李　思　夏忠军
所在院校：湖南理工学院

## 设计说明

选题“山居秋暝”源自唐代著名诗人王维的五言律诗——《山居秋暝》。山作为文人墨客托物言志的载体，体现了一种内心的追求，也体现了中国人对自然与人和谐共生的思考。本案例通过提炼大自然中的“山”元素，运用平面构成的基本要素——点、线、面，精炼地表现出山的形态。以纤细而稠密的线条和平滑方正的块面，展现出山的各种有趣变化形态。观者不仅可以通过视觉直观感受来体验这些变化，还能从意象中体会到深厚的文化底蕴和独特的宁静意境，回味无穷。在传承中式传统家具文化的基础上，作者将提炼自山中的装饰元素与现代家具设计理念相结合，设计出既满足实用性和审美需求，又能表达精神追求的新中式家具。

## Design notes

The topic of "autumn moping in the mountains" comes from the famous Tang Dynasty poet Wang Wei's five-word verse - "Autumn moping in the mountains". Mountain, as the carrier of literati, embodies a kind of inner pursuit. On the expression level, the attachment to landscape reflects the Chinese people's thoughts on the harmonious coexistence of nature and humanity. By refining the "mountain" element in nature, this case uses the basic elements of plane formation —— points, lines, and planes to show the form of mountains. With thin and dense lines and smooth square blocks, it shows all kinds of interesting changes and deformation of the mountains. Viewers can not only experience these changes through visual intuition, but also experience the profound cultural heritage and unique quiet artistic conception from the image, and the aftertaste is endless. On the basis of inheriting the traditional Chinese furniture culture, the author combines the decorative elements extracted from the mountains with the modern furniture design concepts to design new Chinese furniture that not only meets the practical and aesthetic needs, but also expresses the spiritual pursuit.

# 福韵·茶道——陶瓷茶具设计

# FUYUN TEA CEREMONY——Ceramic Tea Set Design

作　　者：薛　江　田明宏
指导老师：邓渭亮　蒋艺芝
所在院校：湘南学院

## 设计说明

该设计以“福禄”为主题。整套茶具采用陶泥材质，通过陶的质感和釉色来体现道教文化中“朴素求真”的理念。简洁的造型和结构设计展现出“大道至简”的设计语言和文化内涵，体现了“天人合一”的哲学思想。

## Design notes

The design takes "Fu Lu" culture as the theme. The whole set of tea sets is made of clay, which reflects the concept of "simplicity and truth" in Taoist culture through the texture and glaze of pottery. The simple shape and structure design show the design language and cultural connotation of "the road is simple" , and embody the philosophical thought of "the unity of heaven and man".

# 户外冒险旅行 GPS 设计

## Outdoor Adventure Travel GPS Design

作　　者：潘文博　李梓仪　查鸿琨
　　　　　周　昱
指导老师：刘　颖
所在院校：中南大学



### 设计说明

户外旅行爱好者在徒步旅行时，常常面临手机 GPS 导航和纸质地图无法满足导航及应急需求的情况，比如手机无信号、电量耗尽，而纸质地图既不能定位也容易损坏。因此，他们需要一款集成安全应急功能的 GPS 导航仪。这款产品融合了手持 GPS、指南针和对讲机的多种功能，专为户外活动设计。该设备能够在无需网络的情况下同步数据、管理路线和航点信息，非常适合登山、徒步、自驾越野、骑行等户外活动。它可以显著提升用户的探索体验，并大幅增强户外冒险者的安全保护。该产品配备应急指南针、手电筒以及手摇发电装置，辅助解决户外安全难题。此外，该产品具备防水、防尘、防摔的“三防”特性，极大地延长了在户外的使用寿命，是每位探险爱好者必备的装备。

### Design notes

When hiking, outdoor travel enthusiasts often face the situation that mobile phone GPS navigation and paper maps cannot meet the navigation and emergency needs, such as mobile phone no signal, power out, and paper maps can not be located and easily damaged. So they needed a GPS navigator with integrated safety and emergency functions. This product combines a variety of functions such as handheld GPS, compass and walkie-talkie, and is designed for outdoor activities. The device can synchronize data and manage route and waypoint information without the need for a network, which is ideal for outdoor adventure activities such as mountaineering, hiking, self-driving off-road and cycling. It can significantly enhance the user's exploration experience and greatly enhance the safety protection of outdoor adventurers. Equipped with an emergency compass, a flashlight and a hand generator, it can solve all aspects of outdoor safety problems. In addition, the product has waterproof, dustproof, anti-fall "three" characteristics, greatly extending the service life of outdoor use, is an essential equipment for every adventure enthusiast.

# 视障学生课桌设计

# Desk Design for Visually Impaired Students

作　　者：戴益全
指导老师：金　鑫
所在院校：广东石油化工学院

## 设计说明

本设计针对的目标人群是高度近视、弱视和盲人学生。在设计上，突破了传统课桌的造型和尺寸限制。专为视障学生设计的课桌也适用于普通学生。课桌的侧面设计有书架以取代传统的桌斗，并适当加大了桌面尺寸。考虑到视障学生上课的需求和心理特点，课桌内部设计了放大显示器和盲文点显器，放大显示器附有盲文提示，同时课桌侧面设置了分类书本的隔板，方便学生使用。此外，本产品采用升降式设计，有助于学生保持最佳的上课姿势。

## Design notes

The target population of this design is students with high myopia, amblyopia and total blindness. In terms of design, it breaks through the shape and size restrictions of traditional desks. And the desks designed for visually impaired students are also suitable for ordinary students. The side of the desk is designed with bookshelves to replace the traditional desk bucket and extend the table size appropriately. Taking into account the needs and psychological characteristics of visually impaired students in class, an enlarged display and a Braille point reader are designed inside the desk. The enlarged display is equipped with a Braille hint, and a partition is set up on the side of the desk to classify books for the convenience of students. In addition, the product is designed with a lifting style, which helps students maintain the best posture when atterding a class.

# BUBU 的爱心世界——海岛主题 IP 形象及其衍生品设计

## BUBU LOVE WORLD——Island Theme IP Image and Derivative Design

作　　者：周美伶　朱启萌　徐耿坚
　　　　　李珂羽　陈炜瀚
指导老师：郭　涵　朱鸿运
所在院校：华南农业大学

### 设计说明

本项目以涠洲岛的布氏鲸为主题，设计了一款 IP 形象。涠洲岛附近的海域是布氏鲸捕食和繁殖的重要地点之一。这一形象通过抽象的设计手法，将鲸鱼的嘴部化作一个爱心形状，既体现了鲸鱼的捕食特性，也象征着对游客的热情欢迎。基于这一形象，我们创作了一系列衍生产品，旨在打造一个属于海岛的独特 IP 形象，同时也宣传和展示当地的生态文化。

### Design notes

This project designed an IP image based on the theme of Brzewalski whale in Weizhou Island. The waters near Weizhou Island are one of the important places for Brykella whales to hunt and breed. Through an abstract design technique, the image turns the whale's mouth into a love heart shape, which not only reflects the whale's hunting characteristics, but also symbolizes the warm welcome to tourists. Based on this image, we created a series of derivative products designed to create a unique IP image of the island, but also to promote and showcase the local ecological culture.

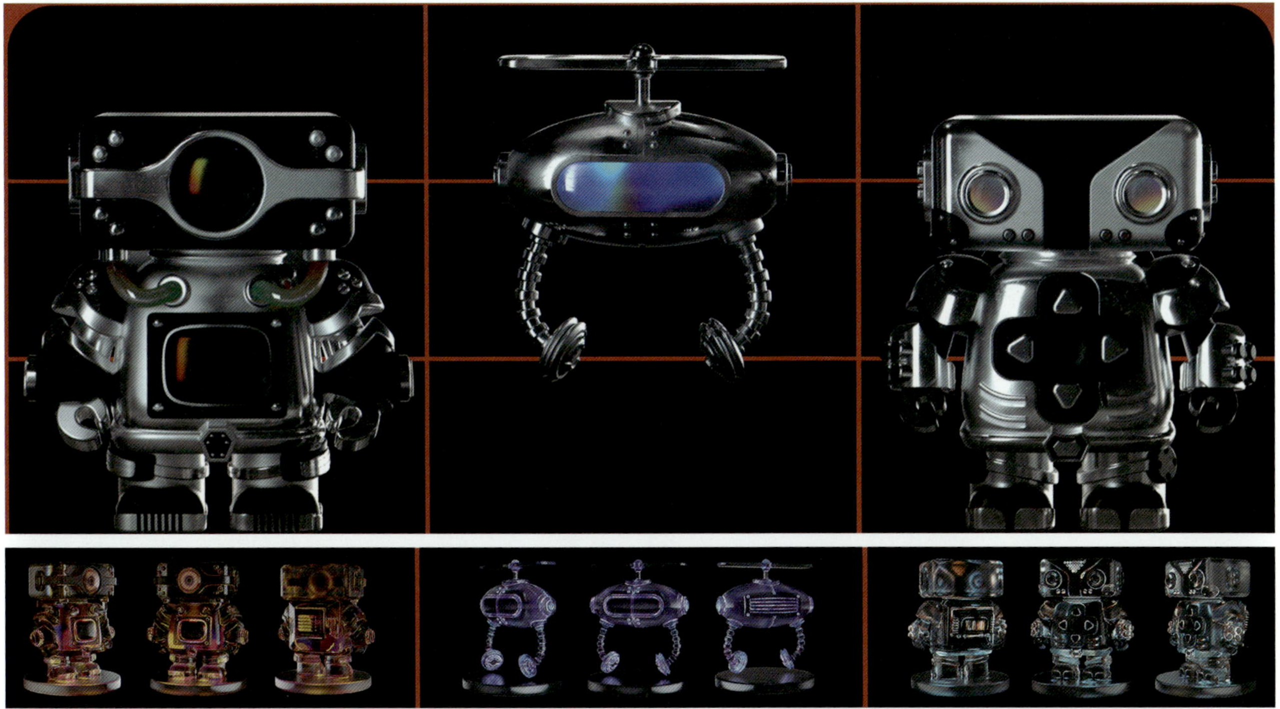

# "TONG 感"——潮玩设计

## EMPATHY —— Fashion Product Design

作　　者：肖海权　庄迪竣　陈政昊
　　　　　马启荣　池国龙
指导老师：张　英
所在院校：广东白云学院

### 设计说明

"TONG 感"潮玩设计，其名称"TONG 感"寓意着童心和共鸣。它既唤起了对童年的怀旧之情，也涵盖了人类的三大感官体验：视觉、触觉和听觉。本系列潮玩产品的设计灵感来源于儿时的老式电子产品，通过这三大感官元素与之相结合，创作出一系列富有童年回忆的潮玩。设计的宗旨是将潮玩具作为载体，与那些镌刻在记忆深处的童年老物件相融合。这样不仅让人们在精神上有所寄托，还能唤起儿时的欢乐时光，帮助缓解生活压力。它鼓励人们在忙碌的生活里暂停脚步，去发现身边的美好事物，从而享受并珍惜当下的美好时光。

### Design notes

"TONG sense" fashion play design, its name "TONG sense" means childlike innocence and resonance. It evokes nostalgia for childhood and encompasses the three major human sensory experiences: sight, touch and hearing. The design inspiration of this series of fashion play products comes from the old electronic products of childhood. Through the combination of these three sensory elements, a series of fashion play rich in childhood memories is created. The purpose of the design is using fashion toys as a carrier, to make integration of those engraved in the memory of the old objects of childhood. This not only gives people spiritual sustains, but also evokes memories of happy times in childhood and helps relieve stress in life. It encourages people to take a break from their busy lives and discover the good things around them, so as to enjoy and cherish the good times in the present.

# 鹿椅

## Deer Chair

作　　者：夏侯莹　李汶琪　莫文杰
朱明欣
指导老师：陈振益
所在院校：五邑大学

## 设计说明

鹿椅的设计融合了鹿的形象与家具的功能性，让人们在家居体验中领略到自然的美，同时推广了人与自然和谐共处的理念。在材料选择上，鹿椅采用高品质的皮革和木材，并根据鹿的自然色彩进行配色，主要采用了充满活力的橙色作为软包的主要色调。工艺方面，使用了热弯技术处理木条，精心切割角度以突出细节，并采用了皮革包裹的技艺。

## Design notes

The design of the deer chair integrates the image of the deer with the functionality of the furniture, so that people can enjoy the beauty of nature in the home, and promote the concept of harmonious coexistence between man and nature. In terms of material selection, the deer chair is made of high-quality leather and wood, and the color scheme is based on the natural color of the deer, mainly using vibrant orange as the main tone of the soft bag. In terms of craftsmanship, the wood strips are treated with hot bending technology, carefully cut angles to highlight details, and the leather wrapping techniques are used.

# 冰山沙发

## Iceberg Module SOFA

作　　者：钟文浩　朱明欣　罗佳仪
　　　　　梁同周　李汶琪
指导老师：陈振益　陈　亮
所在院校：五邑大学

### 设计说明

冰山模块沙发的设计灵感源自于在格陵兰极光下漂浮的冰块，它们在静谧的极光照耀下随着海浪起伏。这款沙发以其完美的造型展现了冰山硬朗的线条感，主要采用了绒布材质，其柔软细腻的触感与坚固的外观形成了鲜明的对比，激发了人们对自然和人类之间关系的思考。通过将自然的元素融入家具设计中，以一种独特的方式促进了人与自然的共生。功能上，冰山模块沙发采用了模块化设计理念。其模块状的造型启发自冰块分裂时形成的裂缝，每个模块都可以单独分离出来，作为一个独立的坐具使用。

### Design notes

The iceberg module sofa is inspired by the ice floating under the Greenland aurora, which rises and falls with the waves under the quiet light of the aurora. Mainly using velvet fabric material, this sofa perfectly shows the hard line sense of iceberg with its shape. its soft and delicate touch and solid appearance form a sharp contrast, inspiring people's thinking about the relationship between nature and human beings.Through the integration of natural elements into the furniture design, It promote the symbiosis between people and nature in a unique way. In terms of function, the iceberg module sofa adopts the modular design concept. Its modular shape is inspired by the cracks when ice cubes break apart, and each module can be detached and used as an independent seat.

# 一童空间

## ADIDIY

作　　者：林志雄
指导老师：陈　旭
所在院校：广州美术学院

### 设计说明

本设计基于趣味性主题、模块化和空间收纳等原则，目的是为儿童打造一款手工体验式学习柜。设计细节如下：整体结构由四个模块组件组成，每个模块的挂件正面装有磁铁，可以吸附在带磁性的板上。柜顶设计成波浪形，便于放置彩笔、白板擦等工具，允许儿童根据需求自定义收纳空间。桌面中央的触摸面采用定制的主题性手工切割板，使儿童能按个人喜好创建自己的“秘密基地”。该设计强调了空间收纳的重要性，帮助整合散乱的文具和手工工具，从而节省家中空间。同时，它也具有教育意义，通过手工体验激发儿童的空间意识，并将整洁有序的收纳习惯融入他们的日常生活中，帮助他们从小培养整理个人物品的良好习惯。

### Design notes

Based on the principles of fun theme, modularity and space storage, the design aims to create a manual experiential learning cabinet for children. The design details are as follows: the overall structure consists of four modular components, each of which has a magnet on the front of the pendant that can be attached to a magnetic plate. The top of the cabinet is designed in a wavy shape to facilitate the storage of tools crayon pen and whiteboard eraser, allowing children to customize the storage space according to their needs. The touch surface in the center of the desktop uses a customized thematic hand cutting board, which allows children to manually "DIY" their own "secret base" according to their personal preferences. The design emphasizes the importance of space storage, helping to consolidate scattered stationery and hand tools to save space in the home. At the same time, it is also educational, stimulating children's spatial awareness through manual experience, and integrating the habit of clean and orderly storage into their daily life, helping them to cultivate the good habit of organizing personal belongings from an early age.

# 三合一的面包包装袋

## A Three In One Brea  Packaging Bag

作　　者：赵昀萌　蔡旭东　尤建丹
高帅佳　柴启璇
指导老师：陆定邦　蒋　雯
所在院校：广东工业大学

### 设计说明

面包店在顾客购买面包时常会提供一次性袋子，这种做法不利于社会的可持续发展。本设计是一款专为面包店设计的多功能环保产品，由环保材料如杜邦纸和无纺布制成。它可用作VIP卡、包装袋和盘子，并可折叠且便携，可作为VIP卡携带，有助于减少面包店使用一次性袋子。顾客购买面包时出示卡片可获得折扣，随后可将卡片展开变成一个环保购物袋，用于装载面包，或者将其折叠成餐盘形状使用。使用后，通过简单的折叠步骤，即可将购物袋恢复为VIP卡的形态。

### Design notes

Bakeries often provide disposable bags when customers buy bread, which is damaging to the sustainable development of society. VGP is a versatile and environmentally friendly product designed for bakeries, made from environmentally friendly materials such as Dupont paper and non-woven fabrics. It can be used as a VIP card, bag and plate. Designed to be foldable and portable, the VGP can be used as a VIP card, helping to reduce the use of disposable bags by bakeries. Customers receive a discount when purchasing bread by presenting the card, which can then be unfolded into an eco-friendly shopping bag for carrying  bread or folded into a serving plate. After use, the shopping bag can be restored to the form of a VIP card through a simple folding step.

# 三十分之七天里的痛感

## Pain In 7 Out of 30 Days

作　　者：刘晓琳
指导老师：何文才　吴越齐
所在院校：广州美术学院

### 设计说明

本系列纤维艺术作品灵感来源于我个人每月生理期的痛经体验。我试图通过材质对比和肌理的变化来表现我所经历的疼痛程度。在作品中，我广泛使用了毛线来体现女性的温柔与感性特质，并辅以手工混合线或玻璃纱来营造疼痛时的撕扯感觉。我希望这一系列作品能让更多人理解和关心身边的女性所承受的痛苦。

### Design notes

This series of fiber art works is inspired by my personal experience with menstrual cramps during my monthly period. I tried to show the level of pain I was experiencing through material contrast and texture changes. In the works, I extensively use wool to reflect the gentle and sensual characteristics of women, and supplemented by hand-mixed thread or glass yarn to create the feeling of tearing in pain. I hope this series of works will make more people understand and care about the suffering of women around them.

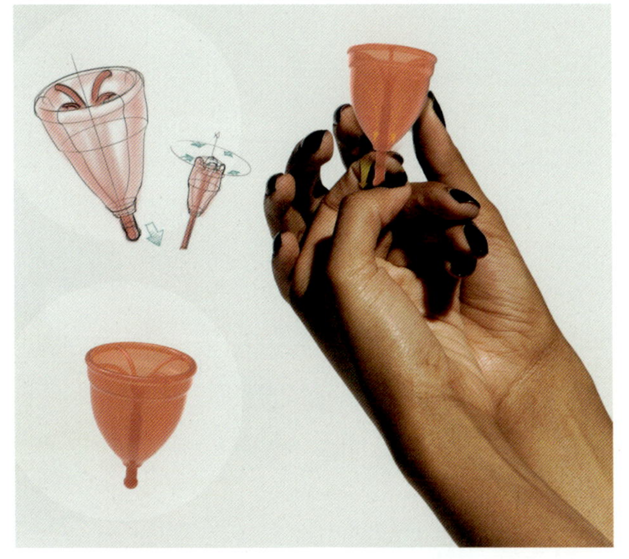

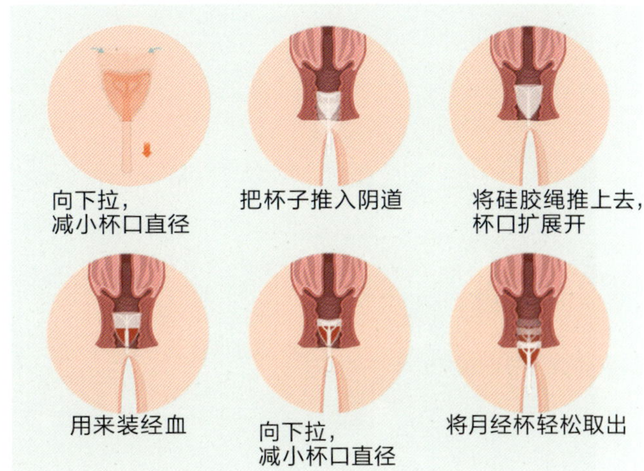




# SMALLER CALIBER CUP ——月经杯创新设计

## SMALLER CALIBER CUP —— Menstrual Cup Design

作　　者：汪启韬　郑小涵　陈文静
指导老师：陈朝杰
所在院校：广东工业大学

## 设计说明

我国女性每年卫生用品的消费量达到1200亿片，一次性用品耗费了大量资源，同时产生的垃圾也会引发一系列环境问题。月经杯是一种放置于阴道中收集经血的器具，其使用年限可长达5~10年，但普及率却相当低。一方面，使用月经杯需要一定的学习成本和技巧；另一方面，有些女性可能会因为担心阴道口疼痛而不敢尝试。我们从雨伞的开合方式中得到灵感，设计了一款与普通月经杯不同的产品，在杯口处有四根硅胶绳相连并汇聚成杯柄，用手指捏住杯柄往下拉，月经杯的杯口便会在拉力作用下往里收缩，直径减小。这样，月经杯的体积在使用前后会发生变化，使得女性使用月经杯变得更简单且无疼痛。

## Design notes

Chinese women's annual consumption of hygiene products reached 120 billion pieces, disposable products consume a lot of resources, while the waste generated will also lead to a series of environmental problems. A menstrual cup is a device placed in the vagina to collect menstrual blood, and its service life can be as long as 5 to 10 years, but the popularity rate is quite low. On the one hand, the use of menstrual cups requires certain learning costs and skills; on the other hand, some women may be afraid to try it for fear of vaginal opening pain. Taking inspiration from the way umbrellas open and close, we designed a product that is different from the ordinary menstrual cup, with four drawstrings connected at the rim of the cup and assembled to form the handle. Just pinch the handle of the cup with your fingers and pull it down, the rim of the menstrual cup will shrink inward under the action of tension, and the diameter will decrease. In this way, the volume of the menstrual cup changes before and after use, making it easier and painless for women to use the menstrual cup.

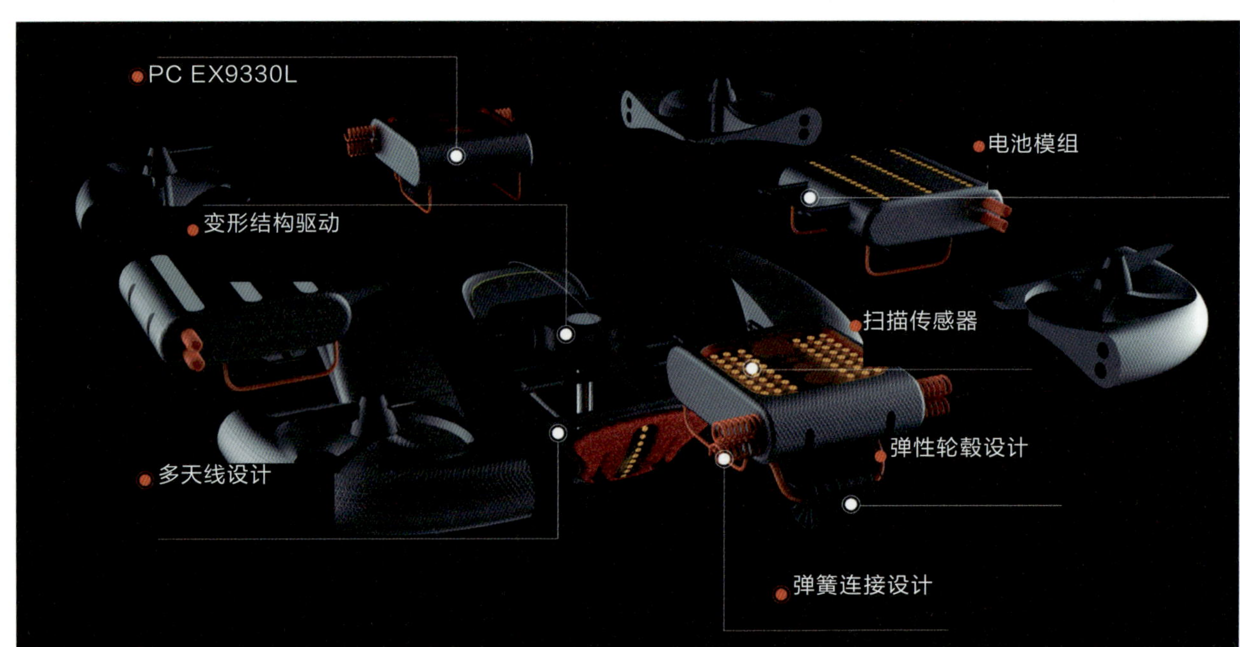


# 迷毂——基于折纸多稳态结构的陆空两用无人机设计

## MI GU——Design of Land Air Dual Use Unmanned Aerial Vehicle Based on Folding Multi Steady State Structure

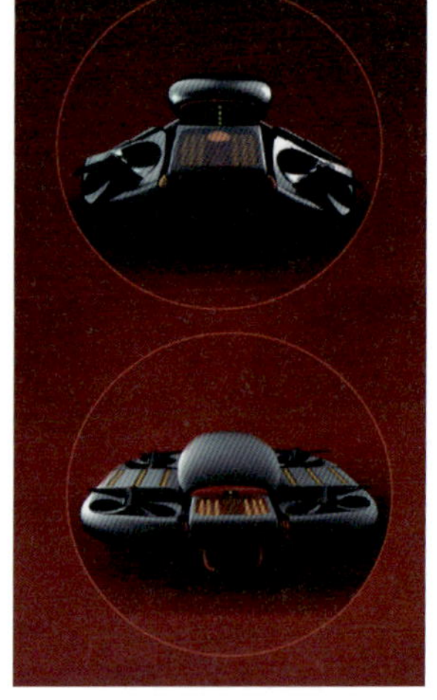

着陆形态

飞行形态

作　　者：杜嘉豪
指导老师：谢　黎
所在院校：东莞理工学院

### 设计说明

本次产品作为矿洞搜寻辅助无人机，具备陆空两用和低空飞行能力，结构稳定且续航性能良好，在不同环境中都能有效地进行搜寻，并扩大搜寻范围，从而更快速地找到被困人员。同时，通过无人机集群中继功能，可实现更稳定的远距离信号通信。相较于传统矿洞机器人，这款陆空两用无人机具有无法比拟的越障能力。

### Design notes

As an auxiliary UAV for mine search, this product has the ability of land and air dual-use and low altitude flight, stable structure and good endurance performance. Using its characteristics, it can effectively search in different environments, and expand the search area, so as to find trapped people more quickly. At the same time, through the UAV cluster relay function, more stable long-distance signal communication can be achieved. Compared with traditional mining robots, this land-air dual-purpose UAV has unparalleled obstacle crossing ability.

# 自动化海产养殖装置设计

## Automatic Mariculture Device Design

装填物料　设定指令　运行状态

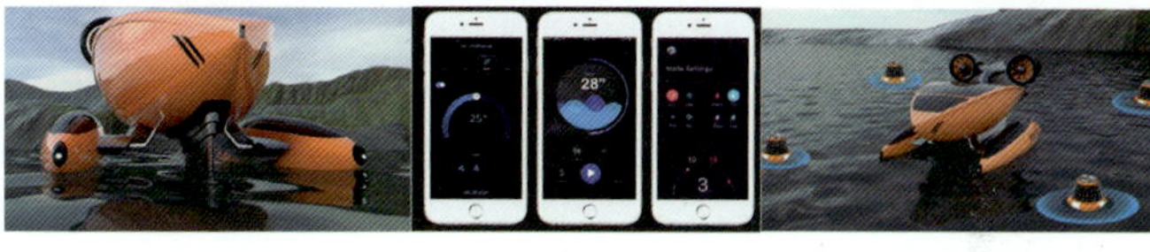

执行指令　数据接收　与制氧机协作

作　　者：李　博
指导老师：杨永福
所在院校：广西大学

### 设计说明

本设计旨在实现以下功能：自动化喂养、精确药物投放、水质检测、环境跟踪监测以及数据共享等。对于渔民来说，该产品可以大幅节省劳动力，提高收益，让养殖过程更加科学和放心。对于科研人员而言，能够实时掌握监测数据，大大降低调研成本，并显著提高科研效率。

### Design notes

The design aims to achieve the following functions: automatic feeding, precise drug delivery, water quality testing, environmental tracking monitoring and data sharing. For fishermen, the product can significantly save labor, improve earnings, and make the farming process more scientific and reassuring. For researchers, real-time monitoring data can be mastered, greatly reducing research costs, and significantly improving research efficiency.

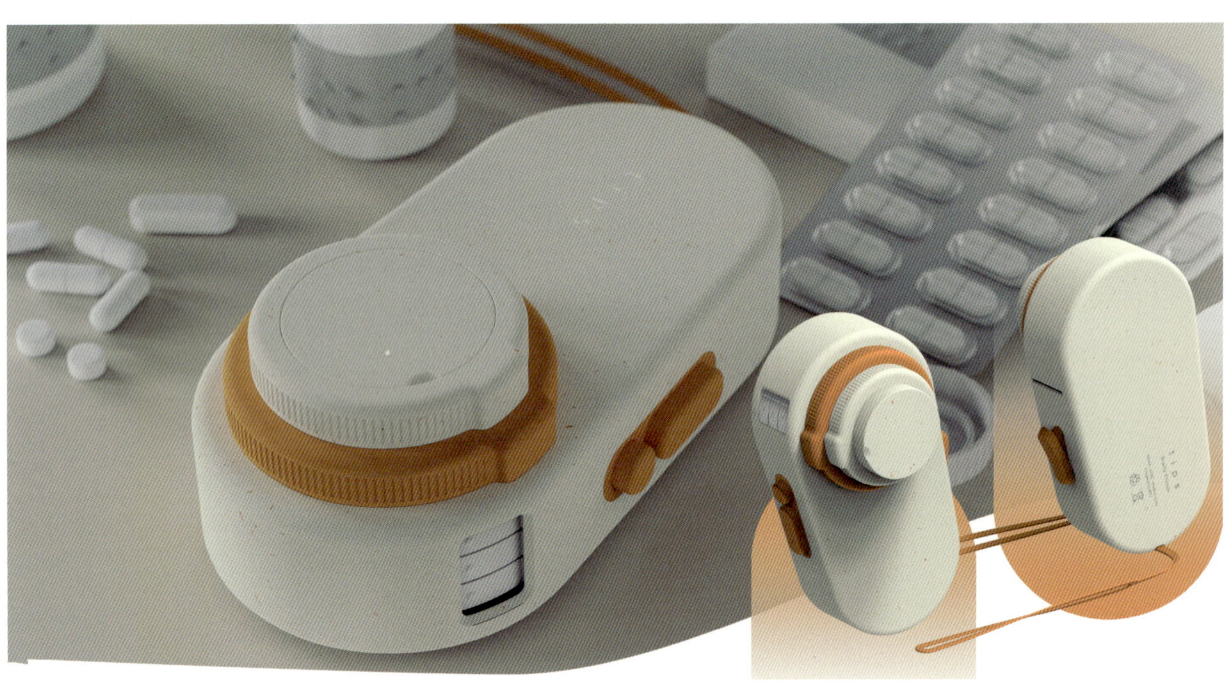

# TIPS——便携式盲文打印机

## TIPS —— Portable Braille Printer

1

胶带进仓示意

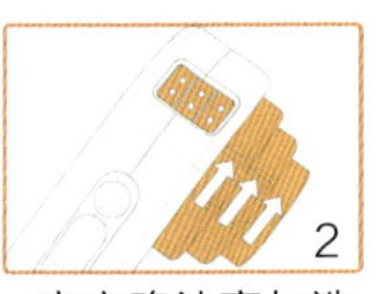

2

盲文确认窗与选择旋钮示意

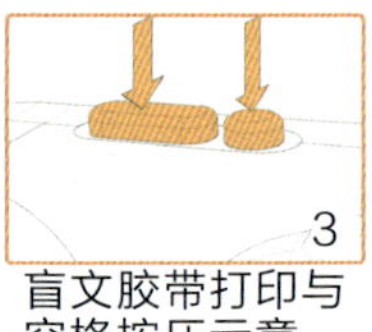

3

盲文胶带打印与空格按压示意

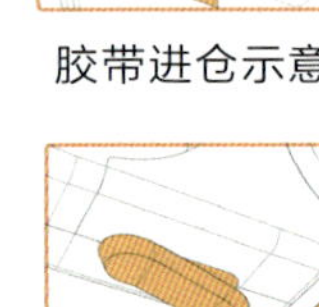

4

盲文胶带切断按压示意

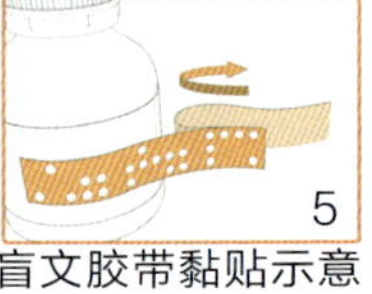

5

盲文胶带黏贴示意

作　　者：唐　帅　张仁颖
指导老师：韦保丞　刘　凯
所在院校：广西科技大学

### 设计说明

Tips 盲文便携式打印机是一款专为视障人士设计的生活标签打印设备。对于视障人士而言，药品、调味料等包装相似但种类繁多的日用品很难通过触摸来快速区分，因此，盲文标签辅助识别显得尤为重要。这款盲文打印机外观小巧，操作便捷。它完全采用机械结构设计，在保证易用性的同时，也降低了电子化带来的高成本和故障率。仅需通过手指简单的确认与按压，便可得到一个具有粘性的塑料盲文标签，从而更好地帮助视障人士优化和标记他们的生活用品。

### Design notes

The Tips Portable Braille Printer is a lifestyle label printing device designed for the visually impaired. For the visually impaired, medicines, seasonings and other daily necessities which similar in packaging but a wide variety of are difficult to distinguish quickly by touch, so Braille label aid recognition is particularly important. This braille printer is compact in appearance and easy to operate. It is completely mechanical structure designed, not only ensuring ease of use, but also reducing the high cost and failure rate brought by electronic. By simply confirming and pressing with your finger, you can get a sticky plastic Braille label that better help the visually impaired optimize and label their daily necessities.

# 大足石刻佛教文化元素桌游玩具设计

## Dazu Stone Carved Buddhist Cultural Elements Board Game Toy Design

作　　者：裴茂童
指导老师：秦　燕
所在院校：重庆理工大学

### 设计说明

经过对现有文创产品特点、桌游市场以及大足历史文化的调研，我们决定将产品定位为线下传统桌游。该游戏以线下经典游戏规则为基础，融入大足宝顶山石刻中的佛教文化元素，并以这些文化内涵和故事情节作为核心来设计游戏规则。在游戏角色的设计上，我们采用大足石刻宝顶山石刻中的佛教文化元素的经典视觉形象，并参考传统壁画的经典配色，如沙黄、赭石、丹青等进行色彩设计。此外，我们选择纸和塑料等性价比高的材料，并运用传统制作工艺，来设计这款经典的线下传统桌游玩具。

### Design notes

After researching the characteristics of the existing cultural and creative products, and the board game market and the history and culture of Dazu, we decided to position the product as a traditional offline board game. The game is based on the offline classic game rules, integrating the Buddhist cultural elements in Dazu Baoding Mountain stone carvings, and using these cultural connotations and storylines as the core to design the game rules. In the design of the game characters, we adopted the classic visual image of Buddhist cultural elements in the Dazu stone carvings of Baoding Mountain, and referred to the classic colors of traditional murals, such as sand yellow, ochre, and Danqing. In addition, we chose cost-effective materials such as paper and plastic, and used traditional manufacturing techniques to design this classic offline traditional board game toy.

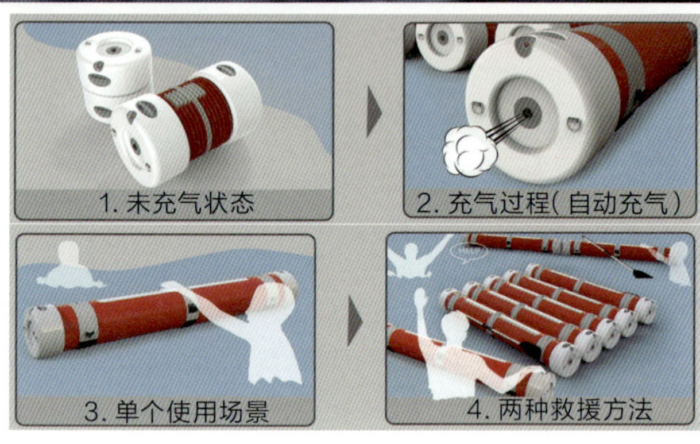


# 基于洪灾自救工具的研究与设计



## Research and Design of Tool Based on Flood Disaster Self Rescue

作　　者：王　茜　郭孚涵　范志娇　刘　洋

指导老师：杨　攀

所在院校：宜宾学院

### 设计说明

这款产品的灵感来自于传统竹筏，其造型经过提炼和简化，以反映竹筏的设计精髓。在产品结构上采用了模块化设计，可拆卸并重新组装，这种结构不仅便于收纳和携带，而且多种组合方式使得使用者在遇到洪水灾害时能够迅速进行自救或互救。功能方面，该产品具备自动充气功能，内置了二氧化碳高压气瓶，用户只需拉开插销，即可刺破气瓶，使气囊迅速充气。

### Design notes

The product is inspired by traditional bamboo rafts, and its shape has been refined and simplified to reflect the essence of bamboo raft design. The modular design is adopted in the product structure, which can be disassembled and reassembled. This structure is not only easy to store and carry, but also has a variety of combination methods which enable users to quickly rescue themselves or each other when they encounter flood disasters. In terms of functions, the product has automatic inflation function, the built-in carbon dioxide high-pressure gas cylinder. The user only needs to pull the latch to puncture the cylinder and inflate the air bag quickly.

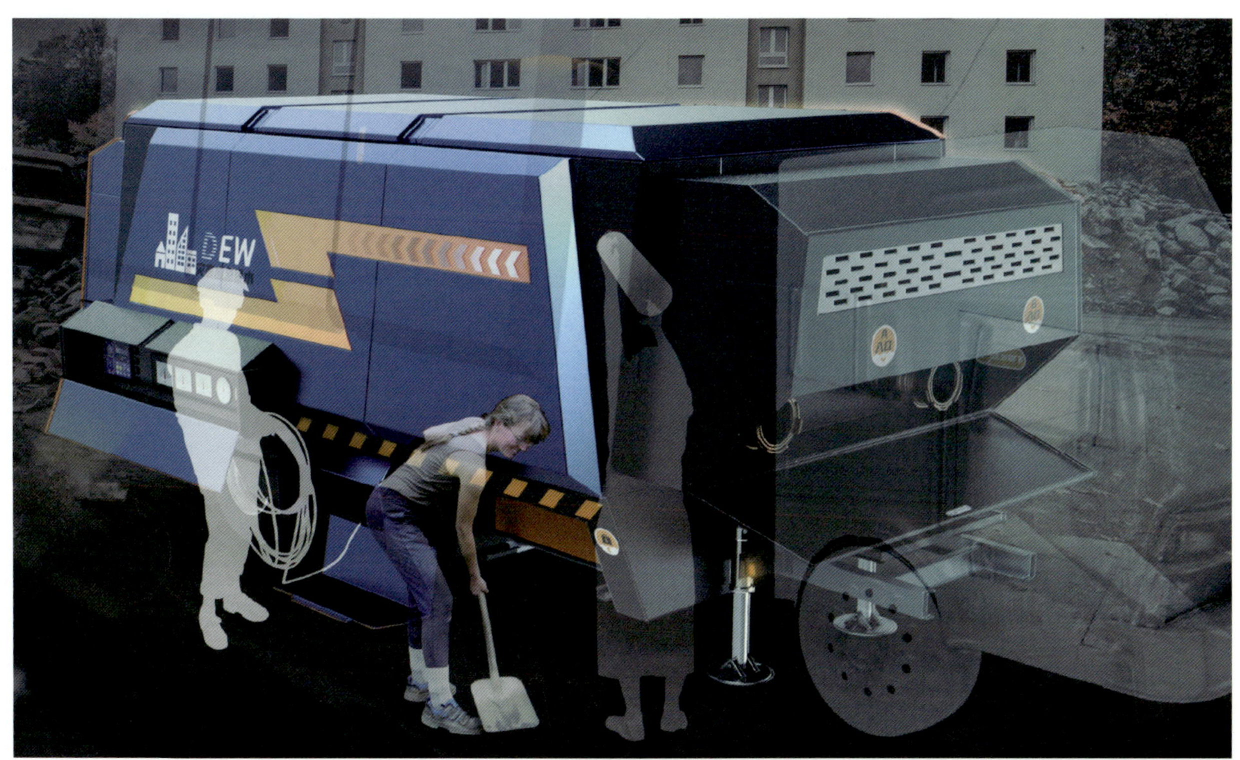




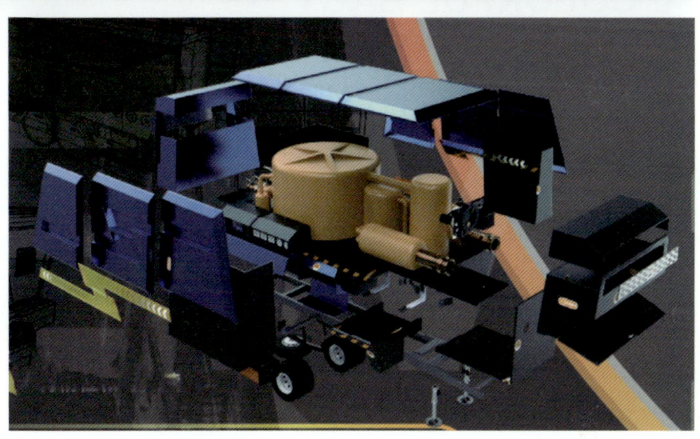

# DEW 地震应急供水设备

## DEW Earthquake Emergency Water Supply Equipment



作　　者：郭策文　宋　李　邱　谣
　　　　　毛钦浩　赵玫橙
指导老师：邓洁茹　刘　源
所在院校：成都工业学院

### 设计说明

DEW 地震应急供水设备是设计用于自然灾害发生期间，当生活用水供应面临威胁时，通过水源的大循环和场地的小循环来提供可持续的临时生活用水设备。该设备能够利用裂隙水、污水、生活废水等多种水源，经过中水回用处理，根据不同的处理等级，将处理后的水输送至避难受灾群众使用。

### Design notes

DEW Seismic Emergency Water Supply Equipment is designed through large circulation of water source and small circulation of the site to provide sustainable temporary domestic water supply during natural disasters when domestic water supply is under threat. The equipment can use fissure water, sewage, domestic waste water and other water sources. According to different treatment levels, the treated water is transported to the shelter for the use of the affected people after the reuse recycled water was treated.

# 彩宝——儿童垃圾分类意识养成的教学产品设计

## COLORFUL BABY —— Teaching Product Design for Children's Awareness of Garbage Sorting



作　　者：刘桂江　卫　哲　高天依
严志民　彭诗文
指导老师：许　佳
所在院校：昆明理工大学

### 设计说明

在20世纪末，美国心理学者马滕斯等人提出了“表象训练”理论。这是一种有效的心理训练方法，它通过使用语言、动画和形象的引导来唤起已有的运动记忆和感知形象。学龄前儿童具有强烈的好奇心、模仿能力、想象力和记忆力，这些特质非常适合应用在表象训练的评估和教学过程中。例如，“彩宝”学龄前儿童垃圾分类教学产品就能够将游戏过程转化为学习体验。该产品顶部配备了识别装置，可以在教师使用的教具中插入相应的识别芯片，这使得儿童在互动时能够重复基础的视觉表象，从而产生无意识和直觉的反应。教师可以通过调整游戏的难易程度来创造一个特定的互动环境，指导学龄前儿童区分不同种类的垃圾，并对垃圾分类的标准有正确的认知。

### Design notes

At the end of the 20th century, American psychologist Martens and others put forward the theory of "representation training". This is an effective mental training method, which evokes the existing motion memory and perceptual image by using language, animation and image guidance. Preschool children have strong curiosity, imitation ability, imagination and memory, which are very suitable for the evaluation and teaching process of image training. For example, "Colorful Baby" preschool children garbage classification teaching product can turn the game process into a learning experience. The product is equipped with a recognition device at the top, and the corresponding recognition chip can be inserted into the teaching aids used by the teacher, which allows the child to repeat the basic visual representation when interacting, resulting in an unconscious and intuitive response. Teachers can create a specific interactive environment by adjusting the difficulty of the game, guide preschoolers to distinguish between different kinds of garbage, and have a correct understanding of the criteria for garbage classification.

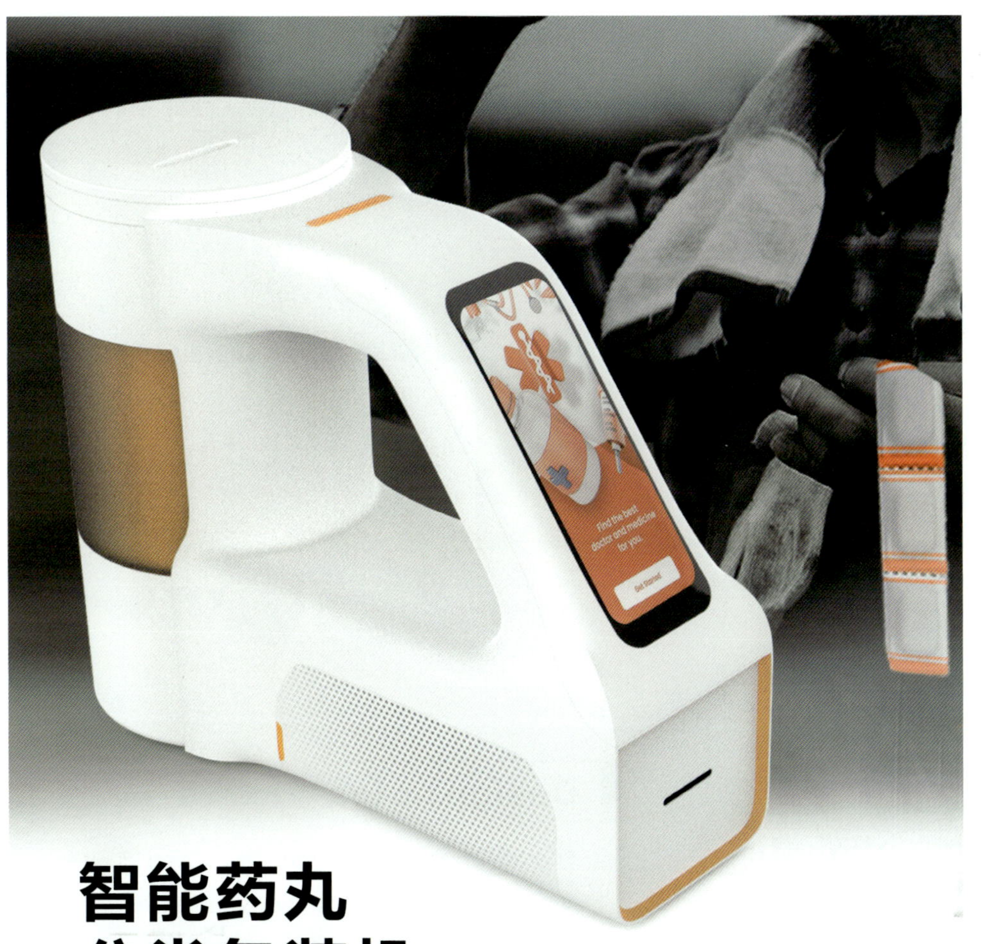

# 智能药丸分类包装机

## Intelligent Pill Sorting and Packaging Machine

作　　者：孙晋琪　高向阳　李晗希
董孟含　牟泽宇
指导老师：许　佳
所在院校：昆明理工大学

### 设计说明

药箱能够通过扫描处方单或药盒上的条码，将药物名称、剂量等信息输入系统，并进行防潮防尘存储。同时，它还能在预设时间提醒用户服药。用户可以选择立即服药，或者选择让系统将当天需要服用的药物进行打包。打包机根据所需剂量将药物打包好，以便外出携带或在环境较差的地方使用，旨在为用户提供一个便捷且卫生的用药体验。

### Design notes

The medicine box can enter the drug name, dosage and other information into the system by scanning the barcode on the prescription sheet or the medicine box, and store it against the moisture and dust. At the same time, it can remind the user to take the medicine at a preset time. Users can choose to take the medication immediately, or choose to have the system pack the medication they need to take that day. The baler packs the medicine according to the required dose for carrying out or use in poor environments, aiming to provide users with a convenient and hygienic medication experience.

# 多功能轮式越野救援森林消防车概念设计

## Concept Design of Multi-Functional Wheeled Off-Road Rescue Forest Firefighting Truck

作　　者：胡　振
指导老师：王　毅
所在院校：西安理工大学

### 设计说明

越野森林消防车具备适应城市道路及户外复杂地形如山路和沙地的强大越野性能，能够最大限度减少地形对消防车快速到达火灾现场的影响，并确保车辆的安全迅速行驶，以便快速展开灭火救援行动。此外，它可以搭载多种灭火装备和工具以及救援设备，以满足不同救援场景下的需求。为适应新时代“智慧消防”的要求，越野森林消防车配备了多种先进电子设备和车载多功能消防无人机。无人机可携带气体分析装置、红外影像采集装置和倾斜摄影装置等多种任务载荷，实现空天一体化的救援系统。这些设备能够快速分析火情，搜索人员，并为消防救援提供准确实时的信息。通过与母车的连接管道，还能提高无人机的续航能力。

### Design notes

The off-road rescue forest firefighting truck has a strong off-road performance to adapt to urban roads and outdoor complex terrain such as mountains and sandy roads, which can minimize the impact of terrain on the firefighting truck's rapid arrival at the fire scene, and ensure the safety and rapid travel of the vehicle in order to quickly launch fire fighting and practice rescue operations. In addition, it can carry a variety of fire fighting equipment and tools as well as rescue equipment to meet the needs of different rescue scenarios. In order to meet the requirements of "smart firefighting" in the new era, the off-road forest firefighting truck is equipped with a variety of advanced electronic equipment and multifunctional on-board firefighting drones. The UAV can carry a variety of mission loads, such as gas analysis device, infrared image acquisition device and tilt camera device, to achieve an integrated rescue system of space and sky. These devices can quickly analyze fires, search for people, and provide accurate and real-time information for fire fighting. Through the connection pipe with the mother vehicle, it can also improve the endurance of the UAV.

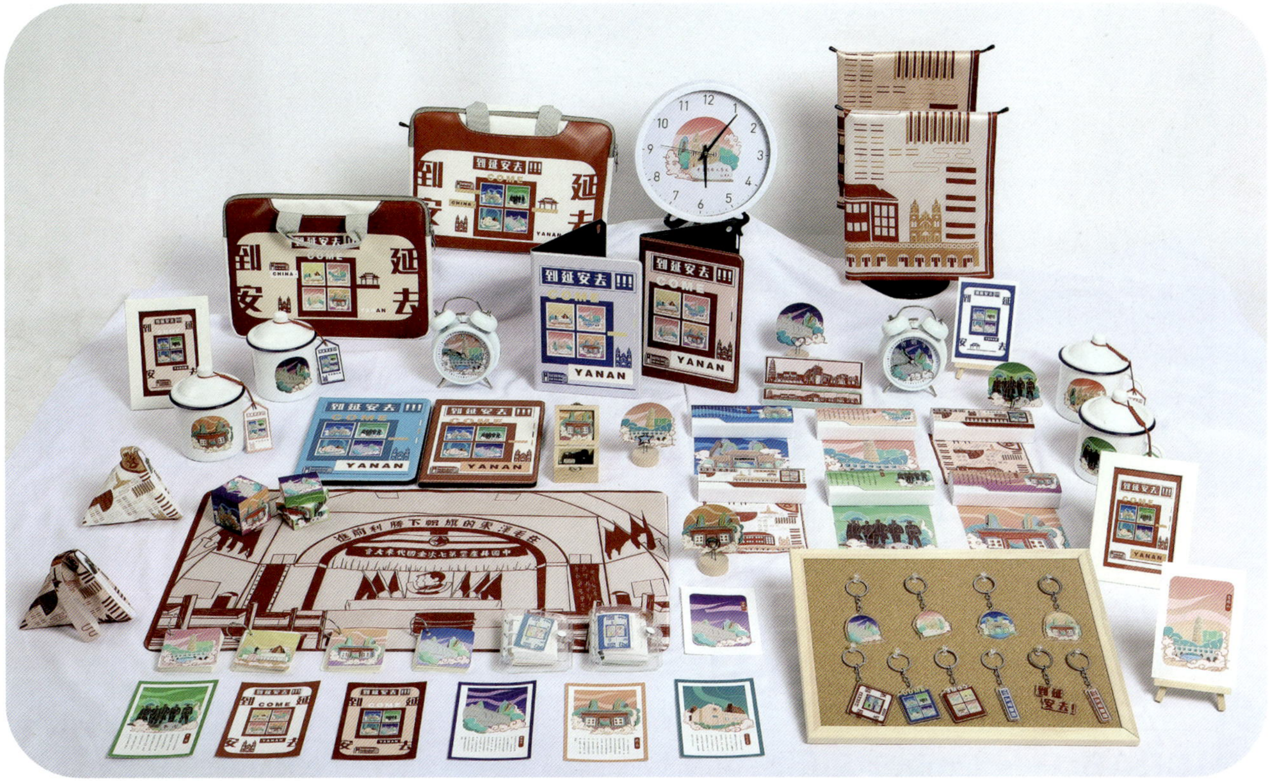



# 到延安去——红色文化创意产品设计

## TO YAN'AN —— Red Cultural Creative Product Design

作　　者：朱　琰　董晨欣　郭文静
　　　　　殷　毅
指导老师：邱春婷
所在院校：西安工程大学

## 设计说明

本作品的设计灵感来源于延安的红色革命建筑，通过插画形式进行创作，并开发出符合现代审美的系列文化创意产品。这不仅能加深人们对延安这座城市的认识和探索，还能有效宣传延安，提高当地文化创意产品的市场价值。在色彩运用上，选择了高饱和度的颜色以增强趣味性。在日常应用上，以生活用品为主要载体，提升了产品的实用性。使用这些文创产品时，人们不仅能够感受到延安的文化底蕴，还能享受到它们的实用价值，从而更有效地传播和弘扬延安精神。

## Design notes

The design inspiration of this work comes from the Red Revolution buildings in Yan 'an, and through the form of illustration, it develops cultural creative products in line with modern aesthetics. This can not only deepen people's understanding and exploration of Yan 'an city, but also effectively promote Yan 'an and improve the market value of local cultural and creative products. In the use of color, choose high saturation color to enhance the interest. In the daily application, using daily necessities as the main carrier, improve the practicality of the product. When using these cultural and creative products, people can not only feel the atmosphere of culture, but also enjoy their practical value, so as to spread and carry forward the Yan 'an spirit more effectively.

# CRAB——下肢外骨骼设计

## CRAB —— Lower Limb Exoskeleton Design

作　　者：史怡铭　景潇蒙　宋　娟
　　　　　黄凯月　闫　敏
指导老师：丁西蓓　苏　胜
所在院校：西安工业大学

### 设计说明

CRAB 下肢外骨骼能显著增强徒步携行者的负重能力。该外骨骼关节设计的关键之处在于结合了螃蟹类生物的关节特点和人体的运动特性，以提升灵活性。膝盖部分的转动结构由两个对称配置的反向半圆形构件组成，当这些半圆相对移动时，它们会在输出端生成特定的运动轨迹。这种设计使得膝盖关节的转动轴实现滚动式的相对运动，有效降低摩擦并确保运动的精确性。通过刚性结构与柔性绑缚相结合的设计，不仅支持穿戴者进行爬行、深蹲、举重等一系列动作，还能减轻因举重物造成的肌肉和骨骼损伤。这样的设计大幅提高了使用者的负重能力，延长了持续机动作业的时间，提升了运动速度，从而有效解决了徒步携行能力不足的问题。

### Design notes

CRAB Lower limb exoskeleton can significantly enhance the load carrying capacity of walkers. The key to the design of the exoskeleton joint is to combine the joint characteristics of the crab and the movement characteristics of the human body to improve flexibility. The rotating structure of the knee part is composed of two reverse semicircular members with symmetrical configuration. When these semicircles move relatively to each other, they generate specific motion trajectories at the output. This design allows the rotating axis of the knee joint to achieve rolling relative motion, effectively reducing friction and ensuring accurate motion. Through the combination of rigid structure and flexible binding design, it not only supports the wearer to crawl, squat, weightlifting and a series of actions, but also reduces the muscle and bone damage caused by heavy lifting. Such a design greatly improves the user's loading capacity, prolongs the continuous mobile operation time, improves the movement speed, and effectively solves the problem of insufficient carrying capacity on foot.

# 疫情环境下的社区防疫产品系统设计

## System Design of Community Epidemic Prevention Products in the Epidemic Environment

作　　者：屈春园　王　晶　郑小琴
指导老师：王　馨
所在院校：西安美术学院

### 设计说明

该设计方案旨在改善社区防疫管理方式，提高工作效率。设计主要由三个部分构成：核酸检测基站、智慧手环和易社区 APP 应用。智慧手环配合易社区 APP，便于社区居民和物业管理者日常使用。通过这一组合，他们可以进行远程交流、实时发布信息以及实现平台化管理。佩戴智慧手环的居民可以展示自己的核酸检测状态和健康情况，这为老年人、儿童和租户等特殊群体提供了无障碍的使用体验。核酸检测基站的设计目的是简化检测流程，它通过集成化的检测设备，对信息可视化、消毒、棉签棒的切除、包装去除、已检测试管的打包和储存等方面进行了优化。居民只需将手环靠近检测基站，即可实现自动数据识别，这不仅提升了居民的检测体验和准确性，也为整个社区带来了更加平台化和科学化的防控管理。

### Design notes

The design scheme aims to improve community epidemic prevention management and improve work efficiency. The design mainly consists of three parts: nucleic acid detection base station, smart bracelet and easy community APP application. Smart bracelet with easy community APP is easy for the community residents and property managers'daily use. Through this combination, they can communicate remotely, publish information in real time, and realize platform management. Residents wearing smart bracelets can display their nucleic acid test status and health conditions, which provides a barrier-free experience for special groups such as the elderly, children and tenants. The nucleic acid detection base station is designed to simplify the detection process and is optimized for information visualization, disinfection, removal of cotton swabs, packaging removal, packaging and the storage of tested tubes through integrated testing equipment. Residents only need to place the bracelet close to the detection base station to realize the automatic data identification, which not only improves the detection experience of residents and increases the accuracy, but also brings to the entire community a better platform and scientific prevention and control management.

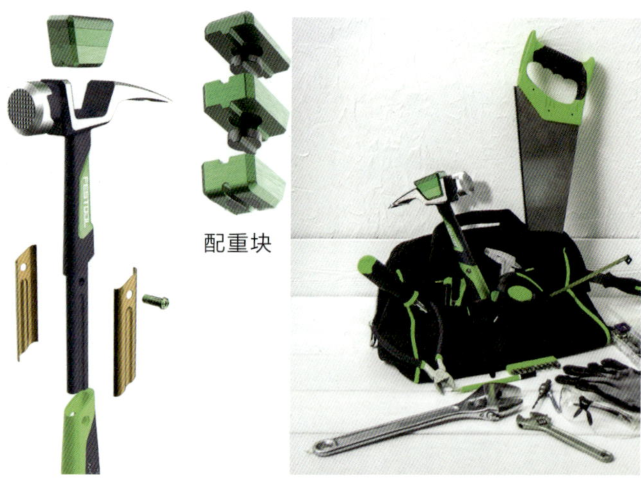


# 自由配重手持工具锤

## Hand Tool Bob-Weight



作　　者：薛　玲　孟怡杉　邹艳妮
指导老师：朱守会　王春霞
所在院校：兰州理工大学

### 设计说明

利用现有品牌产品，结合传统技术并对其进行重新设计，增添新的设计元素，以解决使用锤子时胳膊的疼痛问题，从而达到更佳的使用效果。市场调研揭示了以下问题：钉子种类繁多，工具用途各异；难以确保钉子垂直钉入，不均匀的力道可能导致钉子弯曲；繁多的工具种类给用户带来困扰；握持钉子使用工具时容易造成砸伤或被钉尖扎伤。本产品设计定位于中低端市场，旨在为用户提供优质且舒适的使用体验，突出设计的高品质、安全性和效率，以及符合人体工程学和品牌理念的特点。产品设计的主要创新之处在于改变手柄的形状，使其更加舒适易握，并加入多样的设计元素，允许用户手动调节锤子的配重以适应不同需求。此外，产品的长度与人体小臂长度相近，使操作更安全、便捷。

### Design notes

Using existing brand products with combining traditional technology to re-design a product, adding new design elements in order to solve the problem of pain when hammering, so as to achieve a better result. The market research revealed the following problems: there are many kinds of nails and different kinds of tools.It is difficult to ensure that the nail is driven vertically, and uneven force may cause the nail to bend. A wide variety of tools may bring users trouble. When holding nails and using tools, it is easy to be hit or hurt by the nail tip. The product design is positioned in the middle and low end of the market, aiming to provide users with a high-quality and comfortable using experience, highlighting the high quality, safety and efficiency of the design, as well as the characteristics of ergonomics and brand philosophy. The main innovation of the product design is to change the shape of the handle to make it more comfortable and easy to grip, and with adding a variety of design elements, allowing users to manually adjust the hammer's counterweight to suit different needs. In addition, the length of the product is similar to the length of the human forearm, so the design can better avoid the occurrence of injury in use, making the operation safer and more convenient.

# 消防员自救防护装备设计

## Firefighter Self-Rescue Protective Equipment Design

作　　者：李清月　邢凡凤　耿希媛
郭盼盼　池　钦
指导老师：王　鹏
所在院校：兰州理工大学

### 设计说明

产品设计的初衷是为消防员提供一款智能装备，以便在火灾现场迅速撤离。这款装备包括一个穿戴在消防服外的防护罩，它通过一根可快速回收的绳索与安全区域中的终端相连。当出现紧急情况时，防护面罩上的倒计时功能会提醒消防员做好准备。随后，消防员可以按下装备上的紧急按钮，这时外骨骼装备会从站姿变为蹲姿并锁定所有机械关节点，以确保在拉回过程中结构紧凑，增强集中保护效果。本产品的外部材质坚固，各个可能受到撞击的部位都配备了相应的保护甲，以保护消防员在撤离时避免碰撞伤害。如果消防员失去意识，装备也能感应到人体的生命体征并自动执行上述操作，将消防员拖离火场，确保其在安全区域内得到及时救援。

### Design notes

The original design of the product is to provide firefighters with a smart device to quickly evacuate from the scene of a fire. The device consists of a protective cover worn over a fire suit, which is connected to a terminal in a safe area via a quickly retrievable rope. In the event of an emergency, a countdown function on the protective mask will alert firefighters to be ready. The firefighter can then press the emergency button on the suit, at which point the exoskeleton changes from a standing position to a squatting position and locks all mechanical joints to ensure a compact structure and enhanced centralized protection function during pull-back. The exterior material is strong, and each part that may be impacted is equipped with corresponding protective armor to protect firefighters from collision injuries when evacuating. If the firefighter loses consciousness, the equipment can sense the vital signs of the human body and automatically perform the above operations, dragging the firefighter away from the fire, ensuring that they can be rescued in a safe area.

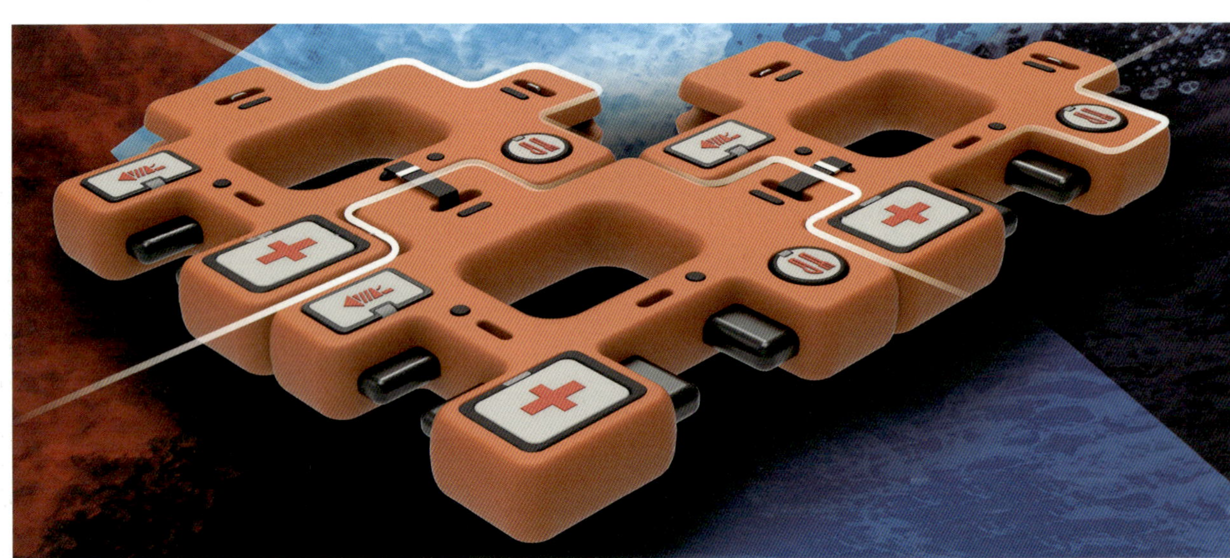

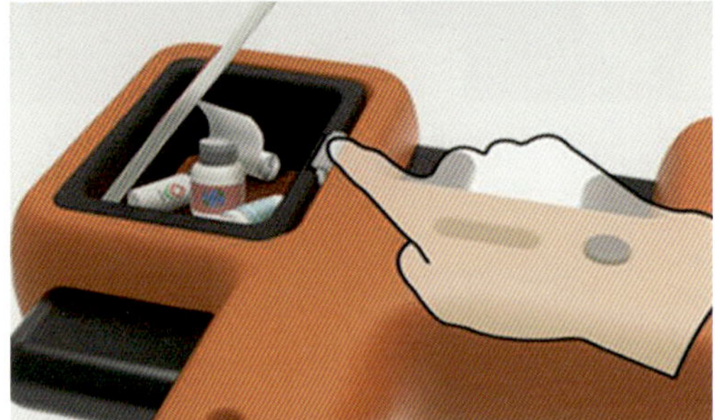

# 救援模块

## Emergency Module

作　　者：王贺祺　马欣怡　杨　宇
　　　　　吕柯飞　张　凡
指导老师：苏建宁　杨文瑾
所在院校：兰州理工大学

### 设计说明

该急救模块是一款旨在用于海上救援和预防儿童常见溺水的水上浮板。在海难发生时，其模块化设计允许快速拼接，从而提升救援效率。此外，它在日常生活中也能有效防止儿童在玩水时发生溺水危险。传统的救生圈在海上难以保持平衡，且在许多事故中可能数量不足，无法满足需求。此外，使用救生圈的落水者往往无法与其他人建立联系。这一独特设计融合了拼图元素和拼接技术，使得在海上事故发生时，可以将它拼接成救生筏，以提高救援的效率和救助的人数。在非紧急情况下，它还可作为普通的游泳辅助设备使用。

### Design notes

The first aid module is a water board designed for sea rescue and prevention of daily drowning of children. In the event of a maritime disaster, its modular design allows for quick splicing, which improves rescue efficiency. In addition, it is also effective in daily life to prevent the risk of drowning when children play in water. Traditional life preservers are difficult to balance at sea and may not sufficient to meet the demand in many accidents. Also, people who fall into the water using life preservers are often unable to connect with others. This unique design incorporates jigsaw elements and splicing technology, making it possible to splice it into a life raft in the event of an accident at sea to improve the efficiency of the rescue and increase the number of people rescued. In non-emergency situations, it can also be used as an ordinary swimming aid.

# 又遇青绿——文房系列用品设计

## MEET GREEN AGAIN——The Design of the Stationery Series for the Study

作　　者：廖文皓　王宇际　王芳林
指导老师：闫文奇　赵志明
所在院校：新疆大学

### 设计说明

本系列产品包括笔、墨、纸、砚、笔山、笔架和笔帘等文房用品，设计灵感源自中国传统山水画，旨在迎合文人雅士的高雅品味。设计主题以青绿山水为主调，借鉴了王希孟《千里江山图》的元素进行产品造型的创新。在创作过程中，我们遵循“减法”原则，即从原画中提炼核心元素，将画中独特的山水色彩和形态融入产品设计之中，这一点构成了本系列产品的独有特色。产品的造型风格采用山水的色彩渐变来表现山峰的起伏与层次感，同时注重扁平化设计的“形”，追求传统文房用品的时尚化表达。通过将传统元素结合到传统用具中，展现了本系列产品独特的传统文化魅力。

### Design notes

This collection includes Chinese writing brush, ink, paper, inkstones, pen hills, pen holders and pen curtains, which are inspired by traditional Chinese landscape painting and designed to cater to the refined tastes of literati. The design theme is green landscape as the main tone, drawing on the elements of Wang Ximeng's "A Thousand Li of Rivers and Mountains" for product modeling innovation. In the process of creation, we follow the principle of "subtraction" , that is, extract the core elements from the original painting, and integrate the unique landscape colors and forms in the painting into the product design, which constitutes the unique characteristics of this series of products. The style of the product uses the gradient of landscape color to express the ups and downs of the mountain peaks and layers, and pays attention to the "shape" of flat tening design, pursuing the fashionable expression of traditional stationery. By combining traditional elements into traditional utensils, the products of this series show the unique charm of traditional culture.

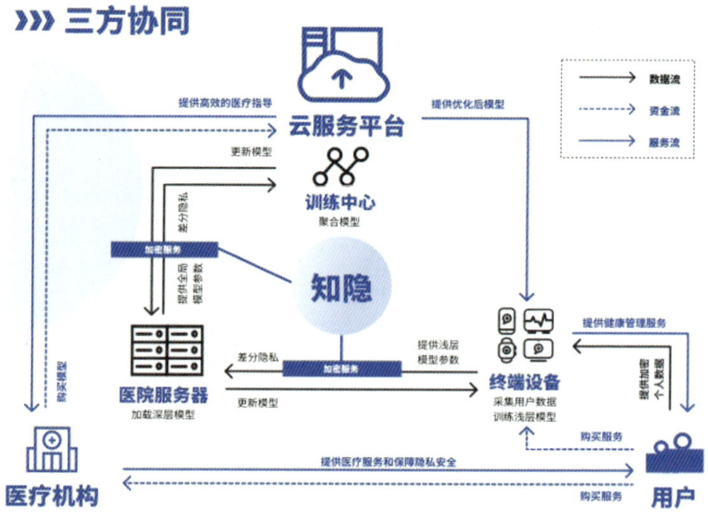

# 知隐：医疗数据隐私保护服务系统设计

## FEEL: Federated Edge Learning System for Medical Data Privacy

作　　者：崔盛兰　吕君妍　张默涵
　　　　　郭烨婷　李碧瑶
指导老师：刘　芳
所在院校：湖南大学

### 设计说明

随着人工智能领域的发展，基于深度学习的 AI 医疗模型已成为帮助医生进行快速、精准诊断的重要工具。然而，这些 AI 模型的训练需要大量的医疗数据，而在数据训练和分析过程中，敏感和隐私数据的安全问题尤其引人关注，特别是对于医疗需求较为频繁的人群，如儿童和老人。因此，我们面临的首要问题是如何在保护敏感数据的同时高效地利用它们来构建智能医疗系统。这个系统能够让用户在确保敏感隐私数据安全的前提下参与到 AI 模型的训练中，以获得高效的 AI 辅助诊疗服务。本系统旨在为智慧医疗领域的用户和医疗机构提供服务，实现医疗行业隐私数据的保护与利用。通过这种方式，本系统能够为用户提供安全、精准、专业的医疗服务，同时守护用户的健康与隐私。

### Design notes

With the development of the field of artificial intelligence, AI medical models based on deep learning have become an important tool to help doctors make fast and accurate diagnoses. However, the training of these AI models requires a large amount of medical data, and the security of sensitive and private data is of particular concern during the data training and analysis process, especially for populations with more frequent medical needs, such as children and the elderly. Therefore, the primary question we face is how to efficiently use sensitive data to build smart healthcare systems while protecting them. Such a system can allow users to participate in the training of AI models while ensuring the security of sensitive private data, so as to obtain efficient AI-assisted medical services. This system is designed to provide services for users and medical institutions in the field of smart medic at treatmentm, protect and utilize private data in the medical industry. In this way, the system can provide users with safe, accurate and professional medical services, while protecting the health and privacy of users.

# 智慧应急消防站人机交互界面设计

## Intelligent Human-Computer Interaction Design of Emergency Firefighting Station

作　　者：杨佳伟
指导老师：谭　浩
所在院校：湖南大学

### 设计说明

在各种突发事件的实际救援过程中，特别是消防应急救援中，依然有很多需要解决的问题，例如：（1）响应时效滞后。（2）救援设施建设不平衡。（3）灾情复杂多样。为解决这些问题，提出了智慧应急消防站的概念。这种消防站能够灵活快速地按需组装，并广泛部署于社区和乡镇，以便在第一时间检测灾情和实施救援。智慧应急消防站的装备能够主动进行状态监测和诊断，支持远程操控，并且在设备管理和使用上不受专业人员限制，实现了面向全民的应急管理理念。智慧应急消防站的人机交互界面设计以运营和维护等环节为中心，注重可视化设计，分为四个主要功能模块：无人机巡检、设备监控、应急救援（接收告警信息和发布救援任务）、宣传管理。设计目标是实现功能结构清晰、界面风格简洁，以此降低使用成本，适应全民参与应急管理的趋势。

### Design notes

In the actual rescue process of various emergencies, especially in fire emergency rescue, there are still many problems to be solved, such as: (1) response time lag. (2) Unbalanced construction of rescue facilities. (3) The disaster situation is complex and diverse. In order to solve these problems, the concept of intelligent emergency fire station is put forward. The stations can be assembled flexibly and quickly on demand, and are widely deployed in communities and towns to detect and respond to disasters in the first place. The equipment of the intelligent emergency firefighting station can actively carry out status monitoring and diagnosis, support remote control, and is not restricted by professionals in equipment management and use. It realized the emergency management concept for the whole people. The human-computer interaction interface design of the intelligent emergency firefighting station is focused on operation and maintenance and pays attention to visual design. According to the workflow, the human-computer interface is divided into four main functional modules: UAV inspection, equipment monitoring, emergency rescue (receiving alarm information and publishing rescue tasks), and publicity management. The design goal is to achieve clear functional structure and simple interface style, so as to reduce the operation cost and adapt to the trend of national participation in emergency management.

a. 画板：辅助画画、作为互动游戏的桌面

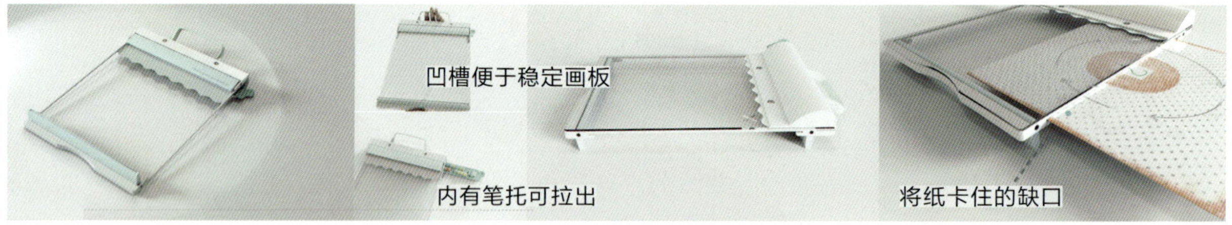


b. 画册：为老人提供可描画的绘画教学与游戏道具

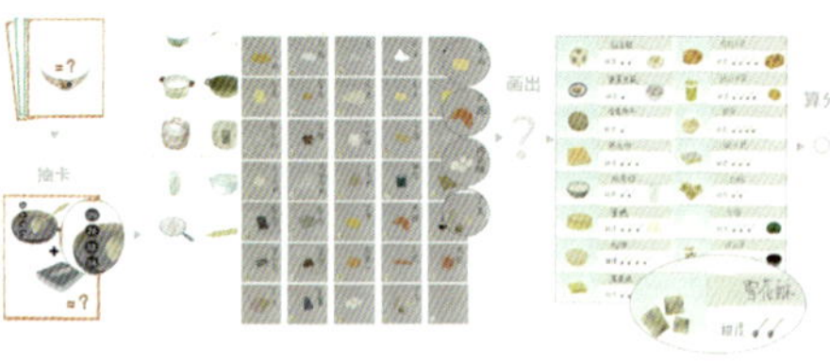

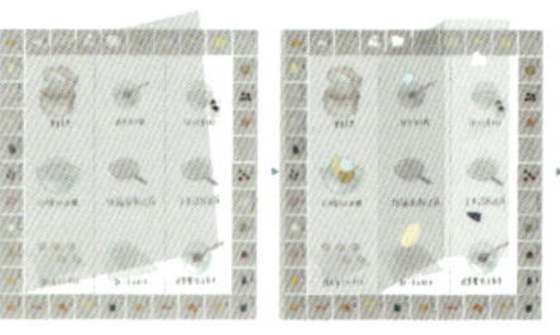

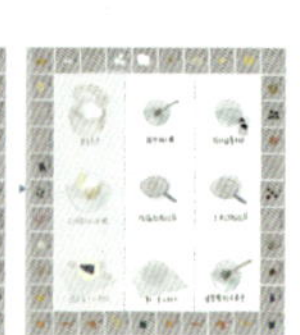

# 以画会友——画板配套设计

## TO MEET FRIENDS THROUGH PAINTING —— Drawing Board Design

作　　者：曾月曦
指导老师：柏　雷
所在院校：南京艺术学院

### 设计说明

本设计专为中高龄老年人群打造一款适用于公共空间的画板以及配套的辅助学习手册。在设计过程中，特别关注老年人的使用场景，将他们的社交特性和生活经验相结合，通过社交互动帮助他们更好地理解周遭的世界。

### Design notes

This design is made to create a drawing board and auxiliary learning manual suitable for public space for middle-aged and elderly people. In the design process, special attention is paid to the use of the elderly, combining their social characteristics and life experience, aiming to help them better understand the world around them through social interaction.

# 木兰说——非遗文化盲盒互动玩法

## THE STORY OF MULAN —— Intangible Cultural Heritage Blind Box Interactive Play

作　　者：余欣宁　费苏儿　徐润禾
指导老师：武奕陈
所在院校：中国美术学院

## 设计说明

非物质文化遗产是社会宝贵的精神瑰宝。本设计旨在运用现代宣传手段，打造面向全民的非遗参与性活动，以促进非遗传承方式的创新与发展。项目从B端出发，即以非物质文化遗产为内容的商家，拉动社会公益组织，通过“点”“线”“面”——“盲盒玩法”“木兰说活动”“全名非遗传承活动”进行文化推广设计，积极履行设计师的社会责任。

受到“古风表演”“国产动画”“国潮糕点”“国风服饰”等社会热点的启发，我们提出了“国潮4.0”这一全新概念。通过结合“国魂”与“潮魂”，设计了非物质文化遗产花木兰主题的盲盒和IP形象。通过在各大平台发布互动短视频，鼓励用户了解文化背景并获取盲盒，从而吸引全民关注非物质文化遗产，共同成为非遗文化的传承者。

## Design notes

The intangible cultural heritage is a valuable spiritual treasure of the society. This design aims to use modern propaganda means to create a participatory activity for the whole people, so as to promote the way of innovation and development of the intangible cultural heritage. The project starts from the B end, that is, businesses with intangible cultural heritage as their content, move social public welfare organizations, and carry out cultural promotion design through "point" , "line" , "face" —— "blind box play" , "Mulan Story activity" , "full name intangible cultural heritage activity" , and actively fulfill the social responsibility of designers.

Inspired by social hot topics such as "ancient style performance" , "Domestic animation" , "National style pastry" and "national style clothing" , we put forward the new concept of "National Style 4.0" . Through the combination of "national soul" and "tide soul" , the blind box and IP image of the intangible cultural heritage Mulan theme are designed. Through the release of interactive short videos on major platforms, users are encouraged to understand the cultural background and obtain blind boxes, so as to attract the attention of the whole people to the intangible cultural heritage and become inheritors of the intangible cultural heritage together.

通过与手机 APP 互联，远程提前开启

到家后即刻享受温暖环境

打开两侧窗叶，可进行衣物烘干

底部可根据场景，更换不同香型香薰

# 暖窗

## Warm Window

作　　者：陈文文　于仕沛　孙　铭
指导老师：严增新
所在院校：中国美术学院



## 设计说明

这是一款融合了新中式设计的塔式取暖器，其侧面展开的延长板仿佛是打开的古典窗扇。“窗户”部分设计有晾衣功能，使得用户在取暖的同时可以烘干衣物。内置的加热元件和环形出风口提升了取暖器的安全性。此外，取暖器底部还配备了香薰模块，用户可以根据个人喜好选择不同的香薰。创新性体现在以下几个方面：1. 结合了暖风机的形式，相对于全屋加热更为高效，同时考虑到了晾衣和安全性，通过内环出风和独立的出风与晾衣模块，确保了电路安全。2. 环形出风设计使风向更加扩散，减少了直接吹风可能带来的不适感。3. 智能化操作，支持手动控制和 APP 远程控制双模式，用户可以提前开启调温，并接收安全隐患的警报提醒。4. 无火香薰模块的设计，能够为整个房间和衣物带来持久的香气。

## Design notes

It's a tower heater that incorporates neo-Chinese design, with an extension panel that unfolds like an open classical window sash on the side. The "window" section is designed with a laundry drying function, allowing users to dry clothes while keeping warm. The built-in heating element and annular air outlet increase the safety of the heater. In addition, the bottom of the heater is also equipped with aromatherapy module, users can choose different aromatherapy flavors according to personal preferences, and enjoy the experience of aroma lingering in the space. This heater not only provides warmth to the home, but also brings a personal and intimate experience. The innovation is embodied in the following aspects:

1. Combined with the form of the heater, it is more efficient than the whole house heating. At the same time, taking into account the drying and safety, the circuit safety is ensured through the inner ring air discharge and independent air discharge and drying module. 2. Annular air design makes the wind more diffuse, reducing the discomfort that may be caused by direct wind. 3. Intelligent operation, support manual control and APP remote control dual mode, users can turn on the temperature adjustment in advance, and receive the alarm reminder of security risks. 4. The fire free aromatherapy module is designed to bring a lasting aroma to the entire room and clothing.

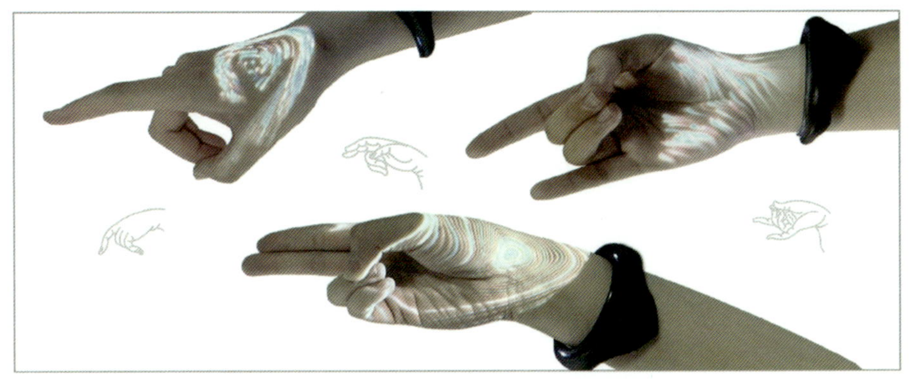

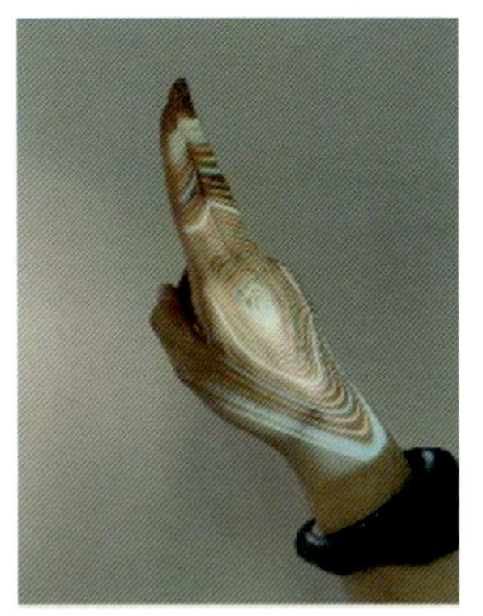



# 光·境——以“敦煌手姿”为交互的投影首饰

## LIGHT REALM —— Interactive Projection Jewelry Based on "Dunhuang Hand Gesture"

作　　者：李兆涵
指导老师：刘震元
所在院校：同济大学

### 设计说明

“光·境”是一款以敦煌手姿为灵感源泉、以光影与手部互动为核心设计理念的投影首饰。它从特定的壁画中提炼出三个元素：轮、触、云，经过数字化处理后，将三种敦煌典型的手姿——持花手、佛手、散花手融入交互体验中。佩戴者通过戴上手钏式的投影配件，可以模仿这些特定的敦煌手姿来解锁和操控光影效果，从而体验到一种非物质的审美感受。

本设计尝试创新性地将可穿戴设备、生成艺术和手势捕捉技术结合在一起，使人们心中的敦煌美景得以感官化和视觉化。这样做不仅增加了视觉上的惊喜和互动的乐趣，还打破了传统“观赏”式首饰展品的体验限制，探索了可穿戴电子设备在首饰领域的应用潜力。我们希望通过不断完善，使这件作品未来能够在博物馆和美术馆等场所得到应用，在实现文化传播的同时，提供寓教于乐的体验。

### Design notes

" Light Realm " is a projection jewelry with Dunhuang hand gestures as its inspiration, the interaction between light and shadow and hands as its core design concept. It extracts three elements from the specific murals: wheel, touch and cloud. After its digitization, three typical Dunhuang hand gestures was integrated——flower holding, Buddha's hand, and flower scattering, into the interactive experience. By wearing a projection accessory in the form of an bracelet, the wearer can imitate these specific Dunhuang hand gestures to unlock and manipulate the effects of light and shadow, thus experiencing an nonmaterial aesthetic feeling.

This design attempts to combine wearable devices, generative art and gesture capture technology in an innovative way, so that the beauty of Dunhuang in people's minds can be sensed and visualized. Doing so not only increases the visual surprise and interactive fun, but also breaks the experience limits of traditional "ornamental" jewelry exhibits and explores the application potential of wearable electronic devices in the field of jewelry. We hope that through continuous improvement, this work can be applied in museums, art galleries and other places in the future, so as to realize cultural communication and provide an educational and entertaining experience.

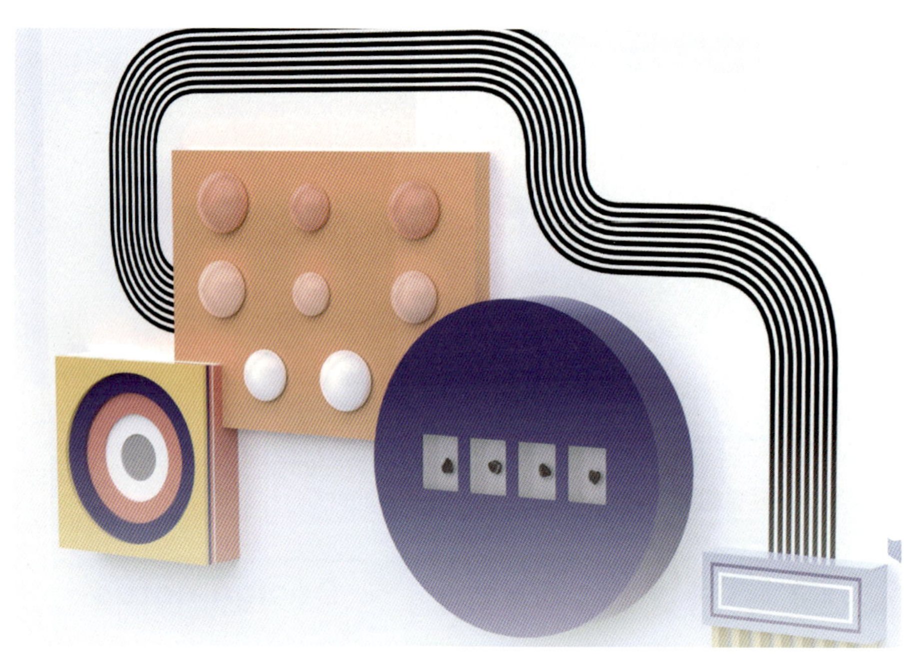

# TOUCH TOUCH——磁流体声音感应装置

## TOUCH TOUCH——Magnetic Fluid Sound Induction Device



作　　者：陈楚怡　吴雅婷
指导老师：周东红
所在院校：中国美术学院

### 设计说明

该装置主要由三个模块组成：距离感应模块、触控感应模块和磁流体模块。它融合了触觉、听觉和视觉三种感官体验，软硬结合的按键设计满足了用户对不同触觉的需求，而钢琴的八音阶和多样的歌曲选择增强了听觉享受。随着音乐节奏跳动的磁流体则提供了视觉上的享受。内部结构以密度板为支架，保证了整体的稳定性；外部用亚克力片覆盖，既确保结构的稳固，又增添了外观的美感。在造型设计上，主要采用了方形和圆形的结合，整体色彩方面，选择了高饱和度的红、黄、蓝三原色，这样的配色不仅中性大方，而且能够营造出活泼和欢快的氛围。

### Design notes

The device is mainly composed of three modules: distance sensing module, touch sensing module and magnetic fluid module. It integrates the three sensory experiences of touch, hearing and vision, the soft and hard combination of key design to meet the needs of users for different senses, while the piano's eight scales and a variety of song choices enhance the auditory enjoyment. Magnetic fluids pulsating to the rhythm of the music provide visual pleasure. The internal structure is supported by the density plate to ensure the overall stability. The exterior is covered with acrylic sheets, which not only ensures the stability of the structure, but also adds the beauty of the appearance. In the shape design, the main use of the combination of square and round, the overall color, the choice of high saturation of red, yellow, blue three primary colors, such a color is not only neutral and liberal, but also to create a lively and cheerful atmosphere.

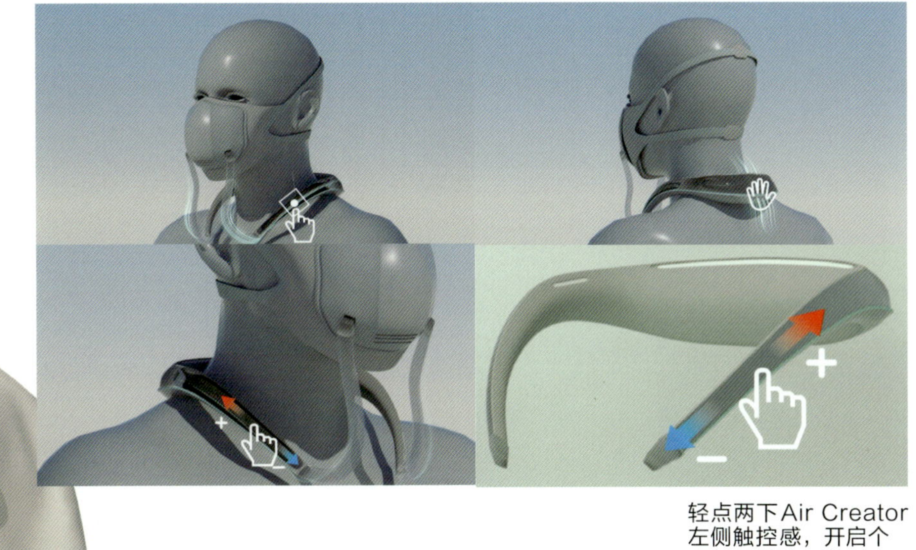

轻点两下Air Creator
左侧触控感，开启个
人新风过滤系统。



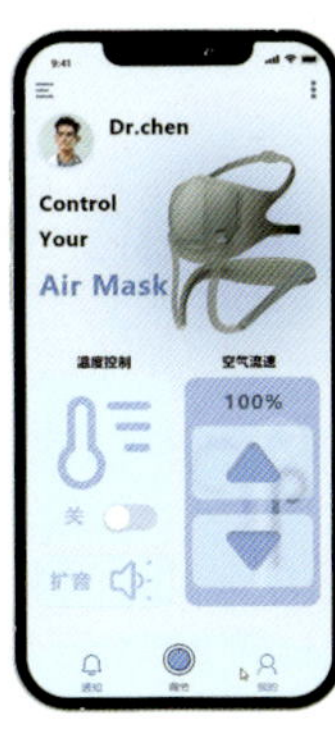


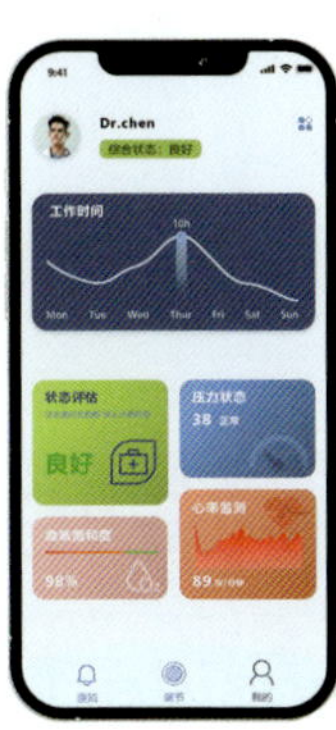

# AIR MASK——智能医用防护面罩

## AIR MASK——Intelligent Medical Protective Mask

作　　者：李高凡　郑翌可　王云淼
　　　　钱　延
指导老师：蒋　晓
所在院校：江南大学

### 设计说明

AIR MASK是一款专为一线抗疫医务人员设计的智能防护面罩，它集成了新风系统、降温功能、病毒防护以及晕厥预警等多种功能，以确保医务人员在工作时的安全与舒适。此外，产品还支持扩音和对讲功能，便于医务人员在工作中进行沟通和即时交流。面罩采用医用级硅胶材质，配合肩带式设计，旨在为长时间工作的医务人员提供舒适的佩戴体验。

### Design notes

AIR MASK is an intelligent protective mask designed for frontline medical personnel. It integrates a variety of functions, such as fresh air system, cooling function, virus protection and syncope warning, to ensure the safety and comfort of medical personnel at work. In addition, the product also supports loudspeaker and intercom functions, which is convenient for medical personnel to communicate and communicate in real time at work. The mask is made of medical-grade silicone material with shoulder strap design to provide a comfortable wearing experience for medical personnel who work for a long time.

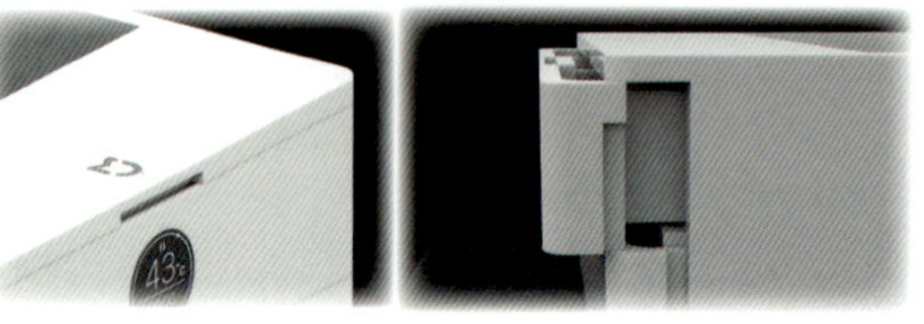

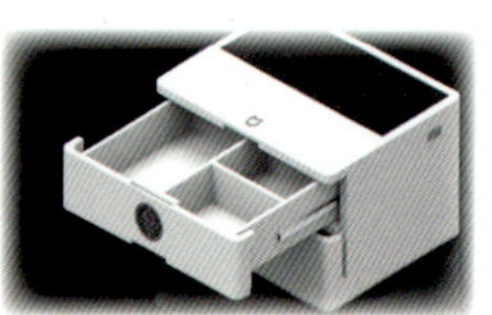

# 智能床头柜

## Smart Bedside Table

作　　者：王昱夫　王添运　黄佳文
　　　　　李　治
指导老师：周　岩
所在院校：哈尔滨工业大学

### 设计说明

本款智能床头柜旨在改善贴身衣物卫生条件和穿着体验问题。内置碳晶加热板和紫外线灯可加热杀菌，拥有定时、即时两种加热模式，满足您的不同需求；简约美观的外观设计、简单易操作的人机交互将优化您的使用体验。您在每个清晨都可以享受这份贴心的温暖！

### Design notes

This smart bedside table is designed to improve the hygiene and wearing experience of intimate clothing. Built-in carbon crystal heating plate and ultraviolet lamp. Take into account heating and sterilization, there are two heating modes to meet your different needs, timed and instant; simple and beautiful appearance design, simple and easy-to-operate human-computer interaction will optimize your experience. You can enjoy this intimate warmth every morning!

# 智能床头柜

## Intelligent Bedside Cupboard

作　　者：王添运　黄佳文　王昱夫
　　　　　李　治　李文希
指导老师：周　岩　朱　磊
所在院校：哈尔滨工业大学

## 设计说明

本智能床头柜优势在于大储存空间搭配合理分区，生活物品井然有序。再加上为贴身衣物专门开发的加热模组，让身体不再受寒冷侵扰。台面顶端配备大功率无线充电模块，不再需要将充电器摆在桌面上，生活更加随意、从容。侧面配有双孔快充插座和USB充电接口。床头从此电力十足。

## Design notes

Large storage space with reasonable zoning makes living items in order. Coupled with the specially developed heating module for closely fitting clothes, your body is no longer disturbed by cold. The top of the cupboard is equipped with a high-power wireless charging module, which eliminates the need to place the charger on the cupboard, making life more casual and leisurely. Two hole quick charging socket and USB charging interface are equipped on the side. The bedside is now fully powered.

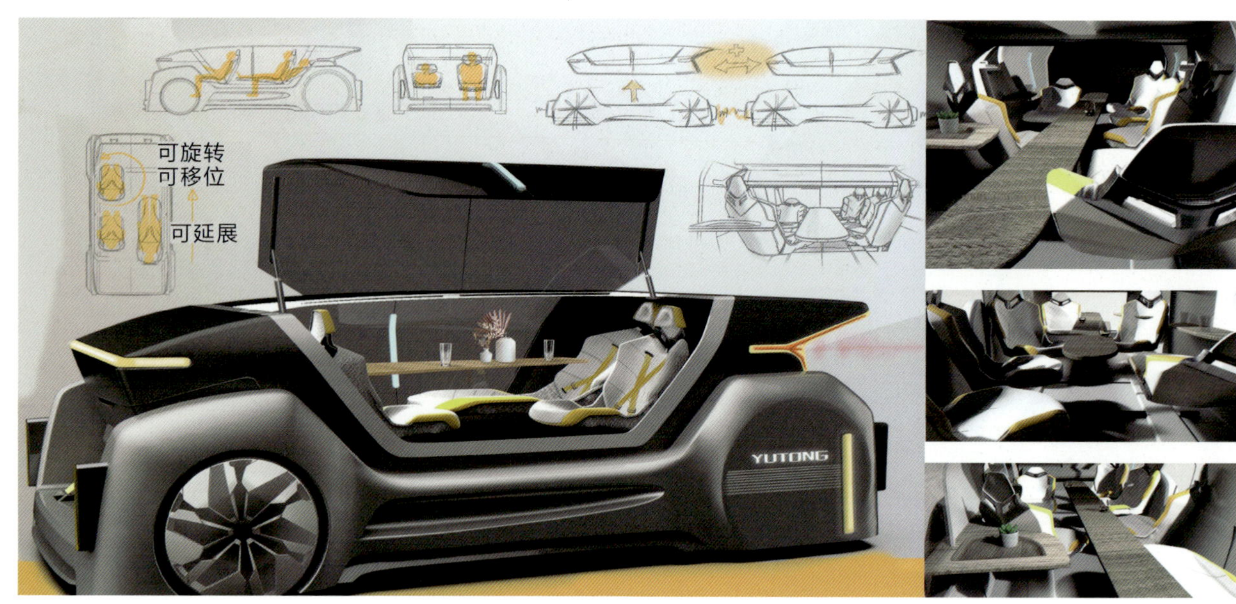


# ONE PLUS ONE ——未来移动娱乐空间

## ONE PLUS ONE ——Future Mobile Entertainment Space



作　　者：勾也平
指导老师：叶俊男
所在院校：华东理工大学

### 设计说明

本设计的核心功能是"可连接"以及"车内自由陈设"，用户可以在打开前后挡板的情况下将两辆车的车厢连接成一个整体，以提供更大的娱乐空间。另外，车内电动桌椅利用电力磁吸固定于底盘上，用户可以必要时调节车内设施的位置来构建符合自己场景需求的车内陈设。座椅还可以360° 旋转或者调节为躺椅，为团体出行创造更舒适的娱乐环境。该概念的核心目的是增加车辆的延展性以及自由度，赋予未来出行更有趣的打开方式。

### Design notes

The core function of the concept is "connectable" and "free interior furnishings" , the user can open the front and rear panels to connect the two car compartments into a whole to provide greater entertainment space; in addition, the car's electric table and chairs using electric magnetism fixed on the chassis, the user can adjust the position of the facilities in the car when necessary to build their own scene for the needs of the car furnishings. The seats can also be rotated 360 ° or be adjusted to reclining chair, to create a more comfortable entertainment environment for group travel. The core purpose of the concept is to increase the extensibility and freedom of the vehicle, giving the future of travel a more interesting way to open.

# 玻璃幕墙清洁机

## Glass Curtain Wall Cleaning Machine

作　　者：许颍铮　单祁琛　宋若宁
指导老师：William Volcoff　陈寿年　田坤坤
所在院校：上海视觉艺术学院

## 设计说明

高层建筑中外墙玻璃容易积灰并且产生污垢影响城市的美观，也会影响室内环境。现有的高空清洁机器需要升降装置，运行轨道单一不灵活，在高空运行也会有风险。由此我们设计了一款清洁机器人，灵活的机械臂和液压控制吸盘可以在玻璃幕墙上自由攀爬，越过障碍物清洁玻璃，使用的清洁材料是工业羊毛毡，与传统的清洁方式相比，具有节水、抛光等特性。通过无线远程操控，我们可以让多台机器人一起全方位清洁，提高效率。

## Design notes

High-rise buildings in the exterior glass is easy to accumulate dust and produce dirt affect the beauty of the city, but also affect the indoor environmental. Existing high-altitude cleaning methods still need lifting devices, which running track is single and inflexible, only applicable to flat glass. The current machine is bulky and risky to run at high altitude, especially the airflow is unstable at high altitude in the city. Thus we designed a cleaning robot, flexible robotic arms and hydraulically controlled suction cups allow it to climb freely on the glass curtain wall to clean the glass over the obstacles. By using the cleaning material industrial wool felt, requires less water consumption, produce a good decontamination effect on the glass surface, also can carry out a polishing effect compared with the traditional cleaning method. Through wireless remote control we can have multiple robots cleaning together in all directions to improve efficiency.

# 荷你心意——智慧公交候车亭设计方案

## LOTUS YOUR HEART —— Overall Facility Design Scheme of Smart Bus Shelter

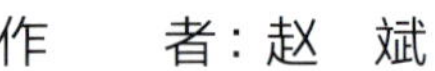

作　　者：赵　斌
指导老师：吴文治　高　瞩
所在院校：上海工程技术大学

### 设计说明

智慧公交候车亭设计方案根据场地特点和设计任务要求，基于设计仿生学提取荷叶的设计元素进行整体设计，在人本主义的设计理念下，探索了大量人工智能的组合模块和运用了大量的可持续设计。候车亭整体设计不仅有着立体律动的美感，在造型的设计上也充分考虑了弱势群体的使用需求。此外，在色彩表现中基于“暖橙”色系的系统设计，在提升候车空间视觉美感上试图打造一个兼具科技智慧与人文关怀的现代化“城市客厅”。

### Design notes

The design is based on the characteristics of the site and the requirements of the design task, and based on design bionics to extract the elements of the lotus leaf for the overall design, also based on high technology to explore a large number of artificial intelligence combination modules and the use of a large number of sustainable design under the humanistic design concept. The overall design of the bus shelter not only has a three-dimensional rhythmic aesthetic, but also takes into account the needs of the underprivileged. In addition, the colour scheme is based on a "warm orange" colour scheme, which enhances the visual aesthetics of the waiting space and attempt to to create a modern "urban living room"  with technological wisdom and humanistic care.

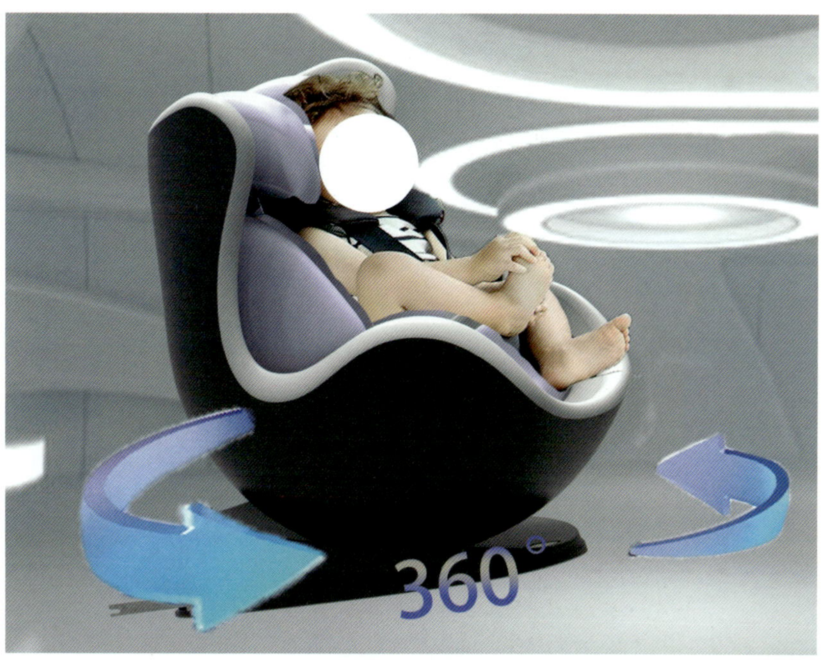

# 智能儿童安全座椅设计

## Intelligent Safety Seat Design for child

作　　者：姚依依
指导老师：宛晓青
所在院校：南京航空航天大学

### 设计说明

智能儿童安全座椅造型灵感来源于蛋壳，具有包裹性、保护性。在结构上进行了创新设计，安全座椅不仅能够在车内使用，而且在户外也可以与推车结合，一举两得。智能化的运用也使用户操作更加便捷，方便安全出行。

### Design notes

The design inspiration for intelligent safety seat for child comes from eggshells, which provide wrapping and protection. Innovation has been carried out in the structure design, making the multifunctional implementation more convenient for users, allowing the safety seat to be used not only in the car, but also in combination with a stroller outdoor, killing two birds with one stone. The application of intelligence also makes it more convenient for users to operate and travel happily.